T0212485

Lecture Notes in Computer Science 9778

Commenced Publication in 1973
Founding and Former Series Editors:
Gerhard Goos, Juris Hartmanis, and Jan van Leeuwen

Editorial Board

David Hutchison
  Lancaster University, Lancaster, UK
Takeo Kanade
  Carnegie Mellon University, Pittsburgh, PA, USA
Josef Kittler
  University of Surrey, Guildford, UK
Jon M. Kleinberg
  Cornell University, Ithaca, NY, USA
Friedemann Mattern
  ETH Zurich, Zürich, Switzerland
John C. Mitchell
  Stanford University, Stanford, CA, USA
Moni Naor
  Weizmann Institute of Science, Rehovot, Israel
C. Pandu Rangan
  Indian Institute of Technology, Madras, India
Bernhard Steffen
  TU Dortmund University, Dortmund, Germany
Demetri Terzopoulos
  University of California, Los Angeles, CA, USA
Doug Tygar
  University of California, Berkeley, CA, USA
Gerhard Weikum
  Max Planck Institute for Informatics, Saarbrücken, Germany

More information about this series at http://www.springer.com/series/7407

Riccardo Dondi · Guillaume Fertin
Giancarlo Mauri (Eds.)

Algorithmic Aspects in Information and Management

11th International Conference, AAIM 2016
Bergamo, Italy, July 18–20, 2016
Proceedings

 Springer

Editors
Riccardo Dondi
Università degli Studi di Bergamo
Bergamo
Italy

Giancarlo Mauri
Università degli Studi di Milano-Bicocca
Milano
Italy

Guillaume Fertin
Université de Nantes
Nantes
France

ISSN 0302-9743   ISSN 1611-3349   (electronic)
Lecture Notes in Computer Science
ISBN 978-3-319-41167-5   ISBN 978-3-319-41168-2   (eBook)
DOI 10.1007/978-3-319-41168-2

Library of Congress Control Number: 2016942022

LNCS Sublibrary: SL1 – Theoretical Computer Science and General Issues

© Springer International Publishing Switzerland 2016
This work is subject to copyright. All rights are reserved by the Publisher, whether the whole or part of the material is concerned, specifically the rights of translation, reprinting, reuse of illustrations, recitation, broadcasting, reproduction on microfilms or in any other physical way, and transmission or information storage and retrieval, electronic adaptation, computer software, or by similar or dissimilar methodology now known or hereafter developed.
The use of general descriptive names, registered names, trademarks, service marks, etc. in this publication does not imply, even in the absence of a specific statement, that such names are exempt from the relevant protective laws and regulations and therefore free for general use.
The publisher, the authors and the editors are safe to assume that the advice and information in this book are believed to be true and accurate at the date of publication. Neither the publisher nor the authors or the editors give a warranty, express or implied, with respect to the material contained herein or for any errors or omissions that may have been made.

Printed on acid-free paper

This Springer imprint is published by Springer Nature
The registered company is Springer International Publishing AG Switzerland

Preface

The papers in this volume were presented at the 11th International Conference on Algorithmic Aspects in Information and Management (AAIM 2016), held during July 18–20, 2016, at the Università degli Studi di Bergamo, Bergamo, Italy. It was the first time the AAIM conference series was held in Europe, the third continent to host AAIM after Asia and America.

The objective of the annual AAIM meetings is (a) to provide an international forum on current trends of research on algorithms, data structures, operation research, and combinatorial optimization and their applications, (b) to stimulate the various fields for which algorithmics can become a crucial enabler, and (c) to strengthen the ties between the Eastern and Western research communities of algorithmics and applications.

Previous AAIM meetings were held in Xian, Hong Kong, Portland, Shanghai, San Francisco, Weihai, Jinhua, Beijing, Dailan, and Vancouver. From the very first AAIM meeting, proceedings have been published in the LNCS series, as volumes 3521, 4041, 4508, 5034, 5564, 6124, 6681, 7285, 7924, and 8546.

The conference program included 18 contributed papers and two invited talks: one by Roberto Grossi from Università di Pisa on "Network Analytics via Pattern Discovery," and the other by Stéphane Vialette from Université Paris-Est on "Unshuffling Words and Permutations." A total of 41 papers were submitted, from 18 different countries (Australia, Canada, China, France, Germany, Greece, Italy, Norway, Oman, Poland, Portugal, Slovakia, Sweden, Taiwan, Tunisia, Turkey, UK, and USA), out of which 18 were accepted, corresponding to an acceptance ratio of 43.9 %. The papers were evaluated by an international Program Committee overseen by the Program Committee co-chairs: Riccardo Dondi, Guillaume Fertin, and Giancarlo Mauri. Each paper was evaluated by at least three Program Committee members, assisted in some cases by additional reviewers. We thank the members of the Program Committee and all the additional external reviewers (whose names are listed later in these proceedings) for their thorough evaluation of the submitted papers and for the fruitful discussion that preceded the final choice of accepted papers. This led to establishing a scientific program of high quality. A special issue of *Theoretical Computer Science* is planned for extended versions of a selection of the papers presented at this year's meeting.

The whole submission and review process was handled electronically via the EasyChair conference system. We thank everyone who contributed to making the meeting a success, the invited speakers, the authors, the Program Committee members, and external reviewers. We thank the Steering Committee for supporting Bergamo as the site for AAIM 2016 and for their help in different issues.

Finally, we thank the Università degli Studi di Bergamo and the Italian Chapter for Theoretical Computer Science for their support, and the local organizers for their assistance.

May 2016

Riccardo Dondi
Guillaume Fertin
Giancarlo Mauri

Organization

Program Committee Co-chairs

Riccardo Dondi   Università degli Studi di Bergamo, Italy
Guillaume Fertin   University of Nantes, France
Giancarlo Mauri   Università degli Studi di Milano-Bicocca, Italy

Program Committee

Francine Blanchet-Sadri   University of North Carolina, USA
Laurent Bulteau   University of Paris-Est Marne-La-Vallée, France
Cedric Chauve   Simon Fraser University, Canada
Zhi-Zhong Chen   Tokyo Denki University, Japan
Marek Chrobak   University of California, Riverside, USA
Ferdinando Cicalese   Università di Verona, Italy
Pierluigi Crescenzi   Università degli Studi di Firenze, Italy
Peter Damaschke   Chalmers University of Technology, Sweden
Bhaskar Dasgupta   University of Illinois at Chicago, USA
Nadia El-Mabrouk   University of Montreal, Canada
Michael Fellows   Charles Darwin University, Australia
Irene Finocchi   Università di Roma La Sapienza, Italy
Pawel Gorecki   University of Warsaw, Poland
Inge Li Gørtz   Technical University of Denmark, Denmark
Frederic Havet   Inria Sophia-Antipolis, France
Danny Hermelin   Ben-Gurion University of the Negev, Israel
Jesper Jansson   Kyoto University, Japan
Minghui Jiang   Utah State University, USA
Christian Komusiewicz   TU Berlin, Germany
Moshe Lewenstein   Bar Ilan University, Israel
Martin Milanic   University of Primorska, Slovenia
Rolf Niedermeier   TU Berlin, Germany
Daniël Paulusma   Durham University, UK
David Peleg   The Weizmann Institute, Israel
Marcin Pilipczuk   University of Warsaw, Poland
Romeo Rizzi   Università di Verona, Italy
Marie-France Sagot   Inria Grenoble Rhône-Alpes and Université de Lyon 1, France
Saket Saurabh   The Institute of Mathematical Sciences, Chennai, India
Marinella Sciortino   Università di Palermo, Italy
Florian Sikora   University of Paris-Dauphine, France

Ioan Todinca   Université d'Orléans, France
Leo van Iersel   Delft University of Technology, The Netherlands
Rossano Venturini   Università di Pisa, Italy
Lusheng Wang   City University of Hong Kong, SAR China
Binhai Zhu   Montana State University, USA

Organizing Committee

Federica Baroni   Università degli Studi di Bergamo, Italy
Paolo Cazzaniga   Università degli Studi di Bergamo, Italy
Mauro Castelli   Universidade Nova de Lisboa, Portugal
Riccardo Dondi   Università degli Studi di Bergamo, Italy
Guillaume Fertin   University of Nantes, France
Marco Lazzari   Università degli Studi di Bergamo, Italy
Giancarlo Mauri   Università degli Studi di Milano-Bicocca, Italy
Italo Zoppis   Università degli Studi di Milano-Bicocca, Italy

Additional Reviewers

Carl Barton   Branko Kavšek   Anthony Przybylski
Robert Bredereck   Manuel Lafond   Gaetano Rossiello
Andrea Cali   Tero Laihonen   Sudeepa Roy
Mauro Castelli   Luca Manzoni   Blerina Sinaimeri
Paolo Cazzaniga   Andrea Marino   Glenn Tesler
Gasper Fijavz   Arnaud Mary   Yuen-Lam Voronin
Vincent Froese   Jérôme Monnot   Haitao Wang
Philippe Gambette   Paolo Penna   Janez Zerovnik
Boris Goldengorin   Yuri Pirola
Frédéric Goualard   Solon Pissis

Invited Talks

Network Analytics via Pattern Discovery

Roberto Grossi

Università di Pisa, Pisa, Italy
grossi@di.unipi.it

Abstract. Social, biological and communication networks of data with a strong linked nature can often be modeled and analyzed as labeled graphs. We describe some new algorithms for pattern discovery in graphs that can be useful for network analytics, focusing on clique enumeration and its applications.

Topics of the talk. The combinatorial approach is complementary to the statistical one, and both are successfully employed in network analytics. We describe some recent work on cliques as special patterns to discover in networks, along with some applications. In the literature, problems related to finding cliques in graphs have a long history because of many applications, such as applied statistics, behavioral and cognitive networks, compiler optimization, complex networks, computational biology, computational geometry, dynamic networks, email networks, financial networks, security, and social networks.

We will start with the edge clique covering problem, which deals with discovering a set of (possibly overlapping) cliques in a given network, such that each edge is part of at least one of these cliques. We address the problem from an alternative perspective reconsidering the quality of the cliques found, and proposing more structured criteria with respect to the traditional measures such as minimum number of cliques.

We will then consider two worst-case efficiency measures when listing patterns in a graph, namely, delay and space, that become relevant for enumeration in massive networks. The delay is the maximum latency between any two consecutively reported solutions. The space is the maximum amount of extra memory that should be allocated to enumerate all the solutions, besides the amount required by the input graph. By optimizing these measures we can provide algorithms which are fast enough to be applied in real examples, where brute-force approaches are too costly.

As an application, we will discuss how to employ a variation of the latter techniques to find large common connected induced subgraph of two input graphs that represent two proteins. Indeed, graph-based methods provide a natural complement to sequence-based methods in bioinformatics and protein modeling, allowing us to identify compound similarity between small molecules, and structural relationships between biological macromolecules that are not spotted by sequence analysis.

Joint work with Alessio Conte, Andrea Marino, Lorenzo Tattini, and Luca Versari

Unshuffling Words and Permutations

Stéphane Vialette

Université Paris-Est, LIGM - UMR CNRS 8049, Champs-sur-Marne, France
vialette@univ-mlv.fr

Given three words u, v_1, and v_2, u is said to be a shuffle of v_1 and v_2 if it can be formed by interleaving the letters from v_1 and v_2 in a way that maintains the left-to-right ordering of the letters from each word. A string is said to be a square for the shuffle product if it is the shuffle of two identical words. In a similar manner, a permutation π is said to be a square if it can be obtained by shuffling two order-isomorphic patterns. The definition is intended to be the natural counterpart to the ordinary shuffle of words and languages.

In this talk recent results in recognising square words and permutations will be surveyed. Additional open problems in the area of recognising square words and permutations will be discussed.

Contents

Item Pricing for Combinatorial Public Projects

Evangelos Markakis[1] and Orestis Telelis[2(✉)]

[1] Department of Informatics, Athens University of Economics and Business,
Athens, Greece
markakis@gmail.com
[2] Department of Digital Systems, University of Piraeus, Piraeus, Greece
telelis@gmail.com

Abstract. We describe and analyze a simple mechanism for the Combinatorial Public Project Problem (CPPP), a prototypical abstract model for decision making by autonomous strategic agents. The problem asks for the selection of k out of m available items, so that the social welfare is maximized. With respect to *truthful* Mechanism Design, the CPPP has been shown to suffer from limited computationally tractable approximability. Instead, we study a non-truthful Item Bidding mechanism, which elicits the agents' preferences through *separate* bids on the items and selects the k items with the highest sums of bids. We pair this outcome determination rule with a payment scheme that determines – for each agent – a separate price for each item in the outcome. For expressive classes of the agents' valuation functions, we establish existence of welfare-optimal pure Nash equilibria and strong equilibria. Subsequently, we derive worst-case upper and lower bounds on the approximation of the optimal welfare achieved at strong equilibrium, and at (mixed) Bayes-Nash equilibrium, under an incomplete information setting. The mechanism retains good stability properties and favors an advantage compared to recent related approaches, given its simple per-item bidding and pricing rules, and its comparable performance with respect to welfare approximation.

1 Introduction

The Combinatorial Public Project Problem (CPPP) was introduced by Papadimitriou, Schapira and Singer in [17], as a foundational abstract model for socially efficient decision making, among strategic agents with *private* combinatorial preferences. In the CPPP, an authority aims at combining k components from a given set of m distinct items, to build a composite service or facility, in the common interest of the participating agents. Hence, the goal is to select a subset of the items that maximizes the Social Welfare, i.e., the sum of the agents' (private) values for the outcome. The agents' private preferences are expressed by combinatorial valuation functions, defined over the set of all subsets of items, $2^{[m]}$. Ultimately, we seek a *mechanism* that determines an outcome along with a payment to be issued by each agent. Apart from its importance as a decision making

This work has been partly supported by the University of Piraeus Research Center.

© Springer International Publishing Switzerland 2016
R. Dondi et al. (Eds.): AAIM 2016, LNCS 9778, pp. 1–13, 2016.
DOI: 10.1007/978-3-319-41168-2_1

problem, the CPPP has served as an example highlighting the computational hardness that emerges under the constraint of incentive compatibility.

In this work, we analyze the performance of a mechanism for the CPPP, which favors simplicity at the expense of incentive compatibility and welfare optimality. It employs an *Item Bidding* interface for eliciting the agents' preferences; instead of reporting their complete valuation functions over outcomes to the mechanism, the agents are invited to *bid* their preference on each distinct item separately. The mechanism employs a simple rule for outcome determination: the k items with the highest sums of bids are included in the outcome (with a deterministic tie-breaking). In effect, under this setting each agent is required to "compress" his (combinatorial) valuation function into an *additive* bid vector. The pricing rule charges every item in the outcome *separately* for each agent, according to an instantiation of Vickrey's 2nd Price rule. Each agent is charged the sum of the computed prices over all items. We refer to this scheme as *Item Pricing*.

We investigate the mechanism's performance at (pure Nash and Bayes-Nash) equilibrium, via a Price of Anarchy analysis. The mechanism may not – in general – produce optimal welfare, and is inevitably vulnerable to manipulation by agents misreporting their values for items. One of the reasons is clearly the restricted expressiveness of the Item Bidding interface. Nevertheless, the study of simple mechanisms for the CPPP is motivated by their appeal for practical applications in multi-agent systems. Regarding truthfulness, a series of works [5,17,20] have established severe computational inapproximability results for tractable truthful mechanisms for the CPPP. Additionally, the existing truthful mechanisms tend to use complex algorithmic schemes for determining outcomes and payments, that restrict their practical applicability and may discourage voluntary participation of agents in the underlying strategic game.

Our mechanism differs from an Item Bidding mechanism studied recently [13], in the definition of the payments rule. Lucier *et al.* studied in [13] an Item Bidding mechanism that uses arguably the simplest Item Pricing rule possible; each agent pays his issued bids for each item in the outcome. This is a "pay-as-bid" scheme akin to the one used in First-Price Auctions. As such, it induces a strategic game that does not have even mixed equilibria, unless discretization of the agents' strategies is used, or carefully crafted tie-breaking. We remedy this issue, through employment of the payments scheme described above, that maintains separate pricing of each item in the outcome, for each agent. We show that it induces pure Nash equilibria for a wide class of valuation functions expressing the agents' preferences, and even strong equilibria (resilient to coordinated deviations of agents), for a smaller class. Our mechanism also exhibits comparable welfare approximation performance to the one of [13], for the studied classes of the agents' preferences. The importance of Item Pricing is highlighted by its compatibility to the Item Bidding interface, which restricts the agents' capability of declaring their values for whole outcomes; it appeals to the natural anticipation that separate bidding should be accompanied by separate pricing.

Contribution. We introduce an Item Bidding mechanism with an Item Pricing payments rule for the CPPP, and analyze its performance at equilibrium. Our work complements and strengthens results of previous works on the subject [13,14] (discussed in Sect. 2). First we prove that the mechanism admits socially optimal pure Nash equilibria, for a wide class of the agents' valuation functions, known as *Fractionally Subadditive* [8] or **XOS** [12] ("Exclusive-OR of Singletons" – see Sect. 3 for precise definition). For a narrower class of valuation functions we show that it admits socially optimal *strong equilibria*, that are resilient to coordinated deviations of subsets of agents.

Subsequently, we investigate the mechanism's performance at ℓ-strong equilibria, that are resilient against coordinated deviations of at most ℓ agents. Our technically most demanding result is an upper bound of $O(\lceil n/\ell \rceil)$ on the ℓ-strong Price of Anarchy, for agents with **XOS** valuation functions, along with a lower bound of $\max\{2, n/\ell\}$. These show that the mechanism's performance improves, as the number of agents that are allowed to coordinate increases, recovering a constant fraction of the optimal welfare at strong equilibrium. Finally, we study an incomplete information setting for our mechanism, where valuation functions of agents are drawn from probability distributions. Using analytical developments of [19,21,22], we prove an upper bound of $O(n)$ on the mechanism's Bayes-Nash Price of Anarchy for agents with **XOS** valuation functions and an upper bound of $O(\lceil n/k \rceil)$ for agents with Unit Demand valuation functions. For comparison, no truthful deterministic CPPP mechanism can approximate the optimal welfare within less than $O(\sqrt{m})$ [17], where m is the number of items, while our bounds are independent of m.

2   Related Work

A significant volume of research, initiated in [6] and followed by several recent works [4,9–11,22], concerns the study of Item Bidding mechanisms for combinatorial and multi-unit auctions. These works concern the performance of *simultaneous* 2nd Price or 1st Price auctions, one per distinct item on sale. Bidders with combinatorial valuation functions over subsets of items issue a separate bid in each single-item auction, simultaneously. Syrgkanis and Tardos developed in [22] a unified *smoothness analysis framework*, for analyzing the performance of Item Bidding auction formats in the incomplete information setting, and of their simultaneous or sequential compositions.

In the context of Item Bidding for the CPPP, our work follows the spirit of [13,14]. For an Item Bidding mechanism with a *"pay-as-bid"* payments rule, Lucier *et al.* establish in [13] a tight upper bound of $\Theta(\log n)$ for the strong Price of Anarchy, when agents have arbitrary valuation functions. Using analytical techniques from [22], they also prove upper bounds of $O(n)$ and $O(\lceil n/k \rceil)$ for the incomplete information setting, when the agents have respectively, *fractionally subadditive*, or unit demand valuation functions. In [14] we studied the performance of Item Bidding paired with Vickrey-Clarke-Groves payments. For the resulting mechanism, we proved almost tight upper bounds of $O(\lceil n/\ell \rceil)$ and

$O\left(\left\lceil \frac{n}{k\cdot\ell} \right\rceil\right)$ for its ℓ-strong Price of Anarchy, when agents have fractionally sub-additive and unit demand valuation functions respectively.

The CPPP was formalized by Papadimitriou, Schapira and Singer in [17]; they proved communication and computational complexity lower bounds of $\Omega(\sqrt{m})$ on the problem's approximability, by deterministic *truthful* mechanisms, when agents have submodular valuation functions. In contrast, the underlying optimization problem for this class of valuation functions is long known to be approximable within a factor $\frac{e}{e-1}$ by the celebrated greedy algorithm of [15]. Schapira and Singer proved in [20] non-constant approximability lower bounds for truthful mechanisms and more general classes of valuation functions and devised a truthful $O(\sqrt{m})$-approximation mechanism. A detailed study of the problem's complexity for several classes of valuation functions appeared in [5]. The only known constant approximation mechanism for the CPPP with submodular valuation functions is randomized and *truthful in expectation* [7].

3 Definitions and Preliminaries

We consider a set $[m] = \{1,\dots,m\}$ of *items* and a set $[n] = \{1,\dots,n\}$ of *agents*. Each agent has *private* combinatorial preferences over $2^{[m]}$, expressed by a non-decreasing valuation function $v_i : 2^{[m]} \mapsto \mathbb{R}^+$, for each $i \in [n]$. Given $k \in \mathbb{Z}^+$, the aim is to choose $X \subseteq [m]$, $|X| = k$, maximizing the *Social Welfare*, $SW(X) = \sum_i v_i(X)$. For succinctness in our presentation, we define:

$$V_I(X) \equiv \sum_{i \in I} v_i(X) \quad \text{and} \quad V_{-I}(X) \equiv \sum_{i \in [n]\setminus I} v_i(X), \quad \text{for any } I \subseteq [n], X \subseteq [m]$$

In effect, $SW(X) = V_{[n]}(X)$; we use $SW(X)$ in this case.

We study an *Item Bidding* mechanism, wherein each agent $i \in [n]$ submits a *separate* bid $b_i(j) \geq 0$ for each item $j \in [m]$, within a bid vector $\mathbf{b}_i = (b_i(1),\dots,b_i(m))$. We use $\mathbf{b}_{-i}$ to denote the profile $(\mathbf{b}_1,\dots,\mathbf{b}_{i-1},\mathbf{b}_{i+1},\dots,\mathbf{b}_n)$ and, for any subset $I \subseteq [n]$, $\mathbf{b}_I$ and $\mathbf{b}_{-I}$ are the bids of all agents in I and of all agents except the ones in I, respectively. We will also use the notation $\mathbf{b}_{-ij}$ to denote the bidding profile that emerges from $\mathbf{b}$, *when the bid $b_i(j)$, of agent i on item j is omitted*. Given a bidding profile $\mathbf{b}$, our mechanism determines an outcome $\mathbb{X}^k(\mathbf{b}) \subseteq [m]$, consisting of the k items with *the highest sums of bids* (where $k < m$), subject to a deterministic tie-breaking rule, for discriminating among two items with equal sums of bids (e.g., smallest index). To describe the payment rule succinctly, let us introduce some notation first. *With reference to a profile $\mathbf{b}$ and its outcome, $\mathbb{X}^k(\mathbf{b})$,* define $B(\mathbb{X}^k(\mathbf{b})) = \sum_i \sum_{j \in \mathbb{X}^k(\mathbf{b})} b_i(j)$. We extend this definition for any subset of agents, $I \subseteq [n]$, as follows:

$$B_I(\mathbb{X}^k(\mathbf{b})) \equiv \sum_{i \in I} \sum_{j \in \mathbb{X}^k(\mathbf{b})} b_i(j) \quad \text{and:} \quad B_{-I}(\mathbb{X}^k(\mathbf{b})) \equiv B_{[n]\setminus I}(\mathbb{X}^k(\mathbf{b}))$$

When I is a singleton, we use simply B_i and B_{-i} respectively. Here, we study an item-based payment which we refer to as *Item Pricing*. To define it, we will use the notation $B_{-ij}(\mathbb{X}^k(\mathbf{b})) \equiv B(\mathbb{X}^k(\mathbf{b})) - b_i(j)$. Then:

$$\textbf{ItemPricing}: \quad p_i(\mathbf{b}) = \sum_{j \in \mathbb{X}^k(\mathbf{b})} \left[B(\mathbb{X}^k(\mathbf{b}_{-ij})) - B_{-ij}(\mathbb{X}^k(\mathbf{b})) \right] \quad (1)$$

where, by definition of $\mathbf{b}_{-ij}$ and $B(\mathbb{X}^k(\cdot))$, we have:

$$B(\mathbb{X}^k(\mathbf{b}_{-ij})) = B_{-i}(\mathbb{X}^k(\mathbf{b}_{-ij})) + \sum_{r \in \mathbb{X}^k(\mathbf{b}_{-ij}) \setminus \{j\}} b_i(r)$$

The pricing defined in (1) can be interpreted as summing up prices, that are computed by instantiations of Vickrey's Second Price rule, per item in the outcome. The *utility* of every agent i under profile $\mathbf{b}$ is: $u_i(\mathbf{b}) = v_i(\mathbb{X}^k(\mathbf{b})) - p_i(\mathbf{b})$. Note that we apply $B(\mathbb{X}^k(\cdot))$ consistently, by summing up bids with reference to both, a profile **and** its outcome. Often, we will need to consider a profile $(\mathbf{b}'_I, \mathbf{b}_{-I})_{-ij}$. We abuse the notation slightly and write $B(\mathbb{X}^k(\mathbf{b}'_I, \mathbf{b}_{-I})_{-ij})$, to avoid an extra pair of parentheses.

Item Pricing treats all k available *"slots"* uniformly; it charges agent i for each such *"slot"*, the externality that i imposes – through his bid on item j – on the agents bidding for the item with the "highest losing" sum of bids, independently of the positional order of j in $\mathbb{X}^k(\mathbf{b})$.

Valuation Functions. We evaluate our mechanism's performance for agents with valuation functions belonging to the rich class of *Fractionally Subadditive* functions [8] – also defined in [12], under the name **XOS**. This class is a strict superset of the widely studied class of *Submodular* functions, which express *decreasing marginal value* of each additional item to an agent. We also consider *Unit Demand* functions (**UD**) and a strict subset of it, referred to as *uniform* **UD** (**uUD**). Formally:

Definition 1. *A function $v : 2^{[m]} \mapsto \mathbb{R}^+$ belongs to the class*

- **XOS**, *if there exists a family of r additive functions, $\alpha_t : [m] \mapsto \mathbb{R}^+, t = 1, \ldots, r$, such that: $v(X) = \max_{t=1,\ldots,r} \alpha_t(X)$, for every $X \subseteq [m]$.*
- **UD**: *$v(X) = \max\{v(\{j\}) \mid j \in X\}$, for every $X \subseteq [m]$.*
- **uUD**, *if v is **UD** and, for every $j \in [m]$, $v(\{j\}) \in \{\nu, 0\}$, for some $\nu > 0$.*

Even for **uUD** valuation functions, the Cppp remains **NP**-hard (Theorem 2.1 of [5]). No tractable truthful $o(\sqrt{m})$-approximation mechanism is known for this class; a hardness result by [5] (cf. Theorem 2.2) suggests an $\Omega(\sqrt{m})$ approximation lower bound, for the category of *Maximum-in-Range mechanisms* [16].

Solution Concepts. We study our mechanism's performance at ℓ–strong equilibrium [2] (for some $\ell \in \{1, \ldots, n\}$), that is, a pure Nash equilibrium that is resilient to *coordinated* deviations of subsets of at most ℓ agents. Formally:

Definition 2. *A bidding profile* **b** *is an* $\ell-$*strong equilibrium if and only if, for every subset* I *of at most* $\ell \geq 1$ *agents and for every joint deviation* $\mathbf{b}'_I$ *of* I, *there exists at least one agent* $i \in I$ *such that* $u_i(\mathbf{b}'_I, \mathbf{b}_{-I}) \leq u_i(\mathbf{b})$.

We use the term *strong equilibrium* when $\ell = n$. To quantify our mechanism's inefficiency, we will derive upper and lower bounds on the $\ell-$*Strong* Price of Anarchy [1], that is, *the worst case ratio of the optimum social welfare over the welfare achieved at* $\ell-$*strong equilibrium*. Following previous works on Item Bidding mechanisms [4, 6, 9, 22], we make a standard *no-overbidding assumption*; that for any subset of k items or less, the sum of bids submitted by each agent for this subset does not exceed his value for it: $\sum_{j \in X} b_i(j) \leq v_i(X)$ for any $X \subseteq [m] : |X| \leq k$. This assumption is justified by the fact that overbidding strategies are weakly dominated in general; they can be made strictly dominated by seamless modifications of the mechanisms, see e.g., [6].

4   Efficient Nash Equilibria

We start with examining the existence of efficient stable configurations for the mechanism. Before proceeding, we take the opportunity to introduce the notion of a *proper deviation*; this notion we shall use also through the next sections.

Definition 3. *Let* **b** *denote a strategy profile,* $I \subseteq [n]$ *any non-empty subset of agents, and* $\hat{\mathbf{b}}_I$ *any joint deviation of agents in* I. *A proper joint deviation* $\mathbf{b}'_I$ *of* I *with respect to* $\hat{\mathbf{b}}_I$ *is defined as (for every* $j \in [m]$):

$$b'_i(j) = \begin{cases} \hat{b}_i(j), \text{ if } j \in \mathbb{X}^k(\hat{\mathbf{b}}_I, \mathbf{b}_{-I}) \\ 0, \qquad\qquad otherwise \end{cases}$$

A *proper* joint deviation $\mathbf{b}'_I$ w.r.t. the joint deviation $\hat{\mathbf{b}}_I$ ensures $\mathbb{X}^k(\mathbf{b}'_I, \mathbf{b}_{-I}) = \mathbb{X}^k(\hat{\mathbf{b}}_I, \mathbf{b}_{-I})$, and that every agent in I bids 0 on every item that is *not* in the outcome $\mathbb{X}^k(\hat{\mathbf{b}}_I, \mathbf{b}_{-I})$. We show first that our mechanism admits socially optimal pure Nash equilibria, for agents with **XOS** valuation functions.

Theorem 1. *For agents with* **XOS** *valuation functions, the Item Bidding mechanism with Item Pricing for the* CPPP *admits welfare-optimal pure equilibria.*

Proof (Sketch). We fix any socially optimal outcome $\mathcal{X}^*$, for which we design a supporting bidding profile, $\mathbf{b}^*$, using the definition of the agents' **XOS** valuation functions. For every agent i we set $b_i^*(j) = 0$, if $j \notin \mathcal{X}^*$; for each item $j \in \mathcal{X}^*$ we define $b_i^*(j)$ so that $\sum_j b_i^*(j) = v_i(\mathcal{X}^*)$. The profile $\mathbf{b}^*$ satisfies $p_i(\mathbf{b}^*) = 0$, for every agent $i \in [n]$. We prove that $\mathbf{b}^*$ is a pure Nash equilibrium, by first assuming that there exists a profitable deviation $\mathbf{b}_i$ for some agent $i \in [n]$, such that $u_i(\mathbf{b}_i, \mathbf{b}^*_{-i}) > u_i(\mathbf{b}^*)$, and subsequently contradicting the optimality of $\mathcal{X}^*$. To this end, we transform $\mathbf{b}_i$ into a *proper deviation* $\mathbf{b}'_i$ first, as described in Definition 3. We argue that $\mathbf{b}'_i$ remains profitable for i, thus, $u_i(\mathbf{b}'_i, \mathbf{b}^*_{-i}) > u_i(\mathbf{b}^*)$. Further analysis of this inequality yields $SW(\mathbb{X}^k(\mathbf{b}'_i, \mathbf{b}^*_{-i})) > SW(\mathbb{X}^k(\mathbf{b}^*)) = SW(\mathcal{X}^*)$, a contradiction.   □

We now move on to show that the existence of welfare-optimal strong equilibria can be guaranteed, for the narrower class of **uUD** valuation functions.

Theorem 2. *The Item Bidding mechanism with Item Pricing for the* CPPP *admits efficient strong equilibria, for agents with* **uUD** *valuation functions.*

Proof (Sketch). Along similar lines as for the proof of Theorem 1, we define a supporting bidding profile, $\mathbf{b}^*$, for a socially optimal outcome $\mathcal{X}^*$. By elaborating on the agents' **uUD** valuation functions, we define $\mathbf{b}^*$ so that $\sum_j b_i^*(j) = v_i(\mathcal{X}^*)$, for every $i \in [n]$ and $b_i^*(j) = 0$, for $j \notin \mathcal{X}^*$. We assume that $\mathbf{b}^*$ is not a strong equilibrium, for the purpose of deriving a contradiction. Then there exists a subset of agents, I, and a joint deviation, $\mathbf{b}_I$, such that $u_i(\mathbf{b}_I, \mathbf{b}_{-I}^*) > u_i(\mathbf{b}^*)$, for every $i \in I$. We obtain a proper joint deviation $\mathbf{b}_I'$ from $\mathbf{b}_I$, for which we ensure that it satisfies an additional property, for each agent $i \in I$: that $B_i(\mathbb{X}^k(\mathbf{b}_I', \mathbf{b}_{-I}^*)) = v_i(\mathbb{X}^k(\mathbf{b}_I', \mathbf{b}_{-I}^*))$. This proper joint deviation remains profitable for all members of I, thus, $p_i(\mathbf{b}_I', \mathbf{b}_{-I}^*) < v_i(\mathbb{X}^k(\mathbf{b}_I', \mathbf{b}_{-I}^*))$ for every agent $i \in I$. This latter inequality we manipulate, via the definition of payments $p_i(\mathbf{b}_I', \mathbf{b}_{-I}^*)$, to show that $SW(\mathbf{b}_I', \mathbf{b}_{-I}^*) > SW(\mathcal{X}^*)$, a contradiction. $\qquad\square$

5 Inefficiency of Strong Equilibria

In this section we consider the mechanism's performance w.r.t. the approximation of the socially optimal welfare, that it achieves at ℓ-strong equilibrium. In particular, we prove the following.

Theorem 3. *The ℓ-strong Price of Anarchy of the Item Bidding Mechanism with Item Pricing for agents with* **XOS** *valuation functions is $O(\lceil n/\ell \rceil)$ and at least $\max\{2, \frac{n}{\ell}\}$.*

Our main and technically most challenging result in this section concerns the proof of the upper bound. We present here the main components of the analysis and how they lead to its establishment. The following lemma formalizes a standard argument for upper bounding the ℓ-strong price of anarchy of strategic games, in the context of proper joint deviations (as given by Definition 3).

Lemma 1. *For any ℓ-strong equilibrium $\mathbf{b}$, any subset of agents $I \subseteq [n]$ of cardinality at most ℓ, and any joint deviation profile $\mathbf{b}_I^*$ for I, there exists a bijection $\pi : \{1, \ldots, |I|\} \mapsto I$, dependent on $\mathbf{b}_I^*$, such that:*

$$u_{\pi(q)}(\mathbf{b}) \geq u_{\pi(q)}(\mathbf{b}_{I_q}', \mathbf{b}_{-I_q}), \quad for \; q = 1, \ldots, |I|$$

where $I_q = \{i \in I : \pi^{-1}(i) \geq q\}$ and $\mathbf{b}_{I_q}'$ is a proper deviation of I_q w.r.t. $\mathbf{b}_{I_q}^*$.

The technical heart of our upper bound's proof is the following lemma which, in combination with Lemma 1 yields the proof of the upper bound of Theorem 3.

Lemma 2. *Let $\mathcal{X}^*$ denote any socially optimal profile. For every profile $\mathbf{b}$, every subset $I \subseteq [n]$ of at most ℓ agents, and for every bijection $\pi : \{1, \ldots, |I|\} \mapsto I$, there exists a joint deviation profile $\mathbf{b}_I^*$ such that:*

$$\sum_{q=1}^{|I|} u_{\pi(q)}(\mathbf{b}'_{I_q}, \mathbf{b}_{-I_q}) \geq \frac{1}{2} V_I(\mathcal{X}^*) - 2SW(\mathbb{X}^k(\mathbf{b})) \tag{2}$$

where $I_q = \{i \in I \mid \pi^{-1}(i) \geq q\}$, for $q = 1, \ldots, |I|$, and $\mathbf{b}'_{I_q}$ is the proper joint deviation of I_q with respect to $\mathbf{b}_{I_q}^*$.

Before explaining the proof of Lemma 2, we show how it is used for Theorem 3.

*Proof (**Upper bound in Theorem** 3).* Let $\mathbf{b}$ denote an ℓ-strong equilibrium profile and consider the joint deviation profile $\mathbf{b}_I^*$ whose existence is asserted by Lemma 2, and apply inequality (2) for the bijection π, prescribed by Lemma 1. Combining the inequalities of Lemma 1 with inequality (2) under π, we obtain:

$$\sum_{i \in I} u_i(\mathbf{b}) \geq \frac{1}{2} V_I(\mathcal{X}^*) - 2SW(\mathbb{X}^k(\mathbf{b}))$$

But, for every agent $i \in I$ we have $v_i(\mathbb{X}^k(\mathbf{b})) \geq u_i(\mathbf{b})$, thus:

$$V_I(\mathbb{X}^k(\mathbf{b})) \geq \frac{1}{2} V_I(\mathcal{X}^*) - 2SW(\mathbb{X}^k(\mathbf{b}))$$

Summing up over at most $\lceil n/\ell \rceil$ *disjoint* subsets of at most ℓ agents each, yields:

$$SW(\mathbb{X}^k(\mathbf{b})) \geq \frac{1}{2} SW(\mathcal{X}^*) - 2\lceil n/\ell \rceil SW(\mathbb{X}^k(\mathbf{b}))$$

which, in turn gives the stated result.

We present the main "ingredients" of the proof of Lemma 2, particularly the definitions of strategies for the agents w.r.t. a socially optimal outcome, and of related constructs, needed for technical derivations. These derivations we describe only briefly, due to lack of space.

*Proof (**of Lemma** 2).* Let $\mathbf{b}$ denote an arbitrary strategy profile, $\mathcal{X}^*$ be a socially optimal outcome and fix any subset $I \subset [n]$ of ℓ agents. We define a joint deviation profile $\mathbf{b}_I^*$ mentioned by Lemma 2, with reference to $\mathcal{X}^*$. For every agent $i \in I$ with a *fractionally subadditive* valuation function v_i, let $\{\alpha_{i,1}, \ldots, \alpha_{i,r_i}\}$ constitute the description of v_i through a set of additive functions (recall Definition 1). Let α^i be any particular additive function in $\{\alpha_{i,1}, \ldots, \alpha_{i,r_i}\}$, such that $v_i(\mathcal{X}^*) = \alpha^i(\mathcal{X}^*)$, thus, $v_i(\mathcal{X}^*) = \sum_{j \in \mathcal{X}^*} \alpha^i(\{j\})$. Let us order the items $j_r^* \in \mathcal{X}^*$ in *non-increasing value* of the sum $\sum_{i \in I} \alpha^i(\{j_r^*\})$, so that $\sum_{i \in I} \alpha^i(\{j_r^*\}) \geq \sum_{i \in I} \alpha^i(\{j_{r+1}^*\})$, for $r = 1, \ldots, k-1$. Then, let $\mathcal{J}^* = \{j_1^*, \ldots, j_{\lceil k/2 \rceil}^*\}$ and define $\mathbf{b}_I^*$ as follows:

$$\text{For every } i \in I : \quad b_i^*(j) = \begin{cases} \alpha^i(\{j\}), & \text{if } j \in \mathcal{J}^* \\ 0, & \text{otherwise.} \end{cases} \tag{3}$$

By definition of $\mathcal{J}^*$, $\mathbf{b}_I^*$ and by definition of **XOS** valuation functions, we have:

$$B_I(\mathbb{X}^k(\mathbf{b}_I^*)) = V_I(\mathbb{X}^k(\mathbf{b}_I^*)) = V_I(\mathcal{J}^*) \geq \sum_{i \in I} \alpha^i(\mathcal{J}^*) \geq \frac{1}{2} V_I(\mathcal{X}^*) \qquad (4)$$

To simplify the notation in our exposition and without loss of generality, we assume that $|I| = \ell$, $I = \{1, 2, \ldots, \ell\}$ and that $\pi(q) = q$; thus the subsets I_q of I referenced in the statement of Lemma 2 are in effect $I_q = \{q, q+1, \ldots, \ell\}$, for $q = 1, \ldots, \ell$ and $\mathbf{b}'_{I_q}$ is a proper joint deviation of I_q w.r.t. $\mathbf{b}_{I_q}^*$. Using (1), we can derive for the utility of any agent $i \in I_q$, for $q = 1, \ldots, \ell$:

$$u_q(\mathbf{b}'_{I_q}, \mathbf{b}_{-I_q}) \geq |\mathbb{X}^k(\mathbf{b}'_{I_q}, \mathbf{b}_{-I_q})| \cdot B(\mathbb{X}^k(\mathbf{b}'_{I_q}, \mathbf{b}_{-I_q}))$$
$$- \sum_{j \in \mathbb{X}^k(\mathbf{b}'_{I_q}, \mathbf{b}_{-I_q})} B(\mathbb{X}^k(\mathbf{b}'_{I_q}, \mathbf{b}_{-I_q})_{-qj}) \qquad (5)$$

To continue our analysis, we find it helpful to partition each (partial) outcome $\mathbb{X}^k(\mathbf{b}'_{I_q}, \mathbf{b}_{-I_q}) \cap \mathcal{J}^*$ with reference to its successor outcome, $\mathbb{X}^k(\mathbf{b}'_{I_{q+1}}, \mathbf{b}_{-I_{q+1}})$:

$$Y_q = \left[\mathbb{X}^k(\mathbf{b}'_{I_q}, \mathbf{b}_{-I_q}) \cap \mathcal{J}^* \right] \setminus \mathbb{X}^k(\mathbf{b}'_{I_{q+1}}, \mathbf{b}_{-I_{q+1}})$$
$$Y_q^* = \left[\mathbb{X}^k(\mathbf{b}'_{I_q}, \mathbf{b}_{-I_q}) \cap \mathcal{J}^* \right] \cap \mathbb{X}^k(\mathbf{b}'_{I_{q+1}}, \mathbf{b}_{-I_{q+1}}) \qquad (6)$$

Y_q is the subset of items from $\mathcal{J}^*$ that are *"evicted"* from the q-th outcome, if agent q abandons I_q and switches to his *"original"* bidding strategy $\mathbf{b}_q$. Y_q^* contains the items from $\mathcal{J}^*$ that *survive* in the $(q+1)$-th outcome on this occasion. Notice that, by definition of $\mathbf{b}'_{I_q}$ as a proper joint deviation of I_q (Definition 3) w.r.t. $\mathbf{b}_I^*$, each agent $i \in I_q$ bids $b_i^*(j)$ on each item $j \in Y_{q-1}^* = \mathbb{X}^k(\mathbf{b}'_{I_q}, \mathbf{b}_{-I_q}) \cap \mathcal{J}^*$ and 0 on the remaining items. We define $Y_0^* = \mathbb{X}^k(\mathbf{b}'_{I_1}, \mathbf{b}_{-I_1}) \cap \mathcal{J}^*$. Then, by definition of $\mathbf{b}'_{I_1} = \mathbf{b}'_I$, we have $|Y_0^*| \leq \lceil k/2 \rceil$ and:

$$Y_q \cup Y_q^* = Y_{q-1}^*, \quad \text{for } q = 1, \ldots, \ell = |I|. \qquad (7)$$

We derive upper bounds for each of the summands in the sum of the right-hand side of (5), which allow us to obtain a new inequality from (5); for $q = 1, \ldots, \ell$ we sum up all such resulting inequalities and, after technical manipulations that involve repetitive application of (7), we obtain:

$$\sum_{q=1}^{\ell} u_q(\mathbf{b}'_{I_q}, \mathbf{b}_{-I_q}) \geq \frac{1}{2} V_I(\mathcal{X}^*) - \sum_{q=1}^{\ell} |Y_q| \cdot \beta_q \qquad (8)$$

where we define: $\beta_q = \max \left\{ \sum_{i \in [n] \setminus I_q} b_i(j) \,\middle|\, j \notin \mathbb{X}^k(\mathbf{b}'_{I_q}, \mathbf{b}_{-I_q}) \right\} \qquad (9)$

Finally, we upper bound $\sum_{q=1}^{\ell} |Y_q| \cdot \beta_q$ by $SW(\mathbb{X}^k(\mathbf{b}))$. $\qquad \square$

We conclude this section with a short note on agents with **UD** valuation functions. An *upper* bound of $O(\lceil n/(k\ell) \rceil)$ and an almost matching lower bound of $\max\{2, \frac{n}{k \cdot \ell}\}$ on the mechanism's ℓ-strong Price of Anarchy follow from Theorems 5 and 6 of [14], concerning the (ℓ-strong) inefficiency of Item Bidding with a Vickrey-Clarke-Groves (VCG) Pricing rule. The reason is that the proofs of these results involve the analysis of bidding profiles for the agents, wherein *every agent issues a non-zero bid on a single item only*. In this case, the VCG-based mechanism of [14] and our Item Pricing rule yield equal payments; thus, the utilities of the agents under the mentioned bidding profiles are also identical and all derivations can be carried in the setting of Item Pricing.

6   Inefficiency Under Incomplete Information

In this section, we examine the performance of the Item Bidding mechanism under the incomplete information setting. Every agent draws his valuation function *independently*, from a distribution $\mathcal{D}_i$, defined over a set of valuation functions $\mathcal{V}_i$. Each agent i is aware of his own distribution $\mathcal{D}_i$ and of the joint distribution $(\times_i \mathcal{D}_i)$. We also allow agents to use mixed strategies. A mixed Bayes strategy, $\mathbb{B}_i$, for agent i, maps each valuation function $v_i \in \mathcal{V}_i$ to a *distribution* $\mathbb{B}_i^{v_i} \equiv \mathbb{B}_i(v_i)$ of bidding strategies, *given v_i*. At Bayes-Nash equilibrium, each agent chooses $\mathbb{B}_i$ so as to maximize his expected utility over the distribution of $\mathbf{v}_{-i}$ – conditionally on v_i – and over the joint distribution of mixed strategies played by all agents.

Several recent works [19, 21, 22] were devoted to the development of a *smoothness analysis framework* – originally conceived by Roughgarden [18] for full information games – for bounding the Price of Anarchy of strategic games induced by auction mechanisms, under incomplete information. The *Bayes-Nash* Price of Anarchy in this setting is the worst-case ratio of the expected socially optimal welfare (over the distribution $\times_i \mathcal{D}_i$) over the expected welfare at Bayes-Nash equilibrium. We adopt a version of the smoothness framework [19] (cf. Definition 3.1), [21] (cf. Definition 3) for the game induced by our Item Bidding mechanism, under incomplete information:

Definition 4. *[19, 21] A Bayesian game induced by a mechanism for the* CPPP *under incomplete information is (λ, μ)-smooth if, for any $\mathbf{v} \in \times_i \mathcal{V}_i$ and for any bidding profile[1] $\mathbf{b}$ there exists a $\mathbf{b}_i^*$ for each agent i, such that:*

$$\sum_{i \in [n]} u_i^{v_i}(\mathbf{b}_i^*, \mathbf{b}_{-i}) \geq \lambda \cdot SW^{\mathbf{v}}(\mathcal{X}^*(\mathbf{v})) - \mu \cdot SW^{\mathbf{v}}(\mathbb{X}^k(\mathbf{b}))$$

where $\mathcal{X}^*(\mathbf{v})$, $u_i^{v_i}(\cdot)$, $SW^{\mathbf{v}}(\cdot)$ are resp. the socially optimal outcome, agents' utilities and the social welfare under the profile $\mathbf{v} \in \times \mathcal{V}_i$ of valuation functions.

---

[1] Following [19, 21], we state the definition w.r.t. pure strategies only for simplicity; however, its consequences are also valid for *mixed* Nash equilibria, as asserted in [21].

Our upper bounds will follow from the main results of [19, 21]:

Theorem 4. *[19, 21] The (mixed) Bayes-Nash Price of Anarchy of a (λ, μ)-smooth Bayesian game is at most $\frac{1+\mu}{\lambda}$.*

In the following Lemma, we identify *deterministic* (pure) bidding strategies for the agents, that fulfill the definition's requirement, for appropriate λ and μ.

Lemma 3. *The Item Bidding mechanism with Item Pricing is:*

- *$(1/2, 2n)$-smooth, for agents with* **XOS** *valuation functions.*
- *$(1, n/k)$-smooth, for agents with* **UD** *valuation functions.*

For **XOS** valuation functions the result follows directly from Lemma 2. By Lemma 3 and Theorem 4:

Corollary 1. *The (mixed) Bayes-Nash Price of Anarchy of the Item Bidding mechanism with Item Pricing is at most: $6n$, for agents with* **XOS** *valuation functions, and $\lceil n/k \rceil$, for agents with* **UD** *valuation functions.*

7   Concluding Remarks

In this work we presented and analyzed an Item Bidding mechanism for the CPPP, incorporating a (per-)Item Pricing payments rule, thus, combining separate bidding on items, with separate pricing. This design differs from the *"pay-as-bid"* Item Bidding mechanism of [13] for the CPPP, as it retains stability properties that are absent from the mechanism of [13].

For agents with **XOS** valuation functions, we established existence of welfare-optimal pure Nash equilibria for our mechanism, and showed that it has ℓ-strong Price of Anarchy at most $O(\lceil n/\ell \rceil)$ and at least $\max\{2, n/\ell\}$. We also discussed how the upper and lower bounds of $O\left(\left\lceil \frac{n}{k \cdot \ell} \right\rceil\right)$ and $\max\{2, \frac{n}{k \cdot \ell}\}$ respectively from [14], for agents with **UD** valuation functions, are valid for our mechanism as well. Under incomplete information, we showed upper bounds of $O(n)$ and $O(\lceil n/k \rceil)$ on the mechanism's Bayes-Nash Price of Anarchy, for agents with **XOS** and **UD** valuation functions respectively. In all, Item Pricing exhibits favorable performance, while retaining optimal stable configurations and separating pricing of *and* bidding on items in contrast to previous works [13, 14].

Let us note that in a recent work [3], Bachrach et al. extended the *smoothness analysis* framework [18, 19, 21, 22] for bounding the strong Price of Anarchy of strategic games, through a definition of *coalitional smoothness* which simply requires the proof of a property reminiscent of (2) in our Lemma 2. However, it requires that the strategy of each agent in any subset of I (see the statement of Lemma 2) is exactly his strategy under $\mathbf{b}_I^*$, whereas we use *proper joint deviations* that may induce a different bid for the same agent in different subsets. In effect, our proof of Lemma 2 is *not* a coalitional smoothness proof in the spirit of [3]. However, the coalitional smoothness notion of [3] may prove a useful tool for reducing the constants involved in our bounds.

References

1. Andelman, N., Feldman, M., Mansour, Y.: Strong price of anarchy. Games Econ. Behav. **65**(2), 289–317 (2009)
2. Aumann, R.: Acceptable points in general cooperative n-person games. In: Tucker, A.W., Luce, R.D. (eds.) Contributions to the Theory of Games, vol. 4. Princeton University Press (1959)
3. Bachrach, Y., Syrgkanis, V., Tardos, É., Vojnović, M.: Strong price of anarchy, utility games and coalitional dynamics. In: Lavi, R. (ed.) SAGT 2014. LNCS, vol. 8768, pp. 218–230. Springer, Heidelberg (2014)
4. Bhawalkar, K., Roughgarden, T.: Welfare guarantees for combinatorial auctions with item bidding. In: Proceedings of the ACM-SIAM Symposium on Disctrete Algorithms (SODA), pp. 700–709 (2011)
5. Buchfuhrer, D., Schapira, M., Singer, Y.: Computation and incentives in combinatorial public projects. In: Proceedings of the ACM Conference on Electronic Commerce (ACM EC), pp. 33–42 (2010)
6. Christodoulou, G., Kovács, A., Schapira, M.: Bayesian combinatorial auctions. In: Aceto, L., Damgård, I., Goldberg, L.A., Halldórsson, M.M., Ingólfsdóttir, A., Walukiewicz, I. (eds.) ICALP 2008, Part I. LNCS, vol. 5125, pp. 820–832. Springer, Heidelberg (2008)
7. Dughmi, S.: A truthful randomized mechanism for combinatorial public projects via convex optimization. In: Proceedings of the ACM Conference on Electronic Commerce (ACM EC), pp. 263–272 (2011)
8. Feige, U.: On maximizing welfare when utility functions are subadditive. SIAM J. Comput. **39**(1), 122–142 (2009)
9. Feldman, M., Fu, H., Gravin, N., Lucier, B.: Simultaneous auctions are (almost) efficient. In: Proceedings of the 45th ACM Symposium on the Theory of Computing (STOC 2013) (2013)
10. Hassidim, A., Kaplan, H., Mansour, Y., Nisan, N.: Non-price equilibria in markets of discrete goods. In: Proceedings of the ACM Conference on Electronic Commerce (EC), pp. 295–296 (2011)
11. de Keijzer, B., Markakis, E., Schäfer, G., Telelis, O.: Inefficiency of standard multiunit auctions. In: Bodlaender, H.L., Italiano, G.F. (eds.) ESA 2013. LNCS, vol. 8125, pp. 385–396. Springer, Heidelberg (2013)
12. Lehmann, B., Lehmann, D.J., Nisan, N.: Combinatorial auctions with decreasing marginal utilities. Games Econ. Behav. **55**(2), 270–296 (2006)
13. Lucier, B., Singer, Y., Syrgkanis, V., Tardos, É.: Equilibrium in combinatorial public projects. In: Chen, Y., Immorlica, N. (eds.) WINE 2013. LNCS, vol. 8289, pp. 347–360. Springer, Heidelberg (2013)
14. Markakis, E., Telelis, O.: Item bidding for combinatorial public projects. In: Proceedings of the 28th Coonference on Artificial Intelligence (AAAI 2014), pp. 749–755 (2014)
15. Nemhauser, G.L., Woolsey, L.A., Fischer, M.L.: An analysis of approximation for maximizing submodular functions – I. Math. Program. **14**, 265–294 (1978)
16. Nisan, N., Ronen, A.: Computationally feasible VCG mechanisms. J. Artif. Intell. Res. **29**, 19–47 (2007)
17. Papadimitriou, C.H., Schapira, M., Singer, Y.: On the hardness of being truthful. In: Proceedings of the Annual IEEE Symposium on Foundations of Computer Science (IEEE FOCS), pp. 250–259 (2008)

18. Roughgarden, T.: Intrinsic robustness of the price of anarchy. In: Proceedings of the ACM Symposium on Theory of Computing (STOC), pp. 513–522 (2009)
19. Roughgarden, T.: The price of anarchy in games of incomplete information. In: Proceedings of the ACM Conference on Electronic Commerce (ACM EC), pp. 862–879. ACM (2012)
20. Schapira, M., Singer, Y.: Inapproximability of combinatorial public projects. In: Papadimitriou, C., Zhang, S. (eds.) WINE 2008. LNCS, vol. 5385, pp. 351–361. Springer, Heidelberg (2008)
21. Syrgkanis, V.: Bayesian Games and the Smoothness Framework. CoRR (ArXiv) abs/1203.5155 (2012)
22. Syrgkanis, V., Tardos, E.: Composable and efficient mechanisms. In: Proceedings of the ACM Symposium on Theory of Computing (ACM STOC), pp. 211–220. ACM (2013)

Norm-Based Locality Measures
of Two-Dimensional Hilbert Curves

H.K. Dai[1]($\boxtimes$) and H.C. Su[2]

[1] Computer Science Department, Oklahoma State University,
Stillwater, OK 74078, USA
dai@cs.okstate.edu
[2] Department of Computer Science, Arkansas State University,
Jonesboro, AR 72401, USA
suh@astate.edu

Abstract. A discrete space-filling curve provides a 1-dimensional indexing or traversal of a multi-dimensional grid space. Applications of space-filling curves include multi-dimensional indexing methods, parallel computing, and image compression. Common goodness-measures for the applicability of space-filling curve families are locality and clustering. Locality reflects proximity preservation that close-by grid points are mapped to close-by indices or vice versa. We present an analytical study on the locality property of the 2-dimensional Hilbert curve family. The underlying locality measure, based on the p-normed metric d_p, is the maximum ratio of $d_p(u, v)^m$ to $d_p(\tilde{u}, \tilde{v})$ over all corresponding point-pairs (u, v) and $(\tilde{u}, \tilde{v})$ in the m-dimensional grid space and 1-dimensional index space, respectively. Our analytical results identify all candidate representative grid-point pairs (realizing the locality-measure values) for all real norm-parameters in the unit interval $[1, 2]$ and grid-orders. Together with the known results for other norm-parameter values, we have almost complete knowledge of the locality measure of 2-dimensional Hilbert curves over the entire spectrum of possible norm-parameter values.

Keywords: Space-filling curves · Hilbert curves · z-order curves · Locality

1 Preliminaries

Discrete space-filling curves have many applications in databases, parallel computation, algorithms, in which linearization techniques of multi-dimensional arrays or grids are needed. Sample applications include heuristics for Hamiltonian traversals, multi-dimensional space-filling indexing methods, image compression, and dynamic unstructured mesh partitioning.

For positive integer n, denote $[n] = \{1, 2, \ldots, n\}$. An m-dimensional (discrete) space-filling curve of length n^m is a bijective mapping $C : [n^m] \to [n]^m$, thus providing a linear indexing/traversal or total ordering of the grid points in $[n]^m$. An m-dimensional grid is said to be of order k if it has side-length $n = 2^k$;

© Springer International Publishing Switzerland 2016
R. Dondi et al. (Eds.): AAIM 2016, LNCS 9778, pp. 14–25, 2016.
DOI: 10.1007/978-3-319-41168-2_2

a space-filling curve has order k if its codomain is a grid of order k. The generation of a sequence of multi-dimensional space-filling curves of successive orders usually follows a recursive framework (on the dimensionality and order), which results in a few classical families, such as Gray-coded curves, Hilbert curves, Peano curves, and z-order curves.

One of the salient characteristics of space-filling curves is their "self-similarity". Denote by H_k^m and Z_k^m an m-dimensional Hilbert and z-order, respectively, space-filling curve of order k. Figure 1 illustrates the recursive constructions of H_k^m and Z_k^m for $m = 2$, and $k = 1, 2$, and $m = 3$, and $k = 1$.

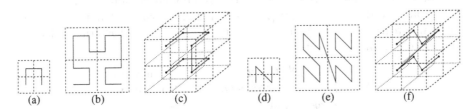

(a)   (b)   (c)   (d)   (e)   (f)

Fig. 1. Recursive constructions of Hilbert and z-order curves of higher order (respectively, H_k^m and Z_k^m) by interconnecting symmetric subcurves, via reflection and/or rotation, of lower order (respectively, H_{k-1}^m and Z_{k-1}^m) along an order-1 subcurve (respectively, H_1^m and Z_1^m): (a) H_1^2; (b) H_2^2; (c) H_1^3; (d) Z_1^2; (e) Z_2^2; (f) Z_1^3.

We measure the applicability of a family of space-filling curves based on: (1) their common structural characteristics that reflect locality and clustering, (2) descriptional simplicity that facilitates their construction and combinatorial analysis in arbitrary dimensions, and (3) computational complexity in the grid space-index space transformation. Locality preservation reflects proximity between the grid points of $[n]^m$, that is, close-by points in $[n]^m$ are mapped to close-by indices/numbers in $[n^m]$, or vice versa. Clustering performance measures the distribution of continuous runs of grid points (clusters) over identically shaped subspaces of $[n]^m$, which can be characterized by the average number of clusters and the average inter-cluster distance (in $[n^m]$) within a subspace.

Empirical and analytical studies of clustering performances of various low-dimensional space-filling curves have been reported in the literature (see [4] and [6] for details). These studies show that the Hilbert and z-order curve families manifest good data clustering properties according to some quality clustering measures, robust mathematical formalism, and viable indexing techniques for querying multi-dimensional data, when compared with other curve families.

The locality preservation of a space-filling curve family is crucial for the efficiency of many indexing schemes, data structures, and algorithms in its applications, for examples, spatial correlation in multi-dimensional indexings, compression in image processing, and communication optimization in mesh-connected parallel computing. To analyze locality, we need to rigorously define its measures that are practical – good bounds (lower and upper) on the locality measure translate into good bounds on the declustering (locality loss) in one space in the presence of locality in the other space.

A few locality measures have been proposed and analyzed for space-filling curves in the literature. Denote by d and d_p the Euclidean metric and p-normed metric (rectilinear metric ($p = 1$) and maximum metric ($p = \infty$)), respectively. Let $\mathcal{C}$ denote a family of m-dimensional curves of successive orders.

We [5] consider a locality measure conditional on a 1-normed distance of δ between points in $[n]^m$:

$$L_\delta(C) = \sum_{i,j \in [n^m] \mid i < j \text{ and } d_1(C(i), C(j)) = \delta} |i - j| \text{ for } C \in \mathcal{C}.$$

They derive exact formulas for L_δ for the Hilbert curve family $\{H_k^m \mid k = 1, 2, \ldots\}$ and z-order curve family $\{Z_k^m \mid k = 1, 2, \ldots\}$ for $m = 2$ and arbitrary δ that is an integral power of 2, and $m = 3$ and $\delta = 1$ (lower-order terms collected in asymptotic form for brevity):

$$L_\delta(H_k^2) = \begin{cases} \frac{17}{2 \cdot 7} \cdot 2^{3k} + O(2^{2k}) & \text{if } \delta = 1 \\ \frac{17}{2 \cdot 7} \cdot 2^{3k + 2 \log \delta} + O(2^{2k + 3 \log \delta}) & \text{otherwise,} \end{cases}$$

$$L_\delta(Z_k^2) = \begin{cases} 2^{3k} + O(2^k) & \text{if } \delta = 1 \\ 2^{3k + 2 \log \delta} + O(2^{2k + 3 \log \delta}) & \text{otherwise;} \end{cases}$$

$$L_1(H_k^3) = \frac{67}{2 \cdot 31} \cdot 2^{5k} + O(2^{3k}) \text{ and } L_1(Z_k^3) = 2^{5k} + O(2^{2k}).$$

With respect to the locality measure L_δ and for sufficiently large k and $\delta \ll 2^k$, the z-order curve family performs better than the Hilbert curve family for $m = 2$ and over the δ-spectrum of integral powers of 2. When $\delta = 2^k$, the domination reverses. The superiority of the z-order curve family persists but declines for $m = 3$ with unit 1-normed distance for L_δ.

For measuring the proximity preservation of close-by points in the indexing space $[n^m]$, Gotsman and Lindenbaum [7] consider the following measures: for $C \in \mathcal{C}$,

$$L_{\min}(C) = \min_{i,j \in [n^m] \mid i < j} \frac{d(C(i), C(j))^m}{|i - j|} \text{ and } L_{\max}(C) = \max_{i,j \in [n^m] \mid i < j} \frac{d(C(i), C(j))^m}{|i - j|}.$$

Alber and Niedermeier [1] generalize $L_{\max}$ to L_p by employing the p-normed metric d_p for real norm-parameter $p \geq 1$ in place of the Euclidean metric d, which is the locality measure studied in our work (and [5]). We summarize below: (1) the representative lower- and upper-bound results and exact formulas for the locality measure L_p of the 2-dimensional Hilbert curve family H_k^2 for various norm-parameter p-values and grid-order k-values, and (2) the contribution of our studies:

1. For $p = 1$: Niedermeier, Reinhardt, and Sanders [8] give a lower bound for $L_1(H_k^2)$: for all $k \geq 1$,

$$L_1(H_k^2) \geq \frac{(3 \cdot 2^{k-1} - 2)^2}{4^{k-1}},$$

and Chochia et al. [3] provide a matching upper bound for $L_1(H_k^2)$ for all $k \geq 2$. We [5] also provide the exact formula for $L_1(H_k^2)$ for all $k \geq 2$.

2. For $p = 2$: Gotsman and Lindenbaum [7] derive a lower and upper bounds for $L_2(H_k^2)$: for all $k \geq 6$,

$$\frac{(2^{k-1} - 1)^2}{\frac{2}{3} \cdot 4^{k-2} + \frac{1}{3}} \leq L_2(H_k^2) \leq 6\frac{2}{3},$$

and Alber and Niedermeier [1] improves the upper bound for $L_2(H_k^2)$: for all $k \geq 1$,

$$L_2(H_k^2) \leq 6\frac{1}{2}.$$

We [5] prove that the lower bound above [7] is the exact formula for $L_2(H_k^2)$: for all $k \geq 5$,

$$L_2(H_k^2) = 6 \cdot \frac{2^{2k-3} - 2^{k-1} + 2^{-1}}{2^{2k-3} + 1}.$$

Bauman [2] obtains a matching lower and upper bounds for $L_2(H_k^2)$ for $k = \infty$:

$$L_2(H_\infty^2) = 6.$$

3. For $2 < p \leq \infty$: Due to the monotonicity of the underlying p-normed metric: for every grid-point pair (v, u), the p-normed metric $d_p(v, u)$ is strictly decreasing in $p \in [1, \infty)$, we [5] prove the same exact formula for $L_p(H_k^2)$ as for the case when $p = 2$:

$$L_p(H_k^2) = 6 \cdot \frac{2^{2k-3} - 2^{k-1} + 2^{-1}}{2^{2k-3} + 1} \text{ for all reals } p \geq 2.$$

When $p = \infty$, Alber and Niedermeier [1] establish a lower and upper bounds for $L_\infty(H_k^2)$, respectively:

$$6(1 - O(2^{-k})) \leq L_\infty(H_k^2) \leq 6\frac{2}{5}.$$

Our proofs of the exact formulas of $L_p(H_k^2)$ for $p \in \{1, 2\}$ in [5] follow a uniform approach: identifying all the representative grid-point pairs, which realize the $L_p(H_k^2)$-value, for each $p \in \{1, 2\}$. The analytical results close the gap between the current best lower and upper bounds with exact formulas for $p \in \{1, 2\}$, and extend to all reals $p \geq 2$. The identifications of candidate representative grid-point pairs rely on sequences of reduction. A reduction of a grid-point pair to another pair is based on the dominance of the underlying locality-measure values of the corresponding grid-point pairs. The geometric characteristics of the underlying p-norms (rectilinear and Euclidean metrics of $p = 1$ and $p = 2$, respectively) help distinguish candidate representative grid-point pairs and verify tedious reductions.

Our study of 2-dimensional curve family H_k^2 is focused on the exact analysis of $L_p(H_k^2)$ for all reals $p \in [1, 2]$. The intrinsic mathematical appeal in completing the computation of $L_p(H_k^2)$ for all possible norm-parameters p is our primary motivation. While the three most obviously important p-values: $\{1, 2, \infty\}$ are intimately related to intuitive concepts, in some cases the structure of applications of the Hilbert curves may suggest a different choice of p-value as the most natural setting for the underlying locality measure.

We present analytical and empirical studies on the locality measure L_p for the 2-dimensional Hilbert curve family for all reals $p \in [1, 2]$. The underlying locality measure L_p, based on the p-normed metric d_p, is the maximum ratio of $d_p(u, v)^m$ to $d_p(\tilde{u}, \tilde{v})$ over all corresponding point-pairs (u, v) and $(\tilde{u}, \tilde{v})$ in the m-dimensional grid space and (1-dimensional) index space, respectively:

1. We identify all the candidate representative grid-point pairs for all norm-parameter p-values in $[1, 2]$ and grid-order k-values. Together with the known results for other norm-parameter values, we have almost complete knowledge of $L_p(H_k^2)$ over the entire spectrum of possible norm-parameter values.
2. Our empirical study, which complements the analytical ones, shows that: (1) The analytical results are consistent with program verification over various norm-parameter p-values and sufficiently large grid-order k-values, and (2) As p increases over the real unit interval $[1, 2]$, the locations of candidate representative grid-point pairs agree with the intuitive interpolation effect over the two delimiting p-values.
3. A practical implication of our results on $L_p(H_k^2)$ is that the exact formulas provide good bounds on measuring the loss in data locality in the index space, while spatial correlation exists in the 2-dimensional grid space.

We present a high-level approach to the main results without any derivations and proofs, supplemented with an empirical study that verifies the analytical results for various p-values and sufficiently large k-values. Complete results: illustrated figures, derivations, and proofs, and verifying computer programs are available from the authors.

2   Analytical Studies of $L_p(H_k^2)$ with $p \in [1, 2]$

For 2-dimensional Hilbert curves, the self-similar structural property guides us to decompose H_k^2 into four identical H_{k-1}^2-subcurves (via reflection and rotation), which are amalgamated together by an H_1^2-curve. Following the linear order along this H_1^2-curve, we denote the four H_{k-1}^2-subcurves (quadrants) as $Q_1(H_k^2)$, $Q_2(H_k^2)$, $Q_3(H_k^2)$, and $Q_4(H_k^2)$. We extend the notion to identify all H_l^m-subcurves of a structured H_k^m for all $l \in [k]$ inductively on the order in an obvious manner.

For a space-filling curve C indexing an m-dimensional grid space, the notation "$v \in C$" refers to "grid point v indexed by C", and $C^{-1}(v)$ gives the index of v in the 1-dimensional index space. The locality measure in our study is, for all reals $p \geq 1$,

$$L_p(C) = \max_{\text{indices } i,j \in [n^m]} \frac{d_p(C(i), C(j))^m}{d_p(i, j)} = \max_{v,u \in C} \frac{d_p(v, u)^m}{|C^{-1}(v) - C^{-1}(u)|}.$$

When $m = 2$, we write $\mathcal{L}_{C,p}(u, v) = \frac{d_p(u,v)^2}{|C^{-1}(v) - C^{-1}(u)|}$.

For subcurves C_1, C_2, C_1', and C_2' of C, a grid-point pair $(v_1, v_2) \in C_1 \times C_2$ is reducible to a grid-point pair $(v_1', v_2') \in C_1' \times C_2'$ if $\mathcal{L}_{C,p}(v_1, v_2) \leq \mathcal{L}_{C,p}(v_1', v_2')$ – denoted by $(v_1, v_2) \preceq (v_1', v_2')$, and subcurve pair $C_1 \times C_2$ is reducible to subcurve pair $C_1' \times C_2'$ if for every $(v_1, v_2) \in C_1 \times C_2$, there exists $(v_1', v_2') \in C_1' \times C_2'$ such that (v_1, v_2) is reducible to (v_1', v_2') – denoted by $C_1 \times C_2 \preceq C_1' \times C_2'$. We define the strict reducibility, denoted by $\prec$, for grid-point pairs and subcurve pairs via the strict inequality of $\mathcal{L}_{C,p}$-values in an obvious manner.

A pair of grid points v and u indexed by C is representative for C with respect to L_p if $\mathcal{L}_{C,p}(v, u) = L_p(C)$, or, equivalently, for all $v', u' \in C$, $(v', u') \preceq (v, u)$. The identifications of candidate representative grid-point pairs for C often involve sequences of reductions – successive considerations of two grid-point pairs and the comparisons of their $\mathcal{L}_{C,p}$-values. Our studies of $L_p(H_k^2)$ cover all norm-parameters $p \geq 1$. However, for all reals $p \in (1, 2)$, the lack of geometric clarity for interpreting L_p-values can adversely increase the complexity: (1) of identifying candidate representative grid-point pairs, and (2) in comparing $\mathcal{L}_{H_k^2, p}$-values for reductions due to the complex interplay of the norm-parameter p-value and grid-order k-value.

2.1   Reductions of Grid-Point Pairs and Subcurve Pairs

For two grid-point pairs (v_1, v_2) and (v_1', v_2') (two subcurve pairs $C_1 \times C_2$ an $C_1' \times C_2'$) of H_k^2, the reduction $(v_1, v_2) \preceq (v_1', v_2')$ ($C_1 \times C_2 \preceq C_1' \times C_2'$, respectively) eliminates (v_1, v_2) ($C_1 \times C_2$, respectively) from the candidacy for representative grid-point pairs. We develop various sufficient conditions for reduction with an example below.

For the grid space $[2^k]^2$ of a 2-dimensional Hilbert curve H_k^2 with a referenced (x, y)-coordinate system (with origin $(1, 1)$) in a canonical orientation (see Fig. 1(a) and (b)), we denote the x- and y-coordinates of a grid point v by $x(v)$ and $y(v)$, respectively.

Lemma 1. *For all norm-parameters $p \in [1, 2]$ and three arbitrary grid points $u, v, v' \in H_k^2$ such that: (1) the sequence of three grid points: (u, v, v') is in indexing order (that is, $(H_k^2)^{-1}(u) \leq (H_k^2)^{-1}(v) \leq (H_k^2)^{-1}(v')$ or $(H_k^2)^{-1}(u) \geq (H_k^2)^{-1}(v) \geq (H_k^2)^{-1}(v'))$, and (2) the two sequences of their x- and y-coordinates: $(x(u), x(v), x(v'))$ and $(y(u), y(v), y(v'))$ have the same monotone property (both increasing or both decreasing), if $|(H_k^2)^{-1}(u) - (H_k^2)^{-1}(v)|(2|x(u) - x(v)||x(v) - x(v')| + |x(v) - x(v')|^2 + 2|y(u) - y(v)||y(v) - y(v')| + |y(v) - y(v')|^2) - |(H_k^2)^{-1}(v) - (H_k^2)^{-1}(v')|(|x(u) - x(v)| + |y(u) - y(v)|)^2 \geq 0 \ (> 0)$, then $(u, v) \preceq (u, v') \ ((u, v) \prec (u, v'))$ via $\mathcal{L}_{H_k^2, p}(u, v) \leq \mathcal{L}_{H_k^2, p}(u, v')$ $(\mathcal{L}_{H_k^2, p}(u, v) < \mathcal{L}_{H_k^2, p}(u, v')$, respectively).*
Note that the sufficient condition for the reduction is independent of the p-value for $\mathcal{L}_{H_k^2, p}$.

For reductions of grid-point pairs, we mostly use various p-independence sufficient conditions as the one in Lemma 1. For reductions of subcurve pairs, simple ones are realized by symmetry arguments with regard to relative subcurve-orientations or succinct geometric interpretations of the $\mathcal{L}_{H_k^2,p}$-computation if possible.

For subcurves in the form of nested subquadrants of H_k^2, we may prove the reduction between subcurve pairs $C_1 \times C_2 \preceq C_1' \times C_2'$ with a divide-and-conquer approach by considering all possible reductions between quadrant-subcurve pairs $Q_{i_1}(C_1) \times Q_{i_2}(C_2)$ (for all $i_1, i_2 \in [4]$) to $Q_{j_1}(C_1') \times Q_{j_2}(C_2')$ (for some $j_1, j_2 \in [4]$). Some reductions of quadrant-subcurve pairs may be resolved by simple symmetry/geometric arguments, while others may entail further reductions of subquadrant-subcurve pairs. These nested reductions generally arrive at some forms of recursive patterns, and mathematical induction is applied to resolve the reductions.

2.2   Identification of Candidate Representative Grid-Point Pairs

The upper-bound argument [5] in establishing the exact formulas for $L_p(H_k^2)$ for $p \in \{1, 2\}$ does not translate into a viable application for $p \in (1, 2)$. For identifying all possible candidate representative grid-point pairs in H_k^2, we consider all grid-point pairs in $Q_i(H_k^2) \times Q_j(H_k^2)$ with $1 \leq i < j \leq 4$ and their possible systematic reductions. Due to a simple reduction ($Q_1(H_k^2) \times Q_4(H_k^2) \preceq Q_2(H_k^2) \times Q_3(H_k^2)$) and geometric symmetry ($Q_2(H_k^2) \times Q_4(H_k^2)$ to $Q_1(H_k^2) \times Q_3(H_k^2)$ and $Q_3(H_k^2) \times Q_4(H_k^2)$ to $Q_1(H_k^2) \times Q_2(H_k^2)$), three cases remain: $Q_1(H_k^2) \times Q_2(H_k^2)$, $Q_1(H_k^2) \times Q_3(H_k^2)$, and $Q_2(H_k^2) \times Q_3(H_k^2)$. An involved analysis of $Q_1(H_k^2) \times Q_3(H_k^2)$ reveals that the quadrant-subcurve pair is void of any candidate representative grid-point pairs.

We summarize the findings below in Theorem 1, in which the sources of (candidate) representative grid-point pairs (named A, B, and C) are illustrated in Fig. 2 and elaborated with (local) (x, y)-coordinates and $\mathcal{L}_{H_k^2,p}$-values in Table 1. For brevity we omit the symmetry ones.

Theorem 1. *Consider the following cases determined by the interplay of the grid-order $k \geq 1$ and norm-parameter $p \in [1, 2]$ of H_k^2:*

1. *Case when $k = 1$:*
   For all $p \in [1, 2)$: One representative grid-point pair with coordinates $((1, 1), (2^k, 2^k))$ and its symmetry.
   For $p = 2$: Three representative grid-point pairs with coordinates $((1, 1), (1, 2^k))$, $((1, 1), (2^k, 2^k))$, and $((1, 2^k), (2^k, 2^k))$, and their symmetries.
2. *Case when $k \in \{2, 3\}$:*
   For all $p \in [1, 2]$: One representative grid-point pair B and its symmetry.
3. *Case when $k = 4$: The p-interval $[1, 2]$ is decomposed into two p-subintervals: $[1, \rho)$ and $(\rho, 2]$, where $\rho \approx 1.825$.*
   For all $p \in [1, \rho)$: One representative grid-point pair B and its symmetry.
   For all $p \in (\rho, 2]$: One representative grid-point pair A and its symmetry.
   For $p = \rho$: Two representative grid-point pairs B and A, and their symmetries.

4. *Case when $k \geq 5$: For all $p \in [1,2]$: $1+(k-2)+(k-4) = 2k-5$ candidate representative grid-point pairs $B, C_1, D_1, C_2, \ldots, C_{k-5}, D_{k-5}, C_{k-4}, D_{k-4}, C_{k-3}, C_{k-2}$, and their symmetries.*

Refined analysis with further reductions eliminates $D_1, \ldots, D_{k-5}$ and D_{k-4}, and their symmetries from the candidacy for representative grid-point pairs.

Theorem 2. *For all grid-orders $k \geq 5$ and norm-parameters $p \in [1,2]$ of H_k^2, the candidate representative grid-point pairs are $B, C_1, C_2, \ldots, C_{k-5}, C_{k-4}, D_{k-4}, C_{k-3}, C_{k-2}$, and their symmetries.*

For all norm-parameters $p \in [1,2]$, there exists a sufficiently large grid-order $k_0 \geq 5$ such that for all grid-orders $k \geq k_0$ of H_k^2, the candidate representative grid-point pairs are $B, C_1, C_2, \ldots, C_{k-5}, C_{k-4}, C_{k-3}, C_{k-2}$, and their symmetries.

Our future work will be focused on establishing analytically the association of representative grid-point pairs in $\{B, C_1, C_2, \ldots, C_{k-5}, C_{k-4}, C_{k-3}, C_{k-2}\}$ with their $\mathcal{L}_{H_k^2, p}$-dominance p-subintervals and relevant grid-orders.

3 Empirical Study on $L_p(H_k^2)$ with $p \in [1,2]$

To complement the analytical results for $L_p(H_k^2)$ for all reals $p \in [1,2]$, we conduct an empirical study on $L_p(H_k^2)$ for all $k \in \{2,3,\ldots,12\}$ and some reals $p \in [1,2]$. We cover the grid space $[2^k]^2$ of a 2-dimensional Hilbert curve H_k^2 in a canonical orientation with Cartesian coordinates: 2^k columns (respectively, rows) indexed by x-coordinates (respectively, y-coordinates) $1, 2, \ldots, 2^k$. For every grid-order $k \in \{2,3,\ldots,12\}$ and real $p \in [1,2]$ with granularity of 0.01 (for $2 \leq k \leq 12$), we locate with computer programs all representative grid-point pairs for H_k^2 with respect to L_p. Figure 2(a) illustrates the three sources $\{A, B, C\}$ of candidate representative grid-point pairs for $k \geq 2$.

Source A identifies the grid-point pair $(u_A, v_A) = ((1, \frac{1}{4} \cdot 2^k + 1), (1, 2^k))$ and its symmetry. The pair (u_A, v_A) serves as the representative grid-point pair "briefly" – for $k = 4$ and $1.83 \leq p \leq 2.00$.

Source B identifies the grid-point pair $(u_B, v_B) = ((2^{k-1}, 1), (1, 2^k))$ and its symmetry. The pair (u_B, v_B) serves as the representative grid-point pair for every $k \in \{2, 3, \ldots, 12\}$ and all reals p of a (shrinking) prefix-interval $[1, \rho_k) \subseteq [1, 2]$ – with ρ_k decreasing as k increases.

Source C identifies a sequence $(C_1, C_2, \ldots, C_{k-2})$ of grid-point pairs:

$$C_t = (u_{C_t}, v_{C_t}) = ((\frac{1}{4} \cdot 2^k + 1, 2^{k-1} + 1), (\frac{3}{4} \cdot 2^k, 2^{k-1} + 2^{k-2-t})),$$

for $t = 1, 2, \ldots, k-2$, and their symmetries, with:

$$x(v_{C_{t+1}}) = x(v_{C_t}) \text{ and } y(v_{C_{t+1}}) - 2^{k-1} = \frac{y(v_{C_t}) - 2^{k-1}}{2},$$

and eventually v_{C_t} converges to $v_{C_{k-2}}$. Note that, for $t = 0$, the grid-point pair $C_0 = (u_{C_0}, v_{C_0}) = ((\frac{1}{4} \cdot 2^k + 1, 2^{k-1} + 1), (\frac{3}{4} \cdot 2^k, 2^{k-1} + 2^{k-2}))$ is not included in C since C_0 can not be a candidate representative grid-point pair (for any k and real $p \in [1, 2]$):

$$\mathcal{L}_{H_k^2, p}(u_B, v_B) = \frac{((2^{k-1} - 1)^p + (2^k - 1)^p)^{\frac{2}{p}}}{2^{2k-2}}$$

$$> \mathcal{L}_{H_k^2, p}(u_{C_0}, v_{C_0}) = \frac{((2^{k-1} - 1)^p + (2^{k-2} - 1)^p)^{\frac{2}{p}}}{\frac{1}{3} \cdot 2^{2k-3} + \frac{1}{3} \cdot 2^{2k-4}}.$$

Empirically, for all $k \in \{5, 6, \ldots, 12\}$ and all reals p of the (growing) suffix-interval $(\rho_k, 2] \subseteq [1, 2]$, all the representative grid-point pairs form a subsequence C' of C composed of: (1) a prefix of C and (2) $(u_{C_{k-2}}, v_{C_{k-2}})$. The suffix-interval $(\rho_k, 2]$ is partitioned into disjoint successive p-subintervals, each of which supports a grid-point pair in the subsequence C' as the representative grid-point pair for H_k^2 (for all reals p of the subinterval). The length of C' (number of all representative grid-point pairs from the source C) should depend on k in general, and on the p-granularity in our empirical setting. Figure 2(b) depicts the sequence of candidate representative grid-point pairs from the source C.

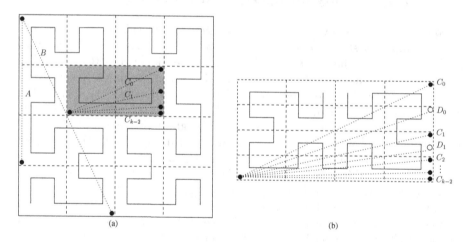

(a)   (b)

Fig. 2. Candidate representative grid-point pairs for H_k^2 with respect to L_p for $k \geq 2$: (a) three sources $\{A, B, C\}$ of candidate representative grid-point pairs; (b) detailed view of the source C.

Table 1 tabulates: (1) for each $k \in \{2, 3, \ldots, 12\}$, the partitioning p-subintervals of $[1, 2]$, and the corresponding representative grid-point pair and its source; and (2) $\mathcal{L}_{H_k^2, p}(u, v)$ ($= L_p(H_k^2)$) for a representative grid-point pair (u, v) in the three sources A, B, and C:

Table 1. Representative grid-point pairs for H_k^2 with respect to L_p for $k \in \{2, 3, \ldots, 12\}$ and $p \in [1.00, 2.00]$ with granularity of 0.01

k	p	(x,y)-coordinates	representative grid-point pair coordinates in terms of k	source
2	$[1.00, 2.00]$	$((2,1),(1,4))$	$((2^{k-1},1),(1,2^k))$	B
3	$[1.00, 2.00]$	$((4,1),(1,8))$	$((2^{k-1},1),(1,2^k))$	B
4	$[1.00, 1.82]$	$((8,1),(1,16))$	$((2^{k-1},1),(1,2^k))$	B
	$[1.83, 2.00]$	$((1,5),(1,16))$	$((1,\frac{1}{4}\cdot 2^k+1),(1,2^k))$	A
5	$[1.00, 1.61]$	$((16,1),(1,32))$	$((2^{k-1},1),(1,2^k))$	B
	$[1.62, 2.00]$	$((9,17),(24,17))$	$((\frac{1}{4}\cdot 2^k+1,2^{k-1}+1),(\frac{3}{4}\cdot 2^k,2^{k-1}+1))$	C_3
6	$[1.00, 1.51]$	$((32,1),(1,64))$	$((2^{k-1},1),(1,2^k))$	B
	$[1.52, 1.55]$	$((17,33),(48,40))$	$((\frac{1}{4}\cdot 2^k+1,2^{k-1}+1),(\frac{3}{4}\cdot 2^k,2^{k-1}+2^{k-3}))$	C_1
	$[1.56, 1.60]$	$((17,33),(48,36))$	$((\frac{1}{4}\cdot 2^k+1,2^{k-1}+1),(\frac{3}{4}\cdot 2^k,2^{k-1}+2^{k-4}))$	C_2
	$[1.61, 2.00]$	$((17,33),(48,33))$	$((\frac{1}{4}\cdot 2^k+1,2^{k-1}+1),(\frac{3}{4}\cdot 2^k,2^{k-1}+1))$	C_4
7	$[1.00, 1.41]$	$((64,1),(1,128))$	$((2^{k-1},1),(1,2^k))$	B
	$[1.42, 1.57]$	$((33,65),(96,80))$	$((\frac{1}{4}\cdot 2^k+1,2^{k-1}+1),(\frac{3}{4}\cdot 2^k,2^{k-1}+2^{k-3}))$	C_1
	$[1.58, 1.66]$	$((33,65),(96,72))$	$((\frac{1}{4}\cdot 2^k+1,2^{k-1}+1),(\frac{3}{4}\cdot 2^k,2^{k-1}+2^{k-4}))$	C_2
	$[1.67, 1.67]$	$((33,65),(96,68))$	$((\frac{1}{4}\cdot 2^k+1,2^{k-1}+1),(\frac{3}{4}\cdot 2^k,2^{k-1}+2^{k-5}))$	C_3
	$[1.68, 2.00]$	$((33,65),(96,65))$	$((\frac{1}{4}\cdot 2^k+1,2^{k-1}+1),(\frac{3}{4}\cdot 2^k,2^{k-1}+1))$	C_5
8	$[1.00, 1.36]$	$((128,1),(1,256))$	$((2^{k-1},1),(1,2^k))$	B
	$[1.37, 1.57]$	$((65,129),(192,160))$	$((\frac{1}{4}\cdot 2^k+1,2^{k-1}+1),(\frac{3}{4}\cdot 2^k,2^{k-1}+2^{k-3}))$	C_1
	$[1.58, 1.68]$	$((65,129),(192,144))$	$((\frac{1}{4}\cdot 2^k+1,2^{k-1}+1),(\frac{3}{4}\cdot 2^k,2^{k-1}+2^{k-4}))$	C_2
	$[1.69, 1.72]$	$((65,129),(192,136))$	$((\frac{1}{4}\cdot 2^k+1,2^{k-1}+1),(\frac{3}{4}\cdot 2^k,2^{k-1}+2^{k-5}))$	C_3
	$[1.73, 2.00]$	$((65,129),(192,129))$	$((\frac{1}{4}\cdot 2^k+1,2^{k-1}+1),(\frac{3}{4}\cdot 2^k,2^{k-1}+1))$	C_6
9	$[1.00, 1.33]$	$((256,1),(1,512))$	$((2^{k-1},1),(1,2^k))$	B
	$[1.34, 1.58]$	$((129,257),(384,320))$	$((\frac{1}{4}\cdot 2^k+1,2^{k-1}+1),(\frac{3}{4}\cdot 2^k,2^{k-1}+2^{k-3}))$	C_1
	$[1.59, 1.69]$	$((129,257),(384,288))$	$((\frac{1}{4}\cdot 2^k+1,2^{k-1}+1),(\frac{3}{4}\cdot 2^k,2^{k-1}+2^{k-4}))$	C_2
	$[1.70, 1.75]$	$((129,257),(384,272))$	$((\frac{1}{4}\cdot 2^k+1,2^{k-1}+1),(\frac{3}{4}\cdot 2^k,2^{k-1}+2^{k-5}))$	C_3
	$[1.76, 1.77]$	$((129,257),(384,264))$	$((\frac{1}{4}\cdot 2^k+1,2^{k-1}+1),(\frac{3}{4}\cdot 2^k,2^{k-1}+2^{k-6}))$	C_4
	$[1.78, 2.00]$	$((129,257),(384,257))$	$((\frac{1}{4}\cdot 2^k+1,2^{k-1}+1),(\frac{3}{4}\cdot 2^k,2^{k-1}+1))$	C_7
10	$[1.00, 1.32]$	$((512,1),(1,1024))$	$((2^{k-1},1),(1,2^k))$	B
	$[1.33, 1.58]$	$((257,513),(768,640))$	$((\frac{1}{4}\cdot 2^k+1,2^{k-1}+1),(\frac{3}{4}\cdot 2^k,2^{k-1}+2^{k-3}))$	C_1
	$[1.59, 1.70]$	$((257,513),(768,576))$	$((\frac{1}{4}\cdot 2^k+1,2^{k-1}+1),(\frac{3}{4}\cdot 2^k,2^{k-1}+2^{k-4}))$	C_2
	$[1.71, 1.76]$	$((257,513),(768,544))$	$((\frac{1}{4}\cdot 2^k+1,2^{k-1}+1),(\frac{3}{4}\cdot 2^k,2^{k-1}+2^{k-5}))$	C_3
	$[1.77, 1.79]$	$((257,513),(768,528))$	$((\frac{1}{4}\cdot 2^k+1,2^{k-1}+1),(\frac{3}{4}\cdot 2^k,2^{k-1}+2^{k-6}))$	C_4
	$[1.80, 1.80]$	$((257,513),(768,520))$	$((\frac{1}{4}\cdot 2^k+1,2^{k-1}+1),(\frac{3}{4}\cdot 2^k,2^{k-1}+2^{k-7}))$	C_5
	$[1.81, 2.00]$	$((257,513),(768,513))$	$((\frac{1}{4}\cdot 2^k+1,2^{k-1}+1),(\frac{3}{4}\cdot 2^k,2^{k-1}+1))$	C_8
11	$[1.00, 1.31]$	$((1024,1),(1,2048))$	$((2^{k-1},1),(1,2^k))$	B
	$[1.32, 1.58]$	$((513,1025),(1536,1280))$	$((\frac{1}{4}\cdot 2^k+1,2^{k-1}+1),(\frac{3}{4}\cdot 2^k,2^{k-1}+2^{k-3}))$	C_1
	$[1.59, 1.70]$	$((513,1025),(1536,1152))$	$((\frac{1}{4}\cdot 2^k+1,2^{k-1}+1),(\frac{3}{4}\cdot 2^k,2^{k-1}+2^{k-4}))$	C_2
	$[1.71, 1.76]$	$((513,1025),(1536,1088))$	$((\frac{1}{4}\cdot 2^k+1,2^{k-1}+1),(\frac{3}{4}\cdot 2^k,2^{k-1}+2^{k-5}))$	C_3
	$[1.77, 1.80]$	$((513,1025),(1536,1056))$	$((\frac{1}{4}\cdot 2^k+1,2^{k-1}+1),(\frac{3}{4}\cdot 2^k,2^{k-1}+2^{k-6}))$	C_4
	$[1.81, 1.82]$	$((513,1025),(1536,1040))$	$((\frac{1}{4}\cdot 2^k+1,2^{k-1}+1),(\frac{3}{4}\cdot 2^k,2^{k-1}+2^{k-7}))$	C_5
	$[1.83, 2.00]$	$((513,1025),(1536,1025))$	$((\frac{1}{4}\cdot 2^k+1,2^{k-1}+1),(\frac{3}{4}\cdot 2^k,2^{k-1}+1))$	C_9
12	$[1.00, 1.31]$	$((2048,1),(1,4096))$	$((2^{k-1},1),(1,2^k))$	B
	$[1.32, 1.58]$	$((1025,2049),(3072,2560))$	$((\frac{1}{4}\cdot 2^k+1,2^{k-1}+1),(\frac{3}{4}\cdot 2^k,2^{k-1}+2^{k-3}))$	C_1
	$[1.59, 1.70]$	$((1025,2049),(3072,2304))$	$((\frac{1}{4}\cdot 2^k+1,2^{k-1}+1),(\frac{3}{4}\cdot 2^k,2^{k-1}+2^{k-4}))$	C_2
	$[1.71, 1.77]$	$((1025,2049),(3072,2176))$	$((\frac{1}{4}\cdot 2^k+1,2^{k-1}+1),(\frac{3}{4}\cdot 2^k,2^{k-1}+2^{k-5}))$	C_3
	$[1.78, 1.81]$	$((1025,2049),(3072,2112))$	$((\frac{1}{4}\cdot 2^k+1,2^{k-1}+1),(\frac{3}{4}\cdot 2^k,2^{k-1}+2^{k-6}))$	C_4
	$[1.82, 1.83]$	$((1025,2049),(3072,2080))$	$((\frac{1}{4}\cdot 2^k+1,2^{k-1}+1),(\frac{3}{4}\cdot 2^k,2^{k-1}+2^{k-7}))$	C_5
	$[1.84, 1.84]$	$((1025,2049),(3072,2064))$	$((\frac{1}{4}\cdot 2^k+1,2^{k-1}+1),(\frac{3}{4}\cdot 2^k,2^{k-1}+2^{k-8}))$	C_6
	$[1.85, 2.00]$	$((1025,2049),(3072,2049))$	$((\frac{1}{4}\cdot 2^k+1,2^{k-1}+1),(\frac{3}{4}\cdot 2^k,2^{k-1}+1))$	C_{10}

$$
\mathcal{L}_{H_k^2,p}(u,v) = \begin{cases} \dfrac{(3\cdot 2^{k-2}-1)^2}{\frac{5}{3}\cdot 2^{2k-4}+\frac{1}{3}} & \text{if } (u,v) \text{ is in } A \\[2ex] \dfrac{((2^{k-1}-1)^p+(2^k-1)^p)^{\frac{2}{p}}}{2^{2k-2}} & \text{if } (u,v) \text{ is in } B \\[2ex] \dfrac{((2^{k-1}-1)^p+(2^{k-2-t}-1)^p)^{\frac{2}{p}}}{\frac{1}{3}\cdot 2^{2k-3}+\frac{1}{3}\cdot 2^{2k-4-2t}} & \text{if } (u,v) = (u_{C_t},v_{C_t}) \text{ in } C, \\ & \text{where } t = 1,2,\ldots,k-2. \end{cases}
$$

Figure 3(a) and (b) show the graphs, using the mathematical software Maple, of the locality measure $\mathcal{L}_{H_k^2,p}(u,v)$ for $k = 4$ and 12, respectively, for all reals $p \in [1,2]$ and all (u,v) in the three sources A, B, and C. Our future work will involve determining, for each k, the dominant functions/measures over successive subintervals of $[1,2]$, whose piece-wise combination yields the (overall) locality measure $L_p(H_k^2)$ for all reals $p \in [1,2]$.

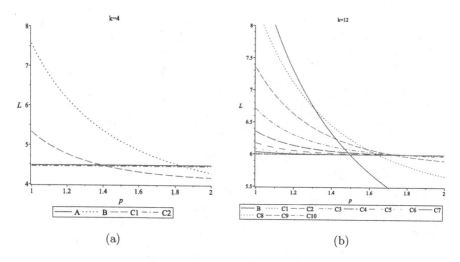

(a)   (b)

Fig. 3. Locality measures corresponding to the grid-point pairs in: (a) A, B, and $C = \{C_2\}$ for $k = 4$ and p-granularity of 0.01; (b) B and $C = \{C_t \mid 1 \leq t \leq k-2\}$ for $k = 12$ and p-granularity of 0.01. (Color figure online)

For the extreme case of $k = 4$ with p-granularity of 0.01, two representative grid-point pairs emerge from the sources B and A over the partitioning subintervals $[1.00, 1.82]$ and $[1.83, 2.00]$, respectively.

For a more general case of $k = 12$ with p-granularity of 0.01, the representative grid-point pairs are from the sources B and C over the partitioning subintervals $[1.00, 1.31]$ and $[1.32, 2.00]$, respectively. Observe that the subsequence C' of all representative grid-point pairs (from the source $C = \{C_t \mid 1 \leq t \leq 10\}$) is $\{C_1, C_2, C_3, C_4, C_5, C_6, C_{10}\}$.

4   Conclusion

Our analytical study of the locality properties of the Hilbert curve family, $\{H_k^2 \mid k = 1,2,\ldots\}$, is based on the locality measure L_p, which is the maximum ratio of $d_p(u,v)^m$ to $d_p(\tilde{u},\tilde{v})$ over all corresponding point-pairs (u,v) and $(\tilde{u},\tilde{v})$ in the m-dimensional grid space and index space, respectively. Our analytical results identify all the candidate representative grid-point pairs of H_k^2 from the three sources A, B, and C (which realize $L_p(H_k^2)$-values) for all norm-parameters $p \in [1,2]$ and grid-orders k, which enable us to have almost complete knowledge of $L_p(H_k^2)$ for all $p \geq 1$ – except for the relation between the candidate grid-point pairs and their dominance p-subintervals. For all real norm-parameters $p \in [1,2]$ with sufficiently small granularity and grid-orders $k \in \{2,3,\ldots,12\}$, our empirical study reveals the three major sources (A, B, and C) of representative grid-point pairs (v,u) that give $\mathcal{L}_{H_k^2,p}(v,u) = L_p(H_k^2)$. The results also suggest that all the representative grid-point pairs of B and C are from B and C', which is a prefix-subsequence of C together with C_{k-2} for some sufficiently large grid-orders $k \in \{5,6,\ldots,12\}$. The study has shed some light on a continuing study of determining the interplay pattern between the norm-parameter p and grid-order k for emerging representative grid-point pairs.

References

1. Alber, J., Niedermeier, R.: On multi-dimensional curves with Hilbert property. Theory Comput. Syst. **33**(4), 295–312 (2000)
2. Bauman, K.E.: The dilation factor of the Peano-Hilbert curve. Math. Notes **80**(5), 609–620 (2006)
3. Chochia, G., Cole, M., Heywood, T.: Implementing the hierarchical PRAM on the 2D mesh: Analyses and experiments. In: Proceedings of the Seventh IEEE Symposium on Parallel and Distributeed Processing, pp. 587–595. IEEE Computer Society, Washington, October 1995
4. Dai, H.K., Su, H.C.: Approximation and analytical studies of inter-clustering performances of space-filling curves. In: Proceedings of the International Conference on Discrete Random Walks (Discrete Mathematics and Theoretical Computer Science, vol. AC (2003)), pp. 53–68, September 2003
5. Dai, H.K., Su, H.C.: On the locality properties of space-filling curves. In: Ibaraki, T., Katoh, N., Ono, H. (eds.) ISAAC 2003. LNCS, vol. 2906, pp. 385–394. Springer, Heidelberg (2003)
6. Dai, H.K., Su, H.C.: Clustering performance of 3-dimensional Hilbert curves. In: Gu, Q., Hell, P., Yang, B. (eds.) AAIM 2014. LNCS, vol. 8546, pp. 299–311. Springer, Heidelberg (2014)
7. Gotsman, C., Lindenbaum, M.: On the metric properties of discrete space-filling curves. IEEE Trans. Image Process. **5**(5), 794–797 (1996)
8. Niedermeier, R., Reinhardt, K., Sanders, P.: Towards optimal locality in mesh-indexings. Discrete Appl. Math. **117**(1–3), 211–237 (2002)

On the Complexity of Clustering with Relaxed Size Constraints

Massimiliano Goldwurm[2], Jianyi Lin[1(✉)], and Francesco Saccà[1]

[1] Dipartimento di Informatica, Università degli Studi di Milano, 20135 Milan, Italy
jianyi.lin@unimi.it
[2] Dipartimento di Matematica, Università degli Studi di Milano, 20133 Milan, Italy
massimiliano.goldwurm@unimi.it

Abstract. We study the computational complexity of the problem of computing an optimal clustering $\{A_1, A_2, ..., A_k\}$ of a set of points assuming that every cluster size $|A_i|$ belongs to a given set M of positive integers. We present a polynomial time algorithm for solving the problem in dimension 1, i.e. when the points are simply rational values, for an arbitrary set M of size constraints, which extends to the ℓ_1-norm an analogous procedure known for the ℓ_2-norm. Moreover, we prove that in the Euclidean plane, i.e. assuming dimension 2 and ℓ_2-norm, the problem is NP-hard even with size constraints set reduced to $M = \{2, 3\}$.

Keywords: Geometric clustering problems · Cluster size constraints · Computational complexity · Constrained k-means

1 Introduction

In the area of unsupervised machine learning and statistical data analysis the clustering methods play an important role with applications in pattern recognition, bioinformatics, signal and image processing, medical diagnostics. Clustering consists in grouping a set of objects into subsets, called clusters, that are maximally homogeneous [5,8]. Partitional or hard clustering requires the subsets to be disjoint and non-empty, and in the usual geometric setting the similarity between objects is measured by distance between points representing the objects [15].

A classical clustering problem is the so-called Euclidean Minimum-Sum-of-Squares [1], Variance-based [10] or k-Means clustering problem: given a finite point set $X \subset \mathbb{R}^d$, find a k-partition $\{A_1, ..., A_k\}$ of X minimizing the sum of weights $W(A_1, ..., A_k) = \sum_i W(A_i) = \sum_i \sum_{x \in A_i} \|x - \mu(A_i)\|^2$ of all clusters, where $\mu(A_i)$ is the sample mean of A_i and $\| \cdot \|$ is the Euclidean norm. This partitional clustering problem is difficult: when d is part of the instance the problem is NP-hard even if the number of clusters is fixed to $k = 2$ [1]; the same occurs for arbitrary k with fixed dimension $d = 2$ [7,16]. Nonetheless, a well-known heuristic for this problem is Lloyd's algorithm [14], also named k-Means Algorithm, which is not guaranteed to converge to the global optimum. This algorithm is usually very fast, but may require exponential time in the worst case [22].

© Springer International Publishing Switzerland 2016
R. Dondi et al. (Eds.): AAIM 2016, LNCS 9778, pp. 26–38, 2016.
DOI: 10.1007/978-3-319-41168-2_3

Often one has some a-priori information on the clusters, that can be incorporated into traditional clustering techniques to increase the clustering performance [2]. Problems that include background information are so-called constrained clustering and can be divided into two classes based on the constraints: instance-level constraints typically define pairs of elements that must be (must-link) or cannot be (cannot-link) in the same cluster [25], and cluster-level constraints prescribe characteristics of each cluster, such as cluster diameter or cluster size [6,21]. In [26] cluster size constraints are used for improving clustering accuracy, for instance allowing one to avoid extremely small or large clusters in standard cluster analysis. In the *size constrained clustering* (SCC) problem, assuming an ℓ_p-norm (we suppose $p \in \mathbb{N}_+$ throughout this work), typically one is given a finite set $X \subset \mathbb{R}^d$ of n points and k positive integers $m_1, ..., m_k$ such that $\sum_i m_i = n$, and searches for a partition $\{A_1, ..., A_k\}$ of X, with $|A_1| = m_1, ..., |A_k| = m_k$, that minimizes the objective function $W(A_1, ..., A_k) = \sum_{i=1}^{k} \sum_{x \in A_i} \|x - c_i\|_p^p$, where each $c_i = \mathrm{argmin}_{c \in \mathbb{R}^d} \sum_{x \in A_i} \|x - c\|_p^p$ is the ℓ_p-centroid of A_i.

For arbitrary $k \in \mathbb{N}$, this problem is NP-hard also in dimension $d = 1$, for any (fixed) ℓ_p-norm, $p \geq 1$; the same negative result holds for arbitrary $d \in \mathbb{N}$ when the number of clusters is fixed to $k = 2$, for every ℓ_p-norm with $p > 1$ [3]. On the contrary, in the case $d = 2 = k$ the problem is solvable in $O(n^2 \log n)$ time assuming Manhattan norm (ℓ_1) and in $O(n \sqrt[3]{m} \log^2 n)$ time with Euclidean norm (ℓ_2) [13], where m is the size of one of the two clusters.

In this work we study a *relaxed version* of the size constrained clustering problem, where the size of each cluster belongs to given set M of integers. We show that in dimension $d = 1$, for an arbitrary (finite) $M \subset \mathbb{N}$, assuming the Manhattan norm, the solution can be obtained in $O(n(ks + n))$ time, where k is the number of clusters and s is the cardinality of M. This extends an analogous algorithm [4] proposed for the Euclidean norm and applied to computational biology problems as a method for identification of promoter regions in genomic sequences. Note instead that, in dimension 1, the SCC problem is NP-hard [3]. On the contrary, in dimension $d = 2$, we prove that even fixing $M = \{2, 3\}$ the problem is NP-hard with Euclidean norm.

2   Problem Definition

In this section we define the problem and fix our notation. Given a positive integer d, for every real $p \geq 1$ and every point $a = (a_1, ..., a_d) \in \mathbb{R}^d$, we denote by $\|a\|_p$ the ℓ_p-norm of a, i.e. $\|a\|_p = (\sum_1^d |a_i|^p)^{1/p}$. Clearly, $\|a\|_2$ and $\|a\|_1$ are the Euclidean and the Manhattan (or Taxicab) norm of a, respectively.

Given a finite set $X \subset \mathbb{R}^d$, a *cluster* of X is a non-empty subset $A \subset X$, while a *clustering* is a partition $\{A_1, ..., A_k\}$ of X in k clusters for some k. Assuming the ℓ_p norm, the *centroid* and the *weight* of a cluster A are the values $C_A \in \mathbb{R}^d$ and $W_p(A) \in \mathbb{R}_+$ defined, respectively, by

$$C_A = \mathrm{argmin}_{c \in \mathbb{R}^d} \sum_{a \in A} \|a - c\|_p^p, \qquad W_p(A) = \sum_{a \in A} \|a - C_A\|_p^p$$

The *weight* of a clustering $\{A_1, ..., A_k\}$ is $W_p(A_1, ..., A_k) = \sum_1^k W_p(A_i)$. We recall that, in case of ℓ_2-norm, the weight of a cluster A can be computed by relation

$$W_2(A) = \frac{1}{|A|} \sum_{(*)} \|a - b\|_2^2 \tag{1}$$

where the sum is extended to all unordered pairs $\{a, b\}$ of distinct elements in A. Moreover, given a set $\mathcal{M} \subset \mathbb{N}$, any clustering $\{A_1, ..., A_k\}$ such that $|A_i| \in \mathcal{M}$ for every $i = 1, \ldots, k$, is called $\mathcal{M}$-*clustering*.

RSC-d Problem (with ℓ_p-norm): Relaxed Size Constrained Clustering in $\mathbb{R}^d$
Given a set $X \subset \mathbb{Q}^d$ of n points, an integer k such that $1 < k < n$ and a finite set $\mathcal{M}$ of positive integers, find an $\mathcal{M}$-clustering $\{A_1, ..., A_k\}$ of X that minimizes $W_p(A_1, ..., A_k)$.[1]

When $\mathcal{M}$ is not included in the instance, but fixed in advance, we call the problem $\mathcal{M}$-**RSC-d** (with ℓ_p-norm). In this work we study these problems in dimension $d = 1, 2$ assuming ℓ_1 and ℓ_2-norm.

3 Dynamic Programming for RSC on the Line

In this section we describe a polynomial-time algorithm for RSC-1 that works assuming either ℓ_1 or ℓ_2-norm. This procedure is based on a dynamic programming technique, in the style of [19], based on the so-called String Property [3, 24]. A simplified version of the procedure in the case of ℓ_2-norm is also presented in [4] and applied to problems of computational biology.

Consider an instance $(X, k, \mathcal{M})$ of RSC-1, where $X = (x_1, x_2, ..., x_n)$ is a sorted sequence of rational numbers, $k \in \{1, \ldots, n-1\}$ and $|\mathcal{M}| = s \le n$. For any $1 \le i \le j \le n$, let $X[i, j]$ be the subsequence $(x_i, x_{i+1}, ..., x_j)$. For a given $p \in \{1, 2\}$, we define the $n \times n$ matrix $U = [U(i, j)]_{i,j=1,...,n}$ by setting $U(i, j) = W_p(X[i, j]) = \sum_{t=i}^{j} |x_t - C_{X[i,j]}|^p$ if $j - i + 1 \in \mathcal{M}$ and $U(i, j) = \infty$ otherwise, that is the weight of cluster $X[i, j]$ when it has admissible size.

Lemma 1. *Given $p \in \{1, 2\}$, for every instance $(X, k, \mathcal{M})$ of RSC-1 with $|X| = n$, matrix U can be computed in $O(n^2)$ time.*

Proof. First, assume $p = 2$. In this case it is easy to check that the weight of any cluster A is $W_2(A) = \sum_{a \in A} a^2 - \frac{1}{|A|}(\sum_{a \in A} a)^2$. Denoting $Q(i) := \sum_{j=1}^{i} x_j^2$ and $S(i) := \sum_{j=1}^{i} x_j$, the finite entries of matrix U reduce to

$$U(i, j) = Q(j) - Q(i-1) - \frac{1}{j - i + 1}(S(j) - S(i-1))^2. \tag{2}$$

The sequences Q and S can be computed in linear time, and thus the computation of (2) requires constant time for each i, j. Hence, matrix U can be computed in $O(n^2)$ time in case $p = 2$.

---

[1] If X does not admit a $\mathcal{M}$-clustering then symbol $\bot$ is returned.

When $p = 1$, the weight $W_1(X[i,j])$ is the sum of the distances between elements and median of $X[i,j]$. Denote $m := (i+j)/2$ and for any cluster $X[i,j]$ set the left and right sums $L(i,j) := \sum_{i \leq h < m} x_h$ and $R(i,j) := \sum_{m < h \leq j} x_h$. It can be checked ([3, Proposition 10]) that $W_1(X[i,j]) = R(i,j) - L(i,j)$. Since $X[i,j]$ is sorted, it can be seen that, for $i < j$,

$$L(i,j) = L(i,j-1) \text{ if } m \in \mathbb{N}, \quad L(i,j) = L(i,j-1) + x_{\lfloor m \rfloor} \text{ otherwise,} \quad (3)$$
$$R(i,j) = R(i,j-1) - x_m + x_j \text{ if } m \in \mathbb{N}, \quad V = R(i,j-1) + x_j \text{ otherwise} \quad (4)$$

and $L(i,i) = R(i,i) = 0$. By means of these recursive formulae the quantities $L(i,j), R(i,j), W_1(X[i,j])$, for all $i \leq j$, can be computed in $O(n^2)$ time, and hence the same holds for determining matrix U when $p = 1$.   □

Now, for every $h \in \{1, \ldots, k\}$ and every $j \in \{1, \ldots, n\}$, let $Z(h,j)$ be the weight of a solution of RSC-1 for the instance $(X[1,j], h, \mathcal{M})$ in case $h \leq j$, while $Z(h,j) = \infty$ if $h > j$. These values can be derived from U.

Proposition 2. *The following properties hold:*
(i) $Z(1,j) = U(1,j)$ for all $j = 1, \ldots, n$;
(ii) $Z(h,j) = \min_{m \in \mathcal{M}} (Z(h-1, j-m) + U(j-m+1, j))$ for all $h = 2, .., k; j = h, \ldots, n$.

Proof. Case *(i)* is obvious. Since $Z(h,j)$ is the weight of the optimal solution for $(X[1,j], h, \mathcal{M})$, the corresponding solution $\{A_1, ..., A_h\}$ satisfies the String Property, i.e. each cluster A_i consists of consecutive points of X [3,24].
Then, its right-most cluster A_h has size $|A_h| = m \in \mathcal{M}$ and weight $W_p(A_h) = W_p(X[j-m+1,j]) = U(j-m+1,j)$.
The other clusters $A_1, ..., A_{h-1}$ form a feasible clustering of RSC-1 for the instance $(X[1, j-m], h-1, \mathcal{M})$, which has minimum weight $W_p(A_1, ..., A_{h-1}) = \sum_1^{h-1} W_p(A_i) = Z(h-1, j-m)$, otherwise it is easy to check that also $\{A_1, ..., A_h\}$ would not be an optimal solution for $(X[1,j], h, \mathcal{M})$.
As a consequence, $Z(h,j) = Z(h-1, j-m) + U(j-m+1, j)$ for some $m \in \mathcal{M}$, and since $Z(h,j)$ has to take the minimum value, property *(ii)* is proved.   □

Relying on the previous proposition we can design an algorithm for RSC-1.

Theorem 3. *For any $p \in \{1,2\}$, RSC-1 with ℓ_p-norm can be solved in $O(n(ks + n))$ time and $O(n^2)$ space.*

Proof. By Lemma 1 we first compute matrix U in $O(n^2)$ time. Then, by means of Proposition 2, matrix $Z = [Z(h,j)]_{h=1,\ldots,k; j=1,\ldots,n}$ can be computed row by row. Each entry requires at most $s = |\mathcal{M}|$ sums and comparisons. The computation is described by the following scheme, where we store in $\ell_{h,j}$ the size of the last cluster of the optimal solution for $(X[1,j], h, \mathcal{M})$, for each pair of indices h, j.

```
begin
  Z := {∞}^{k×n}
  for j = 1, ..., n do  { Z(1,j) := U(1,j)
                        { if U(1,j) ≠ ∞ then ℓ_{1,j} := j
```

```
for h = 2, ..., k do
    for j = h, ..., n do
        m̂ := argmin_{m∈M}{Z(h − 1, j − m) + U(j − m + 1, j)}
        if m̂ is well-defined then
            Z(h, j) := Z(h − 1, j − m̂) + U(j − m̂ + 1, j)
            ℓ_{h,j} := m̂
end
```

Clearly, if $\ell_{k,n}$ is not defined then the symbol $\perp$ is returned since no admissible clustering for $(X, k, \mathcal{M})$ exists. Otherwise, the solution of the problem can be obtained by the following procedure:

```
begin
    j := n
                          ⎧ t_h := ℓ_{h,j}
    for h = k, k − 1, ..., 1 do ⎨ A_h := X[j − t_h + 1, j]
                          ⎩ j := j − t_h
    output {A_1, A_2, ..., A_k}
end
```

The overall time required to compute matrices U and Z is $O(n(ks+n))$. The space necessary to maintain all tables is $O(n^2)$ since $k < n$. $\square$

It is worth noting that the analogous problem, where the size of each cluster is fixed by the instance, is NP-hard even in dimension $d = 1$ for every ℓ_p-norm [3]. This shows that the form of the size constraints for clustering problems is relevant for the existence of polynomial time algorithms.

4 NP-hardness of RSC in the Euclidean Plane

In this section we show that, assuming ℓ_2-norm, the {2, 3}-RSC-2 problem is NP-hard, and therefore RSC-2 also is NP-hard. To this end we introduce a decision version of the problem and describe a polynomial-time reduction from Planar 3-SAT.

Decision {2, 3}-RSC-2 Problem. *Given a point set $X = \{p_1, ..., p_n\} \subset \mathbb{Q}^2$, an integer k, $n/3 \le k \le n/2$, and a rational value $\lambda > 0$ (threshold), decide whether there exists a {2, 3}-clustering $\{A_1, ..., A_k\}$ of X, consisting of k clusters, such that $W_2(A_1, ..., A_k) \le \lambda$.*

Recall that a 3-CNF formula Φ is a boolean formula given by the conjunction of clauses each of which has 3 literals. If V and C are, respectively, the set of variables and the set of clauses of Φ, the *graph of Φ* is defined as the undirected bipartite graph G_Φ such that $V \cup C$ is the family of nodes and $E = \{\{v, c\} : v \in V, c \in C$, and either v or $\bar{v}$ appears in $c\}$ is the set of edges. A formula Φ is said to be *planar* if G_Φ is planar. The Planar 3-SAT problem consists in deciding whether a planar 3-CNF formula Φ is satisfiable.

It is known that Planar 3-SAT is strongly NP-complete [12]. It is also proved that it suffices to consider formulae whose associated graph can be embedded in $\mathbb{R}^2$, with variables arranged on a straight line, and with clauses arranged above and below the straight line [11]. Moreover, the edges between variables and clauses can be drawn in a rectilinear fashion [17].

We also recall that a *box-orthogonal drawing* of a graph G is a planar embedding of G on an integer grid where each vertex is mapped into a (possibly degenerate) rectangle and each edge becomes a path of horizontal or vertical segments of the grid. Rectangles are disjoint and paths do not intersect. Any planar graph of n nodes admits a box-orthogonal drawing computable in $O(n)$ time that uses a $a \times b$ grid, where $a + b \leq 2n$ [9, Theorem 3].

Our goal is to show that Planar 3-SAT is reducible in polynomial time to Decision $\{2, 3\}$-RSC-2. The proof is obtained by adapting the reduction from Planar 3-SAT to an unconstrained version of the k-means problem in the plane, presented in [16]. Here, the main difference is that in our construction we determine directly the rational coordinates of the points given by the reduction, avoiding the approximation of irrational values. Moreover, our reduction does not yield multiple copies of the same point in the plane.

To describe the reduction we show how an arbitrary planar 3-CNF formula Φ, can be associated with an instance (X, k, λ) of the Decision $\{2, 3\}$-RSC-2, computable in polynomial time w.r.t. $|\Phi|$, such that Φ is satisfiable if and only if X admits a partition into k clusters of cardinality 2 or 3, having a total weight at most λ. The definition of such a reduction is split in several phases: the first one computes an embedding of graph G_Φ into a planar integer grid; the others determine the rational coordinates of points in X, and the values k and λ.

The general idea is to build an embedding of G_Φ by representing each clause by a point in the grid, and associating each variable with a cycle on the grid that connects all points of clauses containing the variable. Clearly, these cycles do not overlap, and each clause-point is touched exactly by 3 cycles. Now, the points of X are placed along every cycle, so that there are only 2 optimal $\{2\}$-clusterings for the points of each cycle, which may be associated to the truth assignments of the variable. The satisfiability of each clause will correspond to the possibility of clustering the clause-point with the nearest pair in one of the optimal $\{2\}$-clusterings associated to a variable occurring in the clause.

(1) Embedding of G_Φ into a planar grid

The first phase is described by the following steps, illustrated in Fig. 1.

Step 0. Compute the box-orthogonal drawing D of G_Φ as stated above. We can map any variable into a (non-empty) rectangle and any clause into a vertex of the grid. Moreover, the base of all rectangles can be put on the same horizontal straight line, and the vertices representing clauses above or below such a line.

Step 1. Expand the previous drawing by a factor of 2 and call D_1 the new drawing. This doubles all distances between vertices in D.

Step 2. Shift D_1 half unit upward and rightward and let D_2 be the new drawing. Now, each clause corresponds to a point in the centre of a unit square of the grid, and each path from a rectangle (variable) to a point (clause) crosses just in the middle some unit sides of the grid.

Step 3. Expand all rectangles by half grid unit in all four vertical and horizontal directions, and replace any point (clause) of D_2 by a unit square centred at the same location, erasing the overlapping portion (half unit long) of paths. We call D_3 the new drawing. Now, all rectangles have sides of odd length and no path in D_3 starts from a vertex of a rectangle.

Step 4. Replace every path from a rectangle (variable) to a unit square (clause) by a *strip* of unit width on the grid that cover the same path, erasing the boundary portion of rectangle overlapping the strip. The resulting drawing is called D_4. Now every variable v corresponds to a (sort of) *cycle* on the grid that includes both the residual rectangle representing v and all strips towards the unit squares (clauses) where v occurs, together with one side for each touched square.

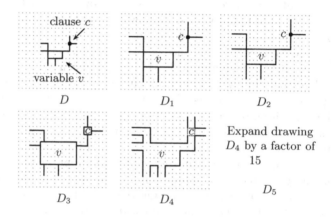

Fig. 1. Main steps of the graph transformations used in the reduction.

Step 5. Expand the previous drawing by a factor of 15. We call D_5 the new drawing. Thus, each clause is now associated with a square on the grid having side of length 15, while the strips described in Step 4 are formed by parallel segments at distance 15 to each other. Moreover, in the following we call *borders* the straight-line segments forming the cycles associated with the variables.

(2) Definition of point set X

Let $V = \{v_1, \ldots, v_n\}$ and $C = \{c_1, \ldots, c_m\}$ be, respectively, the set of variables and the set of clauses of Φ. First, for every $c_j \in C$, X contains a point $z_j \in \mathbb{Q}^2$ located near the centre of the square associated with c_j. The exact position of each z_j is defined by Fig. 2, where the cycles are represented by dashed lines and the sides of the square are removed for sake of simplicity.

Moreover, for every variable $v_i \in V$, X contains a circuit Γ_i of $2L_i$ consecutive points $\{x_{i1}, x_{i2}, \ldots, x_{i(2L_i)}\}$, for a suitable integer L_i. With few exceptions (as in Fig. 2), all $x_{i\ell}$'s lie inside the cycle of drawing D_5 associated with v_i and inside the square associated with the clauses where v_i occurs. The idea is to put almost all points at distance 2 from the borders, setting at distance 5 from each other most consecutive points x_{it}, $x_{i(t+1)}$, as well as points $x_{i(2L_i)}$ and x_{i1}. Hence

$$X = \{z_j \mid j = 1, 2, \ldots, m\} \cup \{x_{i\ell} \mid i = 1, 2, \ldots, n, \; \ell = 1, 2, \ldots, 2L_i\} \quad (5)$$

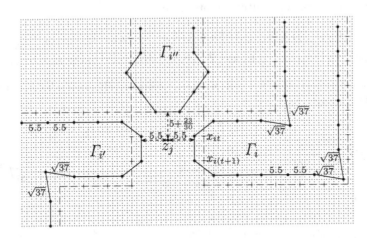

Fig. 2. Points of 3 circuits in the neighbourhood of a clause-point z_j. Edges with length different from 5 are indicated.

The precise position of points $x_{i\ell}$'s is illustrated in Figs. 2 and 3 and is formally defined by conditions (a), (b), (c) given below. Such a position depends on the angles, inside the cycle associated with v_i, formed between two incident borders. Every angle has measure either $\pi/2$ or $3\pi/2$; in the first case we say the angle is *convex*, in the second case we say it is *concave* (e.g., in Fig. 3, angle β is convex, while α is concave).

(a) Near every convex (resp. concave) angle three consecutive points of Γ_i are placed as shown by angles $\beta, \delta, \varepsilon, \zeta, \eta$ (resp. $\alpha, \gamma, \theta, \iota$) in Fig. 3. Note that the second point of the triple always lies on the bisector.

(b) Between any two consecutive angles, the other points of Γ_i are put on a straight-line at distance 2 from the border, so that consecutive points are set at distance 5 from each other, with the exception of two segments of length 4.5 (respectively, 5.5) if both angles are concave (resp., convex). As examples, see in Fig. 3 points between angles β and γ, ι and θ, δ and ε.

(c) If v_i, $v_{i'}$, $v_{i''}$ are the variables occurring in a clause c_j, then near the square of size 15×15 associated with c_j, points of Γ_i, $\Gamma_{i'}$, $\Gamma_{i''}$ are set as defined in Fig. 2. Note that here, all consecutive points are at distance 5 from each other with two exceptions:
   - triple of points close to angles are located according to condition (a);
   - before convex angles, points are located to form two consecutive segments of length 5.5.

(3) Weight of clusters

Note that all pairs of consecutive points in any Γ_i form a segment having one of the following lengths: 4.5, 5, 5.5, $\sqrt{37}$. The weight of the corresponding clusters is easily obtained from Eq. (1): 10.125, 12.5, 15.125, 18.5.

Moreover, every set Γ_i admits only two $\{2\}$-clusterings of minimum weight, consisting of pairs of consecutive points, given by

$$\pi_1(i) = \{\{x_{iu}, x_{i(u+1)}\} \mid u = 1, 3, 5, \ldots, 2L_i - 1\} \qquad \text{and}$$
$$\pi_2(i) = \{\{x_{iu}, x_{i(u+1)}\} \mid u = 2, 4, 6, \ldots, 2L_i - 2\} \cup \{\{x_{i(2L_i)}, x_{i1}\}\}$$

For simplicity, hereafter we call *segment* (respectively, *triangle*) a cluster of cardinality 2 (resp., 3).

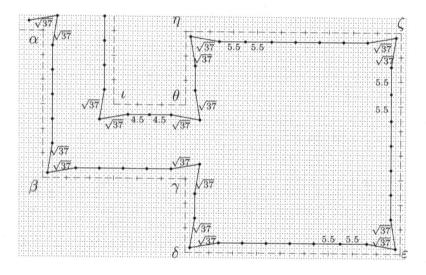

Fig. 3. Points of a circuit Γ_i inside the corresponding rectangle. Segments with length different from 5 are indicated. Note that angles $\alpha, \gamma, \theta, \iota$ are concave, while $\beta, \delta, \varepsilon, \zeta, \eta$ are convex.

Now, consider a clause c_j containing a variable v_i (as a positive or negative literal) and let $x_{it}, x_{i(t+1)}$ be the pair of points of Γ_i nearest to point z_j. We say that z_j *touches* the segment $\{x_{it}, x_{i(t+1)}\}$. Clearly every z_j touches three segments, one for each variable appearing in c_j. Note from Fig. 2, that the distance between z_j and a touched segment is either 5.5 or $5 + \frac{23}{30}$. Then, using Eq. (1), by elementary computation we can determine the weight of any triangle formed by each z_j with its touched segments, as well as the weight of every triangle of consecutive points in any Γ_i. Such a direct computation proves the following property.

Lemma 4. *If point z_j touches a segment $\{x_{it}, x_{i(t+1)}\}$ then the weight of triangle $\{z_j, x_{it}, x_{i(t+1)}\}$ is given by $w = \frac{23402}{675}$, which satisfies $34.66 < w < 34.67$. Moreover, every triangle composed by points of Γ_i has weight greater than w.*

(4) Parity condition

By a suitable choice of the first point x_{i1}, and possibly by adding new points to Γ_i (as explained below), we can assume that the following parity condition holds: in any Γ_i, every segment touched by a point z_j belongs to either $\pi_1(i)$ or $\pi_2(i)$ according to whether v_i or $\overline{v_i}$ appears in c_j, respectively. In order to guarantee this property, slight changes to points of Γ_i near the square including z_j may be necessary, which are illustrated in Fig. 4. This change add two new points (one before and one after the touched segment), and determines 4 more segments of length 4.5, two of which are to be included into $\pi_1(i)$, the others into $\pi_2(i)$. In order to apply this transformation the circuit must contain a rectilinear portion of length at least 30, either horizontal or vertical, as shown in Fig. 4 (left). We may always assume this is satisfied by requiring one more expansion of the initial drawing by a factor of 2 (executing Step 1 twice in the embedding phase).

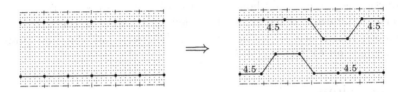

Fig. 4. (Left) 30×15 horizontal strip preserving parity. (Right) 30×15 horizontal strip for changing parity. The vertical case in analogous.

(5) Definition of k and λ

They are given by equalities $k = \sum_1^n L_i$ and

$$\lambda = \frac{5^2}{2}(k - h) + wm + \frac{1}{2}\left[18.5 \cdot s_{\sqrt{37}} + \left(10 + \frac{1}{8}\right) \cdot s_{4.5} + \left(15 + \frac{1}{8}\right) \cdot s_{5.5}\right],$$

where w is defined as in Lemma 4, s_u is the total number of segments of length u in all Γ_i's for $u \in \{\sqrt{37}, 4.5, 5.5\}$, and $h = m + \frac{1}{2}(s_{\sqrt{37}} + s_{4.5} + s_{5.5})$.

It is easy to see that every $\{2, 3\}$-clustering π of X into k clusters must contain exactly m triangles. Indeed, if n_T and n_S denote, respectively, the number of triangles and the number of segments of π, then $|X| = 2n_S + 3n_T = 2k + m$ and $n_S + n_T = k$, which yields $n_T = m$ and $n_S = k - m$. Recall that all triangles in X have weight at least w. Moreover, by construction, π may include at most $s_u/2$ many segments of length u for each $u \in \{\sqrt{37}, 4.5, 5.5\}$ and the remaining $k - h$ cannot have length smaller than 5. This implies $W_2(\pi) \geq \lambda$.

Now, to complete the reduction we verify that Φ is satisfiable if and only if there exists a $\{2, 3\}$-clustering of X of weight at most λ, consisting of k clusters. Suppose Φ is satisfiable and consider a satisfying assignment. For each variable v_i, choose clustering $\pi_2(i)$ or $\pi_1(i)$ according whether its value is 0 or 1, respectively. Since the assignment makes all clauses true, each point z_j can be clustered together with the touched segment in Γ_i, for a variable v_i satisfying clause c_j. By the parity condition, such a touched segment belongs to the chosen clustering (either $\pi_2(i)$ or $\pi_1(i)$). Thus, we obtain m triangles of weight w. The other points in each Γ_i can be clustered as in $\pi_2(i)$ or $\pi_1(i)$ according to the previous choice. This yields a $\{2,3\}$-clustering of X of weight λ having k clusters.

Vice-versa, if there exists a $\{2,3\}$-clustering π of X with k clusters and weight λ, then such a clustering must contain m triangles of weight w. The only way to obtain these triangles is to include each point z_j into a touched segment $\{x_{it}, x_{i(t+1)}\}$. By the parity condition this defines an assignment of values to all variables that makes true each clause of Φ.

Theorem 5. *Assuming ℓ_2-norm, the $\{2,3\}$-RSC-2 problem is strongly NP-hard and it does not admit an FPTAS unless $P = NP$. As a consequence, the same holds in general for RSC-2 problem.*

Proof. The NP-hardness follows from the discussion above. The problem is also strongly NP-hard since the value of all integers in instances (X, k, λ) obtained by the reduction is polynomially bounded w.r.t. $n = |X|$. Moreover, the objective function to minimize is polynomially bounded with respect to the unary size of the instance, and hence, by a classical result [23, Sect. 8.3], the same problem does not admit an FPTAS unless $P = NP$. □

5 Conclusions

In this work, we have studied the clustering problem with relaxed size constraints in dimension 1 and 2 (RSC-1 and RSC-2). First, we have shown a polynomial-time algorithm for RSC-1 in the case of ℓ_1 and ℓ_2-norm. A natural question is whether similar algorithm exists also for ℓ_p-norm with integer $p > 2$. We recall that the clustering in dimension 1 is motivated by bioinformatics applications as illustrated in [4].

Our second result states that $\mathcal{M}$-RSC-2 problem is strongly NP-hard when $\mathcal{M} = \{2,3\}$. Note that with $\mathcal{M} = \{2\}$ the problem reduces to finding a perfect matching of minimum cost in a weighted complete graph, and hence it is solvable in $O(n^3)$ time (even in arbitrary dimension) assuming any ℓ_p-norm, by using classical algorithms [18]. The same occurs when $\mathcal{M} = \{1,2\}$ since this is reducible to finding the minimum cost matching of given cardinality in a weighted graph, which is known to be solvable in polynomial time (see for instance [20, Sect. 3.1.1]). Hence, a natural problem is to determine the sets $\mathcal{M}$ for which the $\mathcal{M}$-RSC-2 problem is NP-hard.

Finally, we conjecture that $\{2,3\}$-RSC-2 remains NP-hard also in case of ℓ_1-norm, by a suitable extension of the proof above.

Acknowledgments. We thank an anonymous referee for his/her useful comments on other problems related to clustering with relaxed size constraints.

References

1. Aloise, D., Deshpande, A., Hansen, P., Popat, P.: NP-hardness of Euclidean sum-of-squares clustering. Mach. Learn. **75**, 245–249 (2009)
2. Basu, S., Davidson, I., Wagstaff, K.: Constrained Clustering: Advances in Algorithms, Theory, and Applications. Chapman and Hall/CRC, Boca Raton (2008)
3. Bertoni, A., Goldwurm, M., Lin, J., Saccà, F.: Size constrained distance clustering: separation properties and some complexity results. Fundamenta Informaticae **115**(1), 125–139 (2012)
4. Bertoni, A., Rè, M., Saccà, F., Valentini, G.: Identification of promoter regions in genomic sequences by 1-dimensional constraint clustering. In: Neural Nets WIRN11, pp. 162–169 (2011)
5. Bishop, C.: Pattern Recognition and Machine Learning. Springer, New York (2006)
6. Bradley, P.S., Bennett, K.P., Demiriz, A.: Constrained K-Means Clustering. Technical report MSR-TR-2000-65, Miscrosoft Research Publication, May 2000
7. Dasgupta, S.: The hardness of k-means clustering. Technical report CS2007-0890, Department of Computer Science and Engineering, University of California, San Diego (2007)
8. Fisher, W.D.: On grouping for maximum homogeneity. J. Am. Stat. Assoc. **53**(284), 789–798 (1958)
9. Fößmeier, U., Kant, G., Kaufmann, M.: 2-Visibility drawings of planar graphs. In: North, S. (ed.) Graph Drawing. LNCS, vol. 1190, pp. 155–168. Springer, Heidelberg (1997)
10. Hasegawa, S., Imai, H., Inaba, M., Katoh, N.: Efficient algorithms for variance-based k-clustering. In: Proceedings of Pacific Graphics 1993, pp. 75–89 (1993)
11. Knuth, D.E., Raghunathan, A.: The problem of compatible representatives. SIAM J. Discrete Math. **5**(3), 422–427 (1992)
12. Lichtenstein, D.: Planar formulae and their uses. SIAM J. Comput. **11**(2), 329–343 (1982)
13. Lin, J., Bertoni, A., Goldwurm, M.: Exact algorithms for size constrained 2-clustering in the plane. Theor. Comput. Sci. **629**, 80–95 (2016)
14. Lloyd, S.: Least squares quantization in PCM. IEEE Trans. Inf. Theor. **28**(2), 129–137 (1982)
15. MacQueen, J.B.: Some methods for classification and analysis of multivariate observations. In: Proceedings of the 5th Berkeley Symposium on Mathematical Statistics and Probability, pp. 281–297 (1967)
16. Mahajan, M., Nimbhorkar, P., Varadarajan, K.: The planar k-means problem is NP-hard. Theor. Comput. Sci. **442**, 13–21 (2012)
17. Mulzer, W., Rote, G.: Minimum-weight triangulation is NP-hard. J. ACM **55**(2), 11 (2008)
18. Papadimitriou, C., Steiglitz, K.: Combinatorial Optimization: Algorithms and Complexity. Dover, New York (1998)
19. Rao, M.R.: Cluster analysis and mathematical programming. J. Am. Stat. Assoc. **66**(335), 622–626 (1971)
20. Stephan, R.: Cardinality constrained combinatorial optimization: complexity and polyhedra. Discrete Optim. **7**(3), 99–113 (2010)
21. Tung, A.K.H., Han, J., Lakshmanan, L.V.S., Ng, R.T.: Constraint-based clustering in large databases. In: Van den Bussche, J., Vianu, V. (eds.) ICDT 2001. LNCS, vol. 1973, pp. 405–419. Springer, Heidelberg (2000)

22. Vattani, A.: k-means requires exponentially many iterations even in the plane. Discrete Comput. Geom. **45**(4), 596–616 (2011)
23. Vazirani, V.: Approximation Algorithms. Springer, Heidelberg (2001)
24. Vinod, H.: Integer programming and the theory of grouping. J. Am. Stat. Assoc. **64**(326), 506–519 (1969)
25. Wagstaff, K., Cardie, C.: Clustering with instance-level constraints. In: Proceedings of the 17th International Conference on Machine Learning, pp. 1103–1110 (2000)
26. Zhu, S., Wang, D., Li, T.: Data clustering with size constraints. Knowl. Based Syst. **23**(8), 883–889 (2010)

Superstring Graph: A New Approach for Genome Assembly

Bastien Cazaux[1,2], Gustavo Sacomoto[3], and Eric Rivals[1,2(✉)]

[1] LIRMM, Université de Montpellier, CNRS UMR 5506, Montpellier, France
{cazaux,rivals}@lirmm.fr
[2] Institut Biologie Computationnelle, Montpellier, France
[3] INRIA Rhône-Alpes and Université Lyon 1, CNRS,
UMR 5558, 69000 Lyon, France
sacomoto@gmail.com

Abstract. With the increasing impact of genomics in life sciences, the inference of high quality, reliable, and complete genome sequences is becoming critical. Genome assembly remains a major bottleneck in bioinformatics: indeed, high throughput sequencing apparatus yield millions of short sequencing reads that need to be merged based on their overlaps. Overlap graph based algorithms were used with the first generation of sequencers, while de Bruijn graph (DBG) based methods were preferred for the second generation. Because the sequencing coverage varies locally along the molecule, state-of-the-art assembly programs now follow an iterative process that requires the construction of de Bruijn graphs of distinct orders (i.e., sizes of the overlaps). The set of resulting sequences, termed unitigs, provide an important improvement compared to single DBG approaches. Here, we present a novel approach based on a digraph, the Superstring Graph, that captures all desired sizes of overlaps at once and allows to discard unreliable overlaps. With a simple algorithm, the Superstring Graph delivers sequences that includes all the unitigs obtained from multiple DBG as substrings. In linear time and space, it combines the efficiency of a greedy approach to the advantages of using a single graph. In summary, we present a first and formal comparison of the output of state-of-the-art genome assemblers.

1 Introduction

Ongoing improvements in DNA sequencing technologies have dramatically increased the throughput of sequencers, thereby authorising the launch of very large genome projects: the 1000 human genomes for studying natural variations [15], the 10 K vertebrate genomes for phylogenomics issues [11] or the 10,000 rice genomes, which aims at getting a genomic overview of all wild and cultivated rice varieties. If getting the collections of raw sequencing reads becomes easier and cheaper, assembling complex eukaryotic genomes remains one of the major practical and theoretical challenges in bioinformatics.

This work is supported by ANR Colib'read (ANR-12-BS02-0008), the Institut de Biologie Computationnelle (ANR-11-BINF-0002).

© Springer International Publishing Switzerland 2016
R. Dondi et al. (Eds.): AAIM 2016, LNCS 9778, pp. 39–52, 2016.
DOI: 10.1007/978-3-319-41168-2_4

With the advent of Next Generation sequencing technologies, most of the sequencing performed yields huge numbers of short reads. For that reason, the de Bruijn Graph (DBG) approach, also termed as Eulerian sequence assembly, has been preferred to the Overlap-Layout-Consensus approach, which resorts to an overlap graph and was used with traditional Sanger sequencing. The DBG encodes each k-mer of the read set as a node and contains an arc from one node to another if the $k + 1$-mer obtained by merging them occurs in at least one read. A path in the DBG represents the sequence obtained by merging the k-mers along it. Many assemblers infer unitigs by traversing non branching paths in the DBG, or contigs if the path chooses some extension when it encounters a branching node. Unitigs, which are the parts of contigs comprised between two branching nodes, represent unambiguous regions of the target genome. However, the choice of the value of k is critical and difficult in practice. Indeed, the density of sequencing reads along the molecule depends on the amount of sequencing and fluctuates for technological and biological reasons. Some regions have low coverage, while others may collapse the expected number of reads several times because they contain genomic repeats.

1.1   Related Works

Recently, some papers have investigated the power of combining the assembly made successively by several DBG of different orders, i.e. varying the order k in a user defined range $[k_{min}, k_{max}]$. The goal is to enable the algorithm to find paths both in tangled and fragmented regions of the graph. More precisely, (1) using larger overlaps to find paths in tangled regions, where repeats in DNA create bubbles and branching nodes because shared k-mers are collapsed in the DBG, and (2) to find paths between connected components with shorter overlaps. The algorithms named IDBA and SPAdes do that by building several DBG with different values of k [1,13]. Their main problem is the necessity to build several DBG. Currently, state-of-the-art methods exploit multiple sizes of overlaps and also impose a constraint on the read coverage (i.e. the density of reads in the sequence region). IDBA [13] builds a DBG of order k_{min}, computes the set of unitigs, merges them with the reads, and iterates this procedure for all k until k_{max} (see Algorithm 2). SPAdes adopts a slightly different algorithm [1]. For all values of k between k_{max} and k_{min}, it computes in parallel the unitigs of each DBG_k, and makes their union. Finally, it builds $DBG_{k_{max}}$ on this set and outputs its unitigs. The result of the two approaches are similar. In practice, building that many $DBGs$ is prohibitive, and hence both IDBA and SPAdes limit themselves to a few (i.e., ≈ 4) values of k.

Boucher et al. propose to extend the BOSS succinct data structure, which succinctly encodes a DBG, to enable the dynamic update from k to $k+1$ [2]. Their practical performance allows to navigate between different orders on bacterial and a Human dataset. However, the question of which size of overlap/order is needed in a given region remains. In other works, we have shown how to build in linear time a DBG of order k from either a Generalised Suffix Tree, a Suffix Array, or a Truncated Suffix Tree of the reads and exhibited an algorithm to update k also in linear time [4].

An alternative and interesting approach, called the manifold de Bruijn graph, which assigns words of arbitrary length to nodes in the graph, was presented in [9]. This perspective is different from the one we propose here.

Formally, the question of assembling strings is modelled as the Shortest Common Superstring, also termed the Shortest Linear Superstring. It requires finding a single superstring containing the input words as substrings. This well-studied problem is known to be NP-hard [7] and APX-hard [12]. Many approximation algorithms have been proposed, which solve a relaxed problem known as the Shortest Cyclic Cover of Strings (SCCS) – see Sect. 2 for a definition of a cyclic cover. SCCS is usually solved in polynomial time with the Hungarian algorithm; we recently exhibited a linear time algorithm for SCCS and introduced the Superstring Graph for this sake [5]. To handle the fact that DNA is double stranded, we have extended this algorithm to the case where either the input word or its reverse complement (in the biological sense) must appear as a substring in the cyclic cover [3].

1.2   Summary of Our Contribution

Let P be a set of words on a finite alphabet. The well-known shortest superstring problems ask for a either cyclic or linear superstring of minimal length, and the Shortest Cyclic Cover problem asks for a collection of cyclic strings of minimal norm (cumulated length). Here, with the Shortest Mixed Cover of Strings or simply Shortest Mixed Cover, we relax the requirements and accept a solution made of a collection of strings that can be linear or cyclic. We introduce a graph that represents all the maximal overlaps between the input words in small space: the Truncated Hierarchical Overlap Graph (THOG). We show first that the Superstring Graph is embedded in the Truncated Hierarchical Overlap Graph of P; and second, that it captures the set of Mixed Covers built by a greedy algorithm that agglomerates words using their largest overlaps ranked in decreasing order.

As mentioned above, Generalised Suffix Tree can also serve to build the DBG of order k for P [4]. However, classical DBG are limited in the size of overlaps, which must be of length $k-1$. This is a strong limitation, and a natural remedy is to consider overlaps of different sizes, by extending the framework of DBG. Current state-of-the-art proposals successively build and explore several DBG to compute unitigs with different overlap sizes [1,13]. Our proposal is to capture multiple overlap sizes in a single graph and to explore its paths to compute unitigs.

Finally, we show that unitigs built with the IDBA approach are substrings of those found in our Superstring Graph. Moreover, we characterise when a unitig from the SG captures an overlap missed in a multiple DBG approach. It can be proven that IDBA solution is contained in the SG solution. A strong point of the SG algorithm is to retain the sensitivity of variable order DBG without building several graphs, which remains computationally prohibitive. Indeed, it is stated in [13] that exploring the whole range of orders $[k_{min}, k_{max}]$ is not feasible on large-scale data. In fact, each iteration in IDBA takes linear time in $||P||$, while the

SG algorithm takes overall linear time in $||P||$. Our contributions are theoretical, but our solution has a linear space complexity (Theorem 3). For simplicity, here we disregard the fact that one usually does not know from which DNA strand the input reads of an assembly problem come from. Hence, both the reads and their reverses complement are considered in assembly problem. However, the approach described in [3] shows that the results developed here can be extended to handle the case of missing information about the DNA strand. Due to space constraints, the proof of Theorems 1, 2, and 3 are omitted here.

1.3   Notation and Basic Definitions

About Strings. We consider two kinds of strings: linear and cyclic strings. For a string s, the length of s is $|s|$. For a linear string s and $i \leq j$ in $\{1, \ldots, |s|\}$, $s[i, j]$ is the linear substring of s beginning at the position i and ending at the position j, $s[i]$ is the substring $s[i, i]$, $s[1, j]$ is a prefix of s and $s[i, |s|]$ is a suffix of s. A prefix (or suffix) s' of s is proper if s' is different of s. For another linear string t, the *maximum overlap* from s to t, denoted by $\text{ov}(s, t)$, is the longest substring that is a proper suffix of s and a proper prefix of t. The prefix from s to t, denoted by $\text{pr}(s, t)$, is such that $s = \text{pr}(s, t)\text{ov}(s, t)$ and the suffix from s to t, denoted by $\text{suf}(s, t)$, is such that $t = \text{ov}(s, t)\text{suf}(s, t)$. The merge of s with t using their maximal overlap is denoted $s \odot t$ and is equal to $\text{pr}(s, t)\text{ov}(s, t)\text{suf}(s, t)$. Since we consider only maximal overlaps, we simply use the term overlap. For simplicity, we denote the concatenation of s with t simply by st.

We say that a linear string w is a substring of a cyclic string c if there exists w_c a linear permutation of c such that w is a substring of w_c^∞ (where $w_c^\infty = w_c w_c \ldots$). To ease distinction between linear and cyclic strings, we will denote a cyclic string c by $\langle c \rangle$.

For a set P of finite strings, we define and denote the *norm* of P by $||P|| := \sum_{w \in P} |w|$. For two strings x and y, we denote by $x \subset_{sub} y$ the fact that x is a substring of y. We denote the *set of factors* of P by $Fact(P) := \{w \mid \exists s_i \in P, w \subset_{sub} s_i\}$. Moreover, for k an integer, we denote by $Fact_k(P)$ the subset of $Fact(P)$ made of strings of length k.

About permutations. Let E be a finite set. A permutation on E is a bijection from E onto itself. Let σ be a permutation on E. The *partition* of E *due to* σ, which is denoted by $Part_\sigma$, is a partition $(E_1, \ldots, E_p)$ of E of maximal cardinality, and such that for any i in $[1, p]$ and for any x of E_i and for any integer k, one has $\sigma^k(x) \in E_i$. Then, one can define p permutations on E, $(\sigma_1, \ldots, \sigma_p)$, such that for any i in $[1, p]$, for any x in E one has $\sigma_i(x) := \sigma(x)$ if $x \in E_i$, and $\sigma_i(x) := x$ otherwise. Then $(\sigma_1, \ldots, \sigma_p)$ is called a *decomposition* of σ in circular permutations.

Throughout the article. let $P := \{s_1, \ldots, s_n\}$ be a set of input words, and $||P||$ denotes the norm of P. Without loss of generality, we always assume that P is factor-free, i.e. for any two strings of P, none is a substring of the other.

2   Permutations and Truncated Hierarchical Overlap Graph

Let $P = \{s_1, \ldots, s_n\}$ be a finite set of linear strings over a finite alphabet. We can define two types of covers:

- a *cyclic cover of strings* of P is a set $C = \{\langle c_1 \rangle, \ldots, \langle c_p \rangle\}$ of cyclic strings such that each string s_i of P is a substring of a $\langle c_j \rangle$ of C, i.e., $s_i \subset_{sub} \langle c_j \rangle$.
- a *mixed cover of strings* of P is a set $C = \{\langle c_1 \rangle, \ldots, \langle c_q \rangle, l_{q+1}, \ldots, l_p\}$ of cyclic and linear strings such that each string s_i of P is a substring of an element of C.

Obviously, one could consider also linear covers of strings. However, by concatenating the strings of a shortest linear cover one gets a shortest linear superstring. Thus, the problem of finding a shortest linear string cover is as hard and as difficult to approximate as the shortest linear superstring problem (NP-hard [7] and APX-hard [12]). Another reason explains our interest in mixed cover of strings: state-of-the-art assemblers like IDBA or SPAdes can yield linear and cyclic strings. Indeed, the de Bruijn Graph may contain an isolated cycle. Hence, their result is indeed a mixed cover of strings. Clearly, a cyclic cover is a mixed cover, and the norm of a shortest cyclic cover of P is at most that of a shortest mixed cover of P. To our knowledge the problem of finding a shortest mixed cover of strings has not yet been studied.

It is known that for each optimal solution and each greedy solution of $SCCS$, there exists a permutation such that this permutation induces this cyclic cover of strings [5]. Figure 1 shows how to build a cyclic cover of strings from a permutation. Indeed, let $P = \{s_1, \ldots, s_n\}$ be a set of strings which is factor-free and let σ be a permutation. We define

$$CC(P, \sigma) = \{circular(P_1, \sigma_1), \ldots, circular(P_m, \sigma_m)\}$$

where the decomposition in circular permutation of σ is $\sigma_1 \ldots \sigma_m$, $Part_\sigma = \{P_1, \ldots, P_m\}$ is such that for any i in $[1, m]$, P_i is the element of $Part_\sigma$ corresponding to σ_i, and for all i between 1 and m where $P_i = \{s_1^i, \ldots, s_{|P_i|}^i\}$:

$$circular(P_i, \sigma_i) := \langle pr(s_1^i, s_{\sigma_i(1)}^i) . pr(s_{\sigma_i(1)}^i, s_{\sigma_i^2(1)}^i) . \ldots . pr(s_{\sigma_i^{|P_i|-1}(1)}^i, s_1^i) \rangle.$$

We denote by $Overlap(CC(P, \sigma))$ the set of overlaps used by the cyclic cover of strings $CC(P, \sigma)$, i.e. $Overlap(CC(P, \sigma)) = \{ov(s_i, s_{\sigma(i)}) \mid \forall i \in \{1, \ldots, n\}\}$.

For any cyclic cover of strings w of P, we can map each word of P on w, and create the permutation σ_w defined so that on the mapping $s_{\sigma_w(i)}$ is just after the string s_i. Hence, we get that $|CC(P, \sigma_w)| \leq |w|$. Indeed, $CC(P, \sigma_w)$ always merges the input words using their maximal overlaps, while w can use any overlap. Thus, we can restrict the problem $SCCS$ to consider only cyclic covers induced by permutations.

Some assemblers consider that a subset of overlaps are unreliable, for example if these are too short [1,13]. In fact they forbid this subset of overlaps. We adapt our definitions to this case and introduce a set F representing the maximal

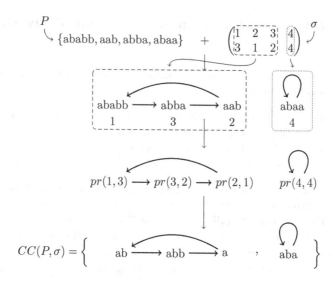

Fig. 1. From a permutation to a cyclic cover. Example with the input set $P :=$ {ababb, aab, abba, abaa}. Instance of a cyclic cover of P obtained with a permutation σ. We obtain the cyclic cover $CC(P,\sigma) = \{\langle ababba \rangle, \langle aba \rangle\}$. (Color figure online)

elements among all forbidden overlaps. All substrings of elements of F will be forbidden. We define variants of shortest cover problems that are constrained by the set of forbidden overlaps (see Definitions 1 and 2).

Definition 1 (Constrained Shortest Cyclic Cover of Strings (CSCCS)).

- **Input:** Two sets of linear strings P and F.
- **Output:** A cyclic cover of strings C induced by a permutation of P such that $Overlap(C) \cap Fact(F) = \emptyset$, which minimises $||C||$.

Note that if we assume two elements x, y of F such that y is a substring of x, $Overlap(C) \cap Fact(\{x\}) = \emptyset$ implies that $Overlap(C) \cap Fact(\{y\}) = \emptyset$. Hence from now on, F is assumed to be factor-free. Unfortunately, some instances of $CSCCS$ lack solutions, and for other instances, the greedy algorithm (Algorithm 1) does not find any solution (see Example 1).

So, we define the following problem which is a relaxed formulation of $CSCCS$.

Definition 2 (Constrained Shortest Mixed Cover of Strings (CSMCS)).

- **Input:** Two sets of linear strings P and F,
- **Output:** A mixed cover of strings C induced by a permutation of P such that $Overlap(C) \cap Fact(F) = \emptyset$, which minimises $||C||$.

We denote by **CMC(P,F)** the set of mixed covers C induced by a permutation of P such that $Overlap(C) \cap Fact(F) = \emptyset$. Let $OPT_{CMC}(P, F)$ be the set of optimal solutions of $CSMCS$ for (P, F).

This time, we can determine easily whether **CMC(P,F)** is empty or not (see Proposition 1 and Example 1). We get the same result as for the greedy solutions of CSMC(P,F) (see Theorem 1 and Example 1).

Proposition 1. *CMC(P,F) is empty if and only if $Fact(F) \cap P$ is not empty.*

Example 1. 1. Let $P = \{abba, baab, bab\}$ and $F = \{b\}$, Then the set of cyclic covers of P constrained by F is empty but $\{\langle abba\rangle, bab\} \in$ **CMC(P,F)**,

2. Let $P = \{abec, bed, cfabe, dgab\}$ and $F = \{b\}$, $OPT_{SCCS}(P) = \{\langle ecfabedgab\rangle\}$; however, the greedy algorithm for $CSCCS$ gives no solution but the greedy algorithm for $CSMCS$ gives $\{\langle cfabe\rangle, bedgab\}$ as a solution.

From now on, we assume that $F \subseteq Fact(P) \setminus P$ and F is factor-free.

Let $P := \{s_1, \ldots, s_n\}$ be a factor-free set of words, and $F \in Fact(P) \setminus P$. For any permutation σ of $\{1, \ldots, |P|\}$, we can obtain a cyclic cover of strings. We ask when such a cyclic cover satisfies the constraint of F, i.e., when it uses a forbidden overlap. For any circular permutation σ_c in a decomposition of σ in circular permutations, we define a set of violations, denoted $Violations(P, F, \sigma_c)$. If this set is empty, the induced cyclic cover is a solution of CSCCS and of CSMCS. If $Violations(P, F, \sigma_c)$ contains only one violation, say i, then the cyclic string can be transformed into a linear string satisfying the constraint of F. The transformation is as follows: the forbidden overlap occurs between s_i and $s_{\sigma_c(i)}$.

One builds the linear word by cutting the word $circular(P, \sigma_c)$ between the words s_i and $s_{\sigma_c(i)}$ to obtain:

$$linear(P, \sigma_c, i) := pr(s_{\sigma_c(i)}, s_{\sigma_c^2(i)})\, pr(s_{\sigma_c^2(i)}, s_{\sigma_c^3(i)}) \, \cdots \, pr(s_{\sigma_c^{n-1}(i)}, s_i)\, s_i$$

Let $F \subseteq Fact(P)$ and σ_c be a circular permutation of $\{1, \ldots, n\}$. We set $Violations(P, F, \sigma_c) := \{i \in \{1, \ldots, n\} \mid \exists f \in F \text{ such that } ov(s_i, s_{\sigma_c(i)}) \subset_{sub} f\}$. Violations are the overlaps used in the cyclic cover (induced by σ) that are substrings of an element of F.

We say that a circular permutation σ_c is *coherent with* (P, F) if and only if $|Violations(P, F, \sigma_c)| \leq 1$. We say that a permutation σ is coherent with (P, F) if each circular permutation in a decomposition of σ in circular permutations is coherent with (P, F). For any circular permutation σ_c that is coherent with (P, F) we define the Mixed Cover (MC) induced by σ_c on (P, F) as

$$MC(P, F, \sigma_c) := \begin{cases} circular(P, \sigma_c) & \text{if } |Violations(P, F, \sigma_c)| = 0, \\ linear(P, \sigma_c, i) & \text{if } Violations(P, F, \sigma_c) = \{i\}. \end{cases}$$

and for any permutation σ coherent with (P, F)

$$MC(P, F, \sigma) := \{MC(P_1, F, \sigma_1), \ldots, MC(P_m, F, \sigma_m)\}$$

where $(\sigma_1, \ldots, \sigma_m)$ is a decomposition of σ in circular permutation, and $Part_\sigma = \{P_1, \ldots, P_m\}$ is such that for any i in $[1, m]$, P_i is the element of $Part_\sigma$ corresponding to σ_i. Let **PMC(P,F)** denote the subset of Mixed Covers induced by a permutation coherent with (P, F). We obtain the following proposition, which means that (1) if there is a solution to CSMC, there also exists one solution induced by a coherent permutation, and (2) an optimal solution is induced by a coherent permutation.

Proposition 2. *Let P be a factor-free set of words and let $F \subseteq Fact(P)$. One has*

1. **$PMC(P,F) = \emptyset$** *if and only if* **$SMC(P,F) = \emptyset$**.
2. **$OPT(P,F) \subseteq PMC(P,F) \subseteq SMC(P,F)$**,

Let us introduce the Truncated Hierarchical Overlap Graph (THOG), which is a generalised version of the Hierarchical Overlap Graph defined in [3].

Let $Ov(P)$ be the set of maximum overlaps from a string of P to another string or the same string of P. Let $Ov^*(P, F)$ be $Ov(P)$ minus the set of all factors of forbidden overlaps; in other words, $Ov^*(P, F) := Ov(P) \setminus Fact(F)$. Now, we define the *Truncated Hierarchical Overlap Graph (THOG)* of (P, F), in which the nodes are either words of P or allowed overlaps between these words, and an arc links a string to the node representing its maximal suffix or the maximal prefix of a string with this string. Two examples of THOG are shown in Fig. 4a and c.

Definition 3. *The Truncated Hierarchical Overlap Graph of (P, F), denoted by $THOG(P, F)$, is the oriented graph $(P \cup Ov^*(P, F), R \cup B)$ where:*

$$R = \{(x, y) \in (P \cup Ov^*(P, F)) \times (P \cup Ov^*(P, F)) \mid y \ longest \ suffix \ of \ x \ in \ P\}$$
$$B = \{(y, x) \in (P \cup Ov^*(P, F)) \times (P \cup Ov^*(P, F)) \mid y \ longest \ prefix \ of \ x \ in \ P\}$$

R denotes the set of red arcs, and B the set of blue arcs.

It is known that overlaps between two strings are explicit nodes in the Generalised Suffix Tree of these words [8]. Hence, all nodes of THOG are explicit nodes of the Generalised Suffix Tree of P. Moreover, a blue arc is a contracted path of edges of the suffix tree, while a red arc is a contracted path of suffix links. Altogether, we can built THOG in linear time.

Proposition 3. *The graph $THOG(P, F)$ can be built in linear time in $\|P\|$.*

Let s_i and s_j be two words of P. We define the RB-path from s_i to s_j, denoted by $RB\text{-}path(s_i, s_j)$, as the path in $THOG(P, F)$ going from s_i to $ov(s_i, s_j)$ using only arcs from R, and then from $ov(s_i, s_j)$ to s_j using only arcs of B. Let σ be a permutation of P coherent with (P, F). Then, for any i between 1 and n, the RB-path from s_i to $s_{\sigma(i)}$ is well defined and exists in $THOG(P, F)$.

THOG construction algorithm We execute Gusfield's algorithm for finding maximal overlap nodes in the Generalised Suffix Tree (GST) of P [8] and along the way we mark the words of F. This gives an explicit list of the THOG nodes. We then perform a depth first traversal of the GST (using the suffix tree arcs) to set all blue arcs of the THOG. Finally, we perform the same using the tree of suffix links to the set of all red arcs. Altogether it takes linear time in the GST of P.

3   Greedy Algorithm and Superstring Graph

Here, we define the Superstring Graph and introduce the greedy algorithm for the problem *Shortest Mixed Cover of Strings* (SMC) of a set P of words. The difference between the norm of the input and the norm of a solution is the compression achieved by this solution. Finally, we show that a Eulerian multi-path of the SG and the associated set of words form a solution of SMC, which approximates the optimal compression by a factor $\frac{1}{2}$, as later shown in Theorem 1.

We define the greedy algorithm for CSMCS (see Algorithm 1).

Algorithm 1. The greedy algorithm for CSMCS

---

1 **Input**: P a set of linear words and $F \subseteq Fact(P)$ **Output**: $C' \in$ **PMC(P,F)**
2 $C := \emptyset$
3 **while** $Ov^*(P, F)$ *is not empty* **do**
4 $\quad$ Select u and v of P which have the longest overlap (u can be equal to v)
5 $\quad$ $P := P \setminus \{u, v\}$
6 $\quad$ **if** $u = v$ *(i.e. $u \odot v$ is cyclic)* **then** $C := C \cup \{u \odot v\}$;
7 $\quad$ **else** $P := P \cup \{u \odot v\}$;
8 **return** $C \cup P$

Let **Greedy(P,F)** denote the set of solutions of algorithm greedy for CSMCS (for simplicity, we say greedy solutions). One has the following theorem, whose third statement gives the $\frac{1}{2}$-approximation ratio of compression of the greedy algorithm. To prove this ratio, one can define a subset system for CSMCS, which turns out to be 2-extendible. The ratio of $\frac{1}{2}$ follows directly from this 2-extendibility [10] (see [6] for details). Note this greedy approximation ratio of $\frac{1}{2}$ for the compression is the same as for the well-studied Shortest Common Superstring problem [6,14]. These considerations support Theorem 1 (omitted proof).

Theorem 1. *Let P be a factor-free set of words and let $F \subseteq Fact(P)$. One has*

1. **Greedy(P,F)** $\subseteq$ **PMC(P,F)** $\cap \{\sigma$ *permutation coherent with* $(P, F)\}$,
2. **Greedy(P,F)** $= \emptyset$ *if and only if* **SMC(P,F)** $= \emptyset$.
3. *Let $w_g \in$ **Greedy(P,F)** and $w_o \in$ **OPT(P,F)**. Then $||P|| - w_g \geq \frac{1}{2}(||P|| - w_o)$.*

The inclusions of set of solutions are illustrated in Fig. 2. Section 3 states how greedy solutions can be found in linear time.

Let σ be a permutation that is coherent with (P, F) and such that $MC(P, F, \sigma)$ is a greedy solution for CSMCS. Let us denote by $G(\sigma)$ the subgraph that consists of the set of RB-paths from s_i to $s_{\sigma(i)}$ for all i in $[1, n]$. As in [5], one can show that any two permutations that are coherent with (P, F) and correspond to a greedy solution for CSMCS, yield the same graph. We call this graph the *Superstring Graph (SG)* and define it as follows. Two examples of superstring graphs are shown in Fig. 4b and d.

Definition 4 (Superstring Graph). *The Superstring Graph of (P, F), denoted $SG(P,F)$, is the graph $G(\sigma)$ where σ is a permutation that is coherent with (P, F) and corresponds to a greedy solution for CSMCS.*

As the Superstring Graph of (P, F) is embedded in THOG of (P, F), it can clearly be built in a time linear in $||P||$.

Theorem 2. *The Superstring Graph of (P, F) can be built in time $O(||P|| + ||F||)$.*

4 Comparing the Superstring Graph with a Multiple Order DBG Approach

The IDBA assembler iteratively builds DBG basically as depicted in Algorithm 2. The only difference concerns the step for removing the so-called short *dead-ends* in the DBG at each iteration. As the name says, a dead-end is a simple path starting after a branching node and ending in a node having a single neighbour. IDBA removes dead-ends shorter than $2k$, which are likely due to nucleotidic errors [13]. Such a dead-end would make up a very short, biologically meaningless, unitig. However, for the simplicity of the proofs, we consider a simplified algorithm without short dead-end removal. As usual, we require that the input set P of words is factor-free.

Algorithm 2. Algorithm IDBA assembler where $DB_m(P, k)$ is the de Bruijn Graph of order k (i.e. , dBG_k^+) where we remove all nodes which represent a k-mer of coverage smaller than m.

---

1 **Input:** A set P of reads factor-free   **Output:** A set $U_{k_{max}}$ of unitigs
2 **for** $k_{min} \leq k \leq k_{max}$ **do**
3 $\quad\mid\quad H_k = DB_m(P, k)$
4 $\quad\mid\quad U_k = $ Unitigs H_k
5 $\quad\llcorner\quad P = P \cup U_k$
6 **return** $U_{k_{max}}$

About IDBA algorithm (Algorithm 2)

Complexity. For each k between k_{min} and k_{max}, the algorithm needs to look at all the strings of the instance, i.e. $||P||$. At the end, the complexity of Algorithm 2 is at least linear in $(k_{max} - k_{min}) \times ||P||$.

Theoretical Solution. Nothing prevents a unitig of $DB_m(P, k)$ from being a cycle. Let w be a string. We denote the cover of w in P by $Cov_P(w) := \{(i, j) \mid \exists r_i \in P \text{ such that } w = r_i[j : j + |w|]\}$. Let $P(k, m)$ denote the set of substrings of P satisfying, for any $w \in P(k, m)$: $|w| \leq k$, and $Cov_P(w) \geq m$, and for all word w' such that w is a proper substring of w', $Cov_P(w') < m$ or $|w'| > k$. Hence, we have that $U_{k_{max}}$, which is a set of cyclic and linear strings, is in fact a mixed cover of string of $P(k_{max} + 1, m)$ with the set of $(k_{min} - 1)$-mers, i.e. $Fact_{k_{min}-1}(P)$, is taken as forbidden overlaps.

We are going to use the Superstring Graph on the Truncated Hierarchical Overlap Graph to build in linear time in $||P||$ an improved mixed cover for the same instance, that is for strings of $P(k_{max} + 1, m)$ with the set of $(k_{min} - 1)$-mers taken as forbidden overlaps. The mixed cover obtained from the superstring graph is smaller in terms of inclusion, of cardinality and of norm than $U_{k_{max}}$ (see Theorem 3).

Let $SG(P, k_{max}, k_{min}, m)$ be the Superstring Graph of $(P(k_{max} + 1, m), Fact_{k_{min}-1}(P))$. A RB-route of a Superstring Graph is a sub-path of a sequence of RB-paths.

Proposition 4. *We can build $SG(P, k_{max}, k_{min}, m)$ in linear time in $||P||$.*

Proof. With the Generalised Suffix Tree of P, we can build $P(k_{max} + 1, m)$ in linear time in the size of P. We can build the Superstring Graph of $(P(k_{max}+1, m), Fact_{k_{min}-1}(P))$ in linear time in $||P(k_{max}+1, m)||$, because in this case, we can determine the nodes of the tree corresponding to the elements of $Fact_{k_{min}-1}(P)$ during the construction of the GST of P without reading these strings. Hence, it improves on the complexity of Theorem 2, and show that one can build the SG in linear time in $||P||$.

Let U_{SG} be the set of labelled maximal RB-routes (u, v) of the $SG(P, k_{max}, k_{min}, m)$ such that $(d_R^{in}(u) = d_B^{in}(u) = 1$ or $d_R^{in}(u) + d_B^{in}(u) \leq 1)$ and $d_R^{out}(v) + d_B^{out}(v) \leq 1$. Here, $d_R^{in}(u)$ denotes the in-degree in number of red arcs of node u in the superstring graph, and $d_R^{out}(u)$ the out-degree of u in number of red arcs. The notation $d_B^{in}(u)$ and $d_B^{out}(u)$ are defined similarly for blue arcs.

Proposition 5. *For all $c \in U_{k_{max}}$, there exists $x \in U_{SG}$ such that $c \subset_{sub} x$.*

Proposition 6. *For all $x \in U_{SG}$, $\exists c_1, \ldots, c_q \in U_{k_{max}}$ such that $x = c_1 \odot \ldots \odot c_q$ and for all i between 1 and $q - 1$, $|ov(c_i, c_{i+1})| \geq k_{min}$.*

Theorem 3. *We can build a mixed cover that includes a solution of Algorithm 2 in time in $O(||P||)$, and in linear space in the size of the de Bruijn Graph of order k_{max} of P.*

Now, we know that the words of U_{SG}, the solution provided by the SG, contains all unitigs of IDBA as substrings. Some words of U_{SG} are exactly equal to some unitigs of IDBA. However, the remaining words of U_{SG}, contain strictly more than one unitig of IDBA as substring. In other words, they elongate the unitigs of IDBA by capturing an overlap missed by IDBA. We formalise this result in the next proposition.

Proposition 7. *(Figure 3) For all $x \in U_{SG}$, $\exists c_1, \ldots, c_q \in U_{k_{max}}$ such that $x = c_1 \odot \ldots \odot c_q$ and for all i between 1 and $q - 1$, there exists $y \in U_{SG}$ such that $\exists c_1'$ and $c_2' \in U_{k_{max}}$ such that $c_1' \odot c_2' \subset_{sub} y$ and $ov(c_i, c_{i+1})$ is a strict prefix of $ov(c_1', c_2')$.*

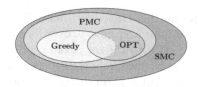
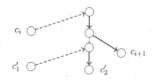

Fig. 2. Inclusions of sets of solutions of the greedy algorithm for CSMCS.

Fig. 3. Illustration of Proposition 7.

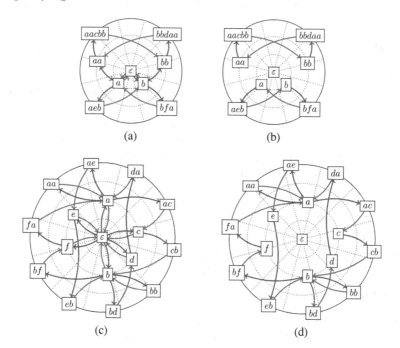

Fig. 4. Examples of truncated hierarchical overlap graphs ((a) and (c)) and of associated Superstring Graphs ((b) and (d)) for $P := \{aacbb, bbdaa, aeb, bfa\}$. (a) and (b) have instance (P, F) with $F := \emptyset$. (c) and (d) have instance $(P(2, 1), F)$ with $F := Fact_1(P)$; F forbids any overlap of length 1 or 0. (Color figure online)

5   Conclusion

State-of-the-art genome assemblers, like IDBA or SPAdes, build multiple DBG with distinct values of k to improve the quality of assembled unitigs. In a formal manner, we compared the result of IDBA with the sequences obtained using the Superstring Graph of an input set P of reads. The SG is a recently introduced digraph with labels on its arcs, which is embedded in a Truncated Hierarchical Overlap Graph (THOG) of P. The SG yields solutions for *Constrained Mixed Cover* that greedily merge the input words using their maximal overlaps; hence, we get a $\frac{1}{2}$-approximation ratio for the compression. We show that the unitigs output by IDBA are always substrings of the sequences assembled with the

SG, and that the converse is false. Indeed, some assembled sequences from the SG extend IDBA unitigs by merging words with smaller overlaps that cannot be incorporated in IDBA. For the first time, a theoretical framework helps to understand and to characterise formally the output of real-world assembly software that adopts a multiple-order de Bruijn graph approach. It also provides a way to improve on their results. Moreover the Superstring Graph offers the possibility to dynamically extend the range of overlap lengths considered without recomputing the unitigs from scratch. It can be adapted to cope with reverse complement of the reads/k-mers using the approach of [3]. The main advantage of the SG, which is linear in the input size, over IDBA is to concentrate all overlaps needed to build a similar assembly in one single graph.

Acknowledgements. We thank the reviewers for their comments and suggestions.

References

1. Bankevich, A., Nurk, S., Antipov, D., Gurevich, A.A., Dvorkin, M., Kulikov, A.S., Lesin, V.M., Nikolenko, S.I., Pham, S., Prjibelski, A.D., Pyshkin, A.V., Sirotkin, A.V., Vyahhi, N., Tesler, G., Alekseyev, M.A., Pevzner, P.A.: SPAdes: a new genome assembly algorithm and its applications to single-cell sequencing. J. Comp. Biol. **19**(5), 455–477 (2012)
2. Boucher, C., Bowe, A., Gagie, T., Puglisi, S.J., Sadakane, K.: Variable-order de bruijn graphs CoRR abs/1411.2718 (2014)
3. Cazaux, B., Cánovas, R., Rivals, E.: Shortest DNA cyclic cover in compressed space. In: Data Compression Conference DCC, pp. 536–545. IEEE Computer Society Press (2016)
4. Cazaux, B., Lecroq, T., Rivals, E.: From indexing data structures to de bruijn graphs. In: Kulikov, A.S., Kuznetsov, S.O., Pevzner, P. (eds.) CPM 2014. LNCS, vol. 8486, pp. 89–99. Springer, Heidelberg (2014)
5. Cazaux, B., Rivals, E.: A linear time algorithm for shortest cyclic cover of strings. J. Discrete Algorithms (2016). doi:10.1016/j.jda.2016.05.001
6. Cazaux, B., Rivals, E.: The power of greedy algorithms for approximating Max-ATSP, cyclic cover, and superstrings. Discrete Appl. Math. (2015). doi:10.1016/j.dam.2015.06.003
7. Gallant, J., Maier, D., Storer, J.A.: On finding minimal length superstrings. J. Comput. Syst. Sci. **20**, 50–58 (1980)
8. Gusfield, D., Landau, G.M., Schieber, B.: An efficient algorithm for the all pairs suffix-prefix problem. Inf. Process. Lett. **41**(4), 181–185 (1992)
9. Lin, Y., Pevzner, P.A.: Manifold de bruijn graphs. In: Brown, D., Morgenstern, B. (eds.) WABI 2014. LNCS, vol. 8701, pp. 296–310. Springer, Heidelberg (2014)
10. Mestre, J.: Greedy in approximation algorithms. In: Azar, Y., Erlebach, T. (eds.) ESA 2006. LNCS, vol. 4168, pp. 528–539. Springer, Heidelberg (2006)
11. G. K. C. of Scientists: Genome 10K a proposal to obtain whole-genome sequence for 10 000 vertebrate species. J. Hered. **100**(6), 659–674 (2009)
12. Ott, S.: Lower bounds for approximating shortest superstrings over an alphabet of size 2. In: Widmayer, P., Neyer, G., Eidenbenz, S. (eds.) WG 1999. LNCS, vol. 1665, pp. 55–64. Springer, Heidelberg (1999)

Efficient FPT Algorithms for (Strict) Compatibility of Unrooted Phylogenetic Trees

Julien Baste[1], Christophe Paul[1], Ignasi Sau[1(✉)], and Celine Scornavacca[2]

[1] CNRS, LIRMM, Université de Montpellier, Montpellier, France
{baste,paul,sau}@lirmm.fr
[2] Institut des Sciences de l'Evolution
(Université de Montpellier, CNRS, IRD, EPHE), Montpellier, France
celine.scornavacca@umontpellier.fr

Abstract. In phylogenetics, a central problem is to infer the evolutionary relationships between a set of species X; these relationships are often depicted via a phylogenetic tree – a tree having its leaves univocally labeled by elements of X and without degree-2 nodes – called the "species tree". One common approach for reconstructing a species tree consists in first constructing several phylogenetic trees from primary data (e.g. DNA sequences originating from some species in X), and then constructing a single phylogenetic tree maximizing the "concordance" with the input trees. The so-obtained tree is our estimation of the species tree and, when the input trees are defined on overlapping – but not identical – sets of labels, is called "supertree". In this paper, we focus on two problems that are central when combining phylogenetic trees into a supertree: the compatibility and the strict compatibility problems for unrooted phylogenetic trees. These problems are strongly related, respectively, to the notions of "containing as a minor" and "containing as a topological minor" in the graph community. Both problems are known to be fixed-parameter tractable in the number of input trees k, by using their expressibility in Monadic Second Order Logic and a reduction to graphs of bounded treewidth. Motivated by the fact that the dependency on k of these algorithms is prohibitively large, we give the first explicit dynamic programming algorithms for solving these problems, both running in time $2^{O(k^2)} \cdot n$, where n is the total size of the input.

Keywords: Phylogenetics · Compatibility · Unrooted phylogenetic trees · Parameterized complexity · FPT algorithm · Dynamic programming

1 Introduction

A central goal in *phylogenetics* is to clarify the relationships of extant species in an evolutionary context. Evolutionary relationships are commonly represented via *phylogenetic trees*, that is, acyclic connected graphs where leaves are univocally labeled by a label set X, and without degree-2 nodes. When a phylogenetic tree is defined on a label set X designating a set of genes issued from a gene family, we

© Springer International Publishing Switzerland 2016
R. Dondi et al. (Eds.): AAIM 2016, LNCS 9778, pp. 53–64, 2016.
DOI: 10.1007/978-3-319-41168-2_5

refer to it as a *gene tree*, while, when X corresponds to a set of extant species, we refer to it as a *species tree*. A gene tree can differ from the species tree depicting the evolution of the species containing the gene for a number of reasons [14]. Thus, a common way to estimate a species tree for a set of species X is to choose *several* gene families that appear in the genome of the species in X, reconstruct a gene tree per each gene family (see [9] for a detailed review of how to infer phylogenetic trees), and finally combine the trees in a unique tree that maximizes the "concordance" with the given gene trees. The rationale underlying this approach is the confidence that, using several genes, the species signal will prevail and emerge from the conflicting gene trees. If the gene trees are all defined on the same label set, we are in the *consensus* setting; otherwise the trees are defined on overlapping – but not identical – sets of labels, and we are in the *supertree* setting. Several consensus and supertree methods exist in the literature (see [2,3,16] for a review), and they differ in the way the concordance is defined.

In this paper, we focus on a problem that arises in the supertree setting: given a set of gene trees $\mathcal{T} = \{T_1, \ldots, T_k\}$ on label sets $\{X_1, \ldots, X_k\}$, respectively, does there exist a species tree on $X := \cup_{i=1}^{k} X_i$ that *displays* all the trees in $\mathcal{T}$? This is the so-called COMPATIBILITY OF UNROOTED PHYLOGENETIC TREES problem. The notion of "displaying" used by the phylogenetic community, which will be formally defined in Sect. 2, coincides with that of "containing as a minor" in the graph community. Another related problem is the STRICT COMPATIBILITY (or AGREEMENT) OF UNROOTED PHYLOGENETIC TREES problem, where the notion of "displaying" is replaced by that of "strictly displaying". This notion, again defined formally in Sect. 2, coincides with that of "containing as a topological minor" in the graph community.

Both problems are polynomial-time solvable when the given gene trees are out-branching (or *rooted* in the phylogenetic literature), or all contain some common label [1,15]. In the general case, both problems are NP-complete [18] and fixed-parameter tractable in the number of trees k [5,17]. The fixed-parameter tractability of these problems has been established via Monadic Second Order Logic (MSOL) together with a reduction to graphs of bounded treewidth. For both problems, it can be checked that the corresponding MSOL formulas [5,17] contain 4 alternate quantifiers, implying by [10] that the dependency on k in the derived algorithms is given by a tower of exponentials of height 4; clearly, this is prohibitively large for practical applications. Therefore, even if the notion of compatibility has been defined quite some time ago [11], at the moment no "reasonable" FPT algorithms exist for these problems, that is, algorithms with running time $f(k) \cdot p(|X|)$, with f a moderately growing function and p a low-degree polynomial. In this paper we fill this lack and we prove the following two theorems.

Theorem 1. *The* COMPATIBILITY OF UNROOTED PHYLOGENETIC TREES *problem can be solved in time* $2^{O(k^2)} \cdot n$, *where k is the number of trees and n is the total size of the input.*

Theorem 2. *The* AGREEMENT OF UNROOTED PHYLOGENETIC TREES *problem can be solved in time* $2^{O(k^2)} \cdot n$, *where k is the number of trees and n is the total size of the input.*

Our approach for proving the two above theorems is to present explicit dynamic programming algorithms on graphs of bounded treewidth. As one could suspect from the fact that the corresponding MSOL formulas are quite involved [5,17], it turns out that our dynamic programming algorithms are quite involved as well, implying that we are required to use a technical data structure.

This paper is organized as follows. In Sect. 2 we provide some preliminaries and we define the problems under study. In Sect. 3 we present our algorithm for the COMPATIBILITY OF UNROOTED PHYLOGENETIC TREES problem. Due to space limitations, its proof of correctness, the analysis of its running time, as well as the entire algorithm for the AGREEMENT OF UNROOTED PHYLOGE-NETIC TREES problem can be found in the full version of this article, available at [arXiv:1604.03008]. Finally, we provide some directions for further research in Sect. 4.

2   Preliminaries

Basic Definitions. Given a positive integer k, we denote by $[k]$ the set of all integers between 1 and k. If S is a set, we denote by 2^S the set of all subsets of S. A *tree* T is an acyclic connected graph. We denote by $V(T)$ its vertex set, by $E(T)$ its edge set, and by $L(T)$ its set of vertices of degree one, called *leaves*. Two trees T and T' are *isomorphic* if there is a bijective function $\alpha : V(T) \cup E(T) \to V(T') \cup E(T')$ such that for every edge $e = \{u, v\} \in E(T)$, $\alpha(e) = \{\alpha(u), \alpha(v)\}$. If T is a tree and S is a subset of $V(T)$, we denote by $T[S]$ the subgraph of T induced by S. *Suppressing* a degree-2 vertex v in a graph G consists in deleting v and adding an edge between the former neighbors of v, if they are not already adjacent. *Identifying* two vertices v and v' of a graph G consists in creating a graph H by removing v and v' and adding a new vertex w such that, for each $u \in V(G) \setminus \{v, v'\}$, there is an edge $\{u, w\}$ in $E(H)$ if and only if $\{u, v\} \in E(G)$ or $\{u, v'\} \in E(G)$. *Contracting* an edge $e = \{u, v\}$ in G consists in identifying u and v. A graph H is a *minor* (resp. *topological minor*) of a graph G if H can be obtained from a subgraph of G by contracting edges (resp. contracting edges with at least one vertex of degree 2). See [8] for more details about the notions of minor and topological minor. If Y is a subset of vertices of a tree T, then $T|_Y$ is the tree obtained from the minimal subtree of T containing Y by suppressing degree-2 vertices. For simplicity, we may sometimes consider the vertices of $T|_Y$ also as vertices of T.

As already mentioned in the introduction, an *unrooted phylogenetic tree* on a label set X is defined as a pair (T, ϕ) with T a tree with no degree-2 vertex along with a bijective function $\phi : L(T) \to X$. We say that a vertex $v \in L(T)$ is *labeled* with label $\phi(v)$. Two unrooted phylogenetic trees (T, ϕ) and (T', ϕ') are *isomorphic* if there exists an isomorphism α from T to T' satisfying that if $v \in L(T)$ then $\phi'(\alpha(v)) = \phi(v)$.

The three graph operations defined above, namely suppressing a vertex, identifying two vertices, and contracting an edge, can be naturally generalized to unrooted phylogenetic trees. In this context, two vertices to be identified are either both unlabeled or both with the same label. In the latter case, the newly created vertex inherits the label of the identified vertices. Finally, contractions in unrooted phylogenetic trees are restricted to edges incident to two unlabeled vertices. In this case, we speak about *upt-contraction*. If (T, ϕ) is an unrooted phylogenetic tree and Y is subset of leaves of $L(T)$, then $(T, \phi)|_Y$ is the unrooted phylogenetic tree $(T|_Y, \phi|_Y)$ where $\phi|_Y$ is the restriction of ϕ to the label set Y.

(Strictly) Compatible Supertree. Let $\mathcal{T} = \{(T_1, \phi_1), (T_2, \phi_2), \ldots, (T_k, \phi_k)\}$ be a collection of unrooted phylogenetic trees, not necessarily on the same label set. We say that an unrooted phylogenetic tree (T, ϕ) is a *compatible supertree* of $\mathcal{T}$ if for every $i \in [k]$, $(T_i, \phi_i) \in \mathcal{T}$ can be obtained from $(T, \phi)|_{L(T_i)}$ by performing upt-contractions. The phylogenetic tree (T, ϕ) is a *strictly compatible supertree* of $\mathcal{T}$ if for every $i \in [k]$, $(T_i, \phi_i) \in \mathcal{T}$ is isomorphic to $(T, \phi)|_{L(T_i)}$. If a collection $\mathcal{T}$ of unrooted phylogenetic trees admits a (strictly) compatible supertree, then we say that $\mathcal{T}$ is *(strictly) compatible*. The two definitions are equivalent when $\mathcal{T}$ contains only binary phylogenetic trees, that is, unrooted trees in which every vertex that is not a leaf has degree 3. Note that, as mentioned in the introduction, the notions of "being a compatible supertree" and "being a strictly compatible supertree" correspond, modulo the conditions on the labels, to the notions of "containing as a minor" and "containing as a topological minor", respectively.

In this paper we consider the following problem:

COMPATIBILITY OF UNROOTED PHYLOGENETIC TREES
Instance: A set $\mathcal{T}$ of k unrooted phylogenetic trees.
Parameter: k.
Question: Does there exist an unrooted phylogenetic tree (T, ϕ) that is a compatible supertree of $\mathcal{T}$?

The AGREEMENT (OR STRICT COMPATIBILITY) OF UNROOTED PHYLOGENETIC TREES problem is defined analogously, just by replacing "compatible supertree" with "strictly compatible supertree". For notational simplicity, we may henceforth drop the function ϕ from an unrooted phylogenetic tree (T, ϕ), and just assume that each leaf of T comes equipped with a label.

Assume that $\widehat{T}$ is a compatible supertree of $\mathcal{T}$. Then, according to the definition of minor, for every $i \in [k]$, every vertex $v \in V(T_i)$ can be mapped to a subtree of $\widehat{T}$, in such a way that the subtrees corresponding to the vertices of the same tree are pairwise disjoint. We call the set of vertices of that subtree the *vertex-model* of v. Observe that by the definition of the upt-contraction operation, the vertex-model of a leaf is a singleton. Hereafter, we denote by $\widehat{\varphi}(v)$ the subset of vertices belonging to the vertex-model of v. Moreover, if $u, v \in V(T_i)$ are two adjacent vertices in T_i, then there is *exactly one* edge in $\widehat{T}$ that connects the vertex-model of u to the vertex-model of v. We call such an edge of $\widehat{T}$ the *edge-model* of $\{u, v\} \in E(T_i)$. Observe that a vertex of $\widehat{T}$ may belong to several vertex-models, but then these vertex-models correspond to vertices from different trees of $\mathcal{T}$. Also, an edge of $\widehat{T}$ may be the edge-model of edges of different trees of $\mathcal{T}$.

Similarly, if $\widehat{T}$ is a strictly compatible supertree of $\mathcal{T}$, then according to the definition of topological minor, for every $i \in [k]$, every vertex $v \in V(T_i)$ can be mapped to a vertex of $\widehat{T}$, called the *vertex-model* of v, in such a way that this mapping is injective when restricted to every $i \in [k]$. In this case, if $u, v \in V(T_i)$ are two adjacent vertices in T_i, then there is exactly one *path* in $\widehat{T}$ that connects the vertex-model of u to the vertex-model of v called the *edge-model* of $\{u, v\} \in E(T_i)$. Similarly to the vertex-models, the edge-models of the same tree need to be pairwise disjoint, except possibly for their endvertices.

Treewidth. A *tree-decomposition* of width w of a graph $G = (V, E)$ is a pair $(\mathsf{T}, \mathcal{B})$, where T is a tree and $\mathcal{B} = \{B_t \mid B_t \subseteq V, t \in V(\mathsf{T})\}$ such that

- $\bigcup_{t \in V(\mathsf{T})} B_t = V$,
- for every edge $\{u, v\} \in E$ there is a $t \in V(\mathsf{T})$ such that $\{u, v\} \subseteq B_t$,
- $B_i \cap B_k \subseteq B_j$ for all $\{i, j, k\} \subseteq V(\mathsf{T})$ such that j lies on the unique path from i to k in T, and
- $\max_{t \in V(\mathsf{T})} |B_t| = w + 1$.

To avoid confusion, we speak about the *nodes* of a tree-decomposition and the *vertices* of a graph. The sets of $\mathcal{B}$ are called *bags*. The *treewidth* of G, denoted by $\mathbf{tw}(G)$, is the smallest integer w such that there is a tree-decomposition of G of width w.

Theorem 3 (Bodlander et al. [4]). *Let G be a graph and k be an integer. In time $2^{O(k)} \cdot n$, we can either decide that $\mathbf{tw}(G) > k$ or construct a tree-decomposition of G of width at most $5k + 4$.*

A tree-decomposition $(\mathsf{T}, \mathcal{B})$ rooted at a distinguished node t_r is *nice* if the following conditions are fulfilled:

- $B_{t_r} = \emptyset$ and this is the only empty bag,
- each node has at most two children,
- for each leaf $t \in V(\mathsf{T})$, $|B_t| = 1$,
- if $t \in V(\mathsf{T})$ has exactly one child t', then either
  - $B_t = B_{t'} \cup \{v\}$ for some $v \notin B_{t'}$ and t is called an *introduce-vertex* node, or
  - $B_t = B_{t'} \setminus \{v\}$ for some $v \in B_{t'}$ and t is called a *forget-vertex* node, or
  - $B_t = B_{t'}$, t is associated with an edge $\{x, y\} \in E(G)$ with $x, y \in B_t$, and t is called an *introduce-edge* node. We add the constraint that each edge of G labels exactly one node of T.
- and if $t \in V(\mathsf{T})$ has exactly two children t' and t'', then $B_t = B_{t'} = B_{t''}$. Then t is called a *join* node.

Note that we follow closely the definition of nice tree-decomposition given in [6], which slightly differs from the usual one [12]. Given a tree-decomposition, then we can build a nice tree-decomposition of G with the same width in polynomial time [6, 12].

Let $(\mathsf{T}, \mathcal{B})$ be a nice tree-decomposition of a graph G. For each node $t \in V(\mathsf{T})$, we define the graph $G_t = (V_t, E_t)$ where V_t is the union of all bags corresponding to the descendant nodes of t, and E_t is the set of all edges introduced by the descendant nodes of t. Observe that the graph G_t may be disconnected.

The Display Graph. Let $\mathcal{T} = \{(T_1, \phi_1), (T_2, \phi_2), \ldots, (T_k, \phi_k)\}$ be a collection of unrooted phylogenetic trees. The *display graph* $D_{\mathcal{T}} = (V_D, E_D)$ of $\mathcal{T}$ is the graph obtained from the disjoint union of the trees in $\mathcal{T}$ by iteratively identifying every pair of labeled vertices with the same label. We denote by L_D the set of vertices of $D_{\mathcal{T}}$ resulting from these identifications. The elements of L_D are called the *labeled vertices*. Observe that every vertex of $V_D \setminus L_D$ (resp. every edge of E_D) is also a vertex (resp. an edge) of some tree $T_i \in \mathcal{T}$. If v is a vertex of L_D, then we will say, with a slight abuse of notation, that v is a vertex of T_i if it results from the identification of some leaf of T_i. Finally, the display graph $D_{\mathcal{T}}$ is equipped with a coloring function $c : V_D \cup E_D \to \{0, \ldots, k\}$ defined as follows. If $v \in L_D$, then we set $c(v) = 0$; if $v \in (V_D \setminus L_D) \cup E_D$ belongs to the tree T_i, we set $c(v) = i$. Observe that if a vertex $v \in L_D$ is incident to an edge e such that $c(e) = i$, then v belongs to T_i. Suppose that $\widehat{T}$ is a (strictly) compatible supertree of $\mathcal{T}$. Then we extend the definition of vertex-model and edge-model for the vertices and edges of the T_i's to the vertices and edges of the display graph $D_{\mathcal{T}}$.

The following theorem provides a bound on the treewith of the display graph of a (strictly) compatible family of unrooted phylogenetic trees:

Theorem 4 (Bryant and Lagergren [5]). *Let* $\mathcal{T} = \{(T_1, \phi_1), (T_2, \phi_2), \ldots, (T_k, \phi_k)\}$ *be a collection of (strictly) compatible unrooted phylogenetic trees, not necessarily on the same label set. The display graph of $\mathcal{T}$ has treewidth at most k.*

3   Compatibility Version

This section provides a proof of Theorem 1. Let $D = (V_D, E_D)$ be the display graph of a collection $\mathcal{T} = \{(T_1, \phi_1), (T_2, \phi_2), \ldots, (T_k, \phi_k)\}$ of unrooted phylogenetic trees, and let $n = |V(D)|$. By Theorems 3 and 4, we may assume that we are given a nice tree-decomposition $(\mathsf{T}, \mathcal{B})$ of D of width at most $5k + 4$, as otherwise we can safely conclude that $\mathcal{T}$ is not compatible. Let t_r be the root of T, and recall that $B_{t_r} = \emptyset$.

Our objective is to build a compatible supertree $\widehat{T}$ of $\mathcal{T}$, if such exists. (We would like to note that there could exist an exponential number of compatible supertrees; we are just interested in constructing *one* of them.) As it is usually the case of dynamic programming algorithms on tree-decompositions, for building $\widehat{T}$ we process $(\mathsf{T}, \mathcal{B})$ in a bottom-up way from the leaves to the root, where we will eventually decide whether a solution exists or not. We first describe the data structure used by the algorithm along with a succinct intuition behind the defined objects, and then we proceed to the description of the dynamic programming algorithm itself.

Description of the Data Structure. Before defining the dynamic-programming table associated with every node t of $(\mathsf{T}, \mathcal{B})$, we need a few more definitions.

Definition 1. *Given a node t of $(\mathsf{T}, \mathcal{B})$, its graph $G_t = (V_t, E_t)$, and a subset $Z \subseteq V_t$, a (Z, t)-supertree is a tuple $\mathfrak{T} = (T, \varphi, \psi, \rho)$ such that*

- *T is a tree containing at most $|B_t| + |Z|$ vertices,*

- $\varphi : Z \to 2^{V(T)}$, *called the* vertex-model function, *associates every* $v \in Z$ *with a subset* $\varphi(v)$ *such that*
  - $T[\varphi(v)]$ *is connected and if* v *is a labeled vertex, then* $|\varphi(v)| = 1$, *and*
  - *if* u *and* v *are two vertices of* Z *such that* $c(u) = c(v)$, *then* $\varphi(u) \cap \varphi(v) = \emptyset$,
- $\psi : E(T) \to 2^{[k]}$, *called the* edge-model function, *associates a subset of colors with every edge of* T, *and*
- $\rho : Z \to V(T)$, *called the* vertex-representative function, *selects, for each vertex* $v \in Z$, *a* representative $\rho(v)$ *in the vertex-model* $\varphi(v) \subseteq V(T)$.

Moreover, we say that a (Z, t)-*supertree* (T, φ, ψ, ρ) *is* valid *if*

- *for every* $\{u, v\} \in E_t$ *such that* $u, v \in Z$, *then the unique edge* e *between* $\varphi(u)$ *and* $\varphi(v)$ *exists in* T *and satisfies* $c(\{u, v\}) \in \psi(e)$.

For a node t *of* $(\mathsf{T}, \mathcal{B})$, *we define a* B_t-*supertree as a* (B_t, t)-*supertree and a* V_t-*supertree as a* (V_t, t)-*supertree.*

To give some intuition on why (Z, t)-supertrees capture partial solutions of our problem, let us assume that $\widehat{T}$ is a compatible supertree of $\mathcal{T}$ and consider a node t of $(\mathsf{T}, \mathcal{B})$. Then we can define a B_t-supertree $\mathfrak{T} = (T, \varphi, \psi, \rho)$ as follows:

- For every vertex $v \in B_t$, $\rho(v)$ can be chosen as any element in the set $\widehat{\varphi}(v)$,
- $T = \widehat{T}|_Y$, where $Y = \bigcup_{v \in B_t} \rho(v)$,
- for every vertex $v \in B_t$, $\varphi(v) = V(T) \cap \widehat{\varphi}(v)$, where $\widehat{\varphi}(v)$ is the vertex-model of v in $\widehat{T}$, and
- for every edge $e \in E(T)$, $i \in \psi(e)$ if there exist an edge $\{u, v\} \in E_t$, with $c(\{u, v\}) = i$, and an edge $f \in E(\widehat{T})$ such that f is incident to a vertex of $\widehat{\varphi}(u)$ and to a vertex of $\widehat{\varphi}(v)$, and f is on the unique path in $\widehat{T}$ between the vertices incident to e.

The edge-model function ψ introduced in Definition 1 allows to keep track, for every edge $e \in E(T)$, of the set of trees in $\mathcal{T}$ containing an edge having e as an edge-model. Observe that the size of a vertex-model $\widehat{\varphi}(v)$ in $\widehat{T}$ of some vertex $v \in V_D$ may depend on n (so, a priori, we may need to consider a number of vertex-models of size exponential in n). We overcome this problem via the vertex-representative function ρ, which allows us to store a tree T of size at most $2k$. This tree T captures how the vertex-models in $\widehat{T}$ "project" to the current bag, namely B_t, of the tree-decomposition of the display graph.

Before we describe the information stored at each node of the tree-decomposition, we need three more definitions.

Definition 2. *A tuple* $\mathfrak{T}_s = (T_s, \varphi_s, \psi_s, \rho_s)$ *is called a* shadow B_t-*supertree if there exists a* B_t-*supertree* $\mathfrak{T} = (T, \varphi, \psi, \rho)$ *such that*

- T_s *is a tree obtained from* T *by subdividing every edge once, called* shadow tree. *The new vertices are called* shadow vertices *and denoted by* $S(T_s)$, *while the original ones, that is,* $V(T_s) \setminus S(T_s)$, *are denoted by* $O(T_s)$,

- *for every* $v \in B_t$, $\varphi_s(v)$ *is a subset of* $V(T_s)$ *such that* $T_s[\varphi_s(v)]$ *is connected and such that* $\varphi(v) = \varphi_s(v) \cap O(T_s)$, *where we licitly consider the vertices in* $\varphi(v)$ *as a subset of* $O(T_s)$. *Furthermore, if* $u, v \in B_t$ *with* $c(u) = c(v)$, *then* $\varphi_s(u) \cap \varphi_s(v) = \emptyset$,
- $\psi_s : E(T_s) \rightarrow 2^{[k]}$ *such that for every* $s \in S(T_s)$, *if* x *and* y *are the neighbors of* s *in* T_s, *then* $\psi_s(\{x, s\}) = \psi_s(\{s, y\}) = \psi(\{x, y\})$, *and*
- $\rho_s : B_t \rightarrow V(T_s)$ *such that for every* $v \in B_t$, $\rho_s(v) = \rho(v)$.

We say that $\mathfrak{T}_s$ is a *shadow* of $\mathfrak{T}$. Note that $\mathfrak{T}$ may have more than one shadow satisfying Definition 2.

Definition 3. *Let* $\mathfrak{T} = (T, \varphi, \psi, \rho)$ *be a* (Z, t)-*supertree. The restriction of* $\mathfrak{T}$ *to a subset of vertices* $Y \subseteq V_t$ *is defined as the* (Y, t)-*supertree* $\mathfrak{T}|_Y = (\tilde{T}, \tilde{\varphi}, \tilde{\psi}, \tilde{\rho})$, *where*

- $\tilde{T} = T|_Z$, *where* $Z = \{\rho(v) \mid v \in Y\}$,
- *for every* $v \in Y$, $\tilde{\varphi}(v) = \varphi(v) \cap V(T|_Y)$,
- *for every* $e \in E(\tilde{T})$, $\tilde{\psi}(e) = \bigcup_{f \in E(P_e)} \psi(f)$, *where* P_e *is the unique path in* T *between the vertices incident to* e, *and*
- *for every* $v \in Y$, $\tilde{\rho}(v) = \rho(v)$.

If $\mathfrak{T}$ *is a* (Z, t)-*supertree and* $B_t \subseteq Z$, *we define a* shadow restriction *of* $\mathfrak{T}$ *to* B_t *as a shadow of* $\mathfrak{T}|_{B_t}$, *and we denote it by* $\mathfrak{T}|^s_{B_t}$.

Definition 4. *Two* (Z, t)-*supertrees* $\mathfrak{T} = (T, \varphi, \psi, \rho)$ *and* $\mathfrak{T}' = (T', \varphi', \psi', \rho')$ *are* equivalent, *and we denote it by* $\mathfrak{T} \simeq \mathfrak{T}'$, *if there exists an isomorphism* α *from* T *to* T' *such that*

- $\forall v \in Z, \forall a \in \varphi(v), \alpha(a) \in \varphi'(v)$,
- $\forall e \in E(T), \psi(e) = \psi'(\alpha(e))$, *and*
- $\forall v \in Z, \alpha(\rho(v)) = \rho'(v)$.

Every node t of $(\mathsf{T}, \mathcal{B})$ is associated with a set $\mathcal{R}_t$ of pairs $(\mathfrak{T}, \gamma)$, called *colored shadow* B_t-*supertrees*, where $\mathfrak{T} = (T, \varphi, \psi, \rho)$ is a shadow B_t-supertree and $\gamma : V(T) \rightarrow 2^{[k]}$ is the so-called *coloring function*. The dynamic programming algorithm will maintain the following invariant:

Invariant 1. *A colored shadow* B_t-*supertree* $(\mathfrak{T} = (T, \varphi, \psi, \rho), \gamma)$ *belongs to* $\mathcal{R}_t$ *if and only if there exists a valid* V_t-*supertree* $\mathfrak{T}_{\mathsf{ps}} = (T_{\mathsf{ps}}, \varphi_{\mathsf{ps}}, \psi_{\mathsf{ps}}, \rho_{\mathsf{ps}})$ *such that*

(1) $\mathfrak{T} \simeq \mathfrak{T}_{\mathsf{ps}}|^s_{B_t}$,
(2) *for every* $a \in V(T)$, *a color* $i \in \gamma(a)$ *if and only if there exists* $u \in V_t$ *with* $c(u) = i$ *such that* $a \in \varphi_{\mathsf{ps}}(u)$, *and*
(3) *for every* $z \in S(T)$ *with neighbors* x *and* y *in* $V(T)$, *a color* $i \in \gamma(z)$ *if there exists* $u \in V_t$ *with* $c(u) = i$ *and* $x, y \notin \varphi_{\mathsf{ps}}(u)$ *such that the unique path between* x *and* y *in* T_{ps} *uses at least one vertex of* $\varphi_{\mathsf{ps}}(u)$.

Intuitively, condition **(2)** of Invariant 1 guarantees that for every vertex $v \in V(T)$, we can recover the set of trees for which v has already appeared in a vertex-model of a vertex of $V_t \setminus B_t$. On the other hand, condition **(3)** of Invariant 1 is useful for the following reason. When a vertex is forgotten in the tree-decomposition, we need to keep track of its "trace", in the sense that the colors given to the corresponding shadow vertex guarantee that the algorithm will construct vertex-models appropriately. If γ is a coloring function satisfying conditions (2) and (3), we say that γ is *consistent* with $\mathfrak{T}_{\mathsf{ps}}$.

For $Z = \emptyset$, we denote by $\oslash$ the unique colored shadow (Z, t)-supertree. From the above description, it follows that the collection $\mathcal{T}$ is compatible if and only if $\oslash \in \mathcal{R}_{t_r}$. Indeed, for $t = t_r$ we have that $B_{t_r} = \emptyset$ and $V_{t_r} = V_D$. In that case, the only condition imposed by Invariant 1 is the existence of a valid V_D-supertree. Then, by Definition 1, the existence of such a supertree is equivalent to the existence of a compatible supertree $\widehat{T}$ of $\mathcal{T}$ in which the vertex-models and edge-models are given by the functions φ and ψ, respectively. Finally, note that the first condition of Definition 1, namely that $|\widehat{T}| \leq |B_{t_r}| + |V_{t_r}| = |V_D|$, is not a restriction on the set of solutions, as we may clearly assume that the size of a compatible supertree is always at most the size of the display graph.

Description of the Dynamic Programming Algorithm. Let $(\mathsf{T}, \mathcal{B})$ be a nice tree-decomposition of the display graph D of $\mathcal{T}$. We proceed to describe how to compute the set $\mathcal{R}_t$ for every node $t \in \mathsf{T}$. For that, we will assume inductively that, for every descendant t' of t, we have at hand the set $\mathcal{R}_{t'}$ that has been correctly built. We distinguish several cases depending on the type of node t:

1. **t is a leaf with $B_t = \{v\}$:** $\mathcal{R}_t = \{((T, \varphi, \psi, \rho), \gamma)\}$, where T is a tree with only one vertex a, $\rho(v) = a$, $\varphi(v) = \{a\}$, $\psi : \emptyset \to 2^{[k]}$, and $\gamma(a) = \{c(v)\}$.

2. **t is an introduce-vertex node such that the introduced vertex v is unlabeled:** For every element $(\mathfrak{T}' = (T', \varphi', \psi', \rho'), \gamma')$ of $\mathcal{R}_{t'}$, we add to $\mathcal{R}_t$ the elements of the form $(\mathfrak{T} = (T, \varphi, \psi, \rho), \gamma)$ that can be built according to one of the following four cases. For all of them, we define the vertex-representative function such that $\rho(v) = a$ for some vertex $a \in V(T)$, and for every $u \in B_{t'}$, $\rho(u) = \rho'(u)$. The different cases depend on this vertex a.

   (i) $\rho(v) = a$ **such that** $a \in V(T')$ **and** $c(v) \notin \gamma'(a)$. See Fig. 1(i) for an example. We define $T = T'$. Let us define φ, ψ, and γ.

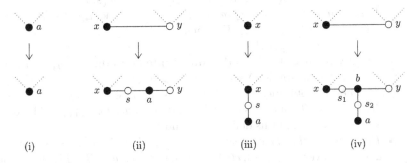

Fig. 1. The four possible cases (i–iv) in the dynamic programming algorithm. The configurations above correspond to T', while the ones below correspond to T. Full dots correspond to vertices in $O(T)$, the other ones being in $S(T)$.

- *Definition of the vertex-model function:* $T[\varphi(v)]$ is connected, contains a, and for every $z \in \varphi(v)$, $c(v) \notin \gamma'(z)$. For every $u \in B_{t'}$, $\varphi(u) = \varphi'(u)$.
- *Definition of the edge-model function:* For every $e \in E(T)$, $\psi(e) = \psi'(e)$.
- *Definition of the coloring function:* For every $z \in V(T)$, $\gamma(z) = \gamma'(z) \cup \{c(v) \mid z \in \varphi(v)\}$.

(ii) $\rho(v) = a$ **and** a **subdivides an edge** $\{x, y\}$ **of** T' **with** $c(v) \notin \psi'(\{x,y\})$. See Fig. 1(ii) for an example. Since T' is a shadow tree, assume w.l.o.g. that $x \in O(T')$ and $y \in S(T')$. Then T is obtained from T' by removing the edge $\{x, y\}$, adding two vertices $a \in O(T)$ and $s \in S(T)$ and three edges $\{x, s\}$, $\{s, a\}$, and $\{a, y\}$. Let us define φ, ψ, and γ.

- *Definition of the vertex-model function:* $T[\varphi(v)]$ is connected, contains a, and for each $z \in \varphi(v)$, $c(v) \notin \gamma'(z)$. For each $u \in B_{t'}$, $T[\varphi(u)]$ is connected, $\varphi'(u) \subseteq \varphi(u) \subseteq \varphi'(u) \cup \{a\} \cup S(T)$, and if u is unlabeled, then $\varphi(u) = \varphi'(u)$. For each $u, u' \in B_t$ with $c(u) = c(u')$, $\varphi(u) \cap \varphi(u') = \emptyset$.
- *Definition of the edge-model function:* For each $e \in E(T) \setminus \{\{x, s\}, \{s, a\}, \{a, y\}\}$, $\psi(e) = \psi'(e)$. Also, $\psi(\{x, s\}) = \psi(\{s, a\}) = \psi(\{a, y\}) = \psi'(\{x, y\})$.
- *Definition of the coloring function:* For each $z \in O(T) \setminus \{a\}$, $\gamma(z) = \gamma'(z) \cup \{c(v) \mid z \in \varphi(v)\}$. $\gamma(a) = \{i \mid \exists u \in B_t : c(u) = i \text{ and } a \in \varphi(u)\} \cup \psi'(\{x, y\})$. For each $z \in S(T')$, $\gamma(z) = \gamma'(z) \cup \{i \mid \exists u \in B_t : c(u) = i \text{ and } z \in \varphi(u)\}$. Finally, $\gamma(s) = \{i \mid \exists u \in B_t : c(u) = i \text{ and } s \in \varphi(u)\} \cup \psi'(\{x, y\})$.

(iii) $\rho(v) = a$ **with** $a \notin V(T')$ **and** a **is connected to a vertex** $x \in V(T')$. See Fig. 1(iii) for an example. T is obtained from T' by adding two vertices $a \in O(T)$ and $s \in S(T)$ and two edges $\{a, s\}$ and $\{s, x\}$. Let us define φ, ψ, and γ.

- *Definition of the vertex-model function:* $T[\varphi(v)]$ is connected, contains a, and for each $z \in \varphi(v)$, $c(v) \notin \gamma'(z)$. For each $u \in B_{t'}$, $T[\varphi(u)]$ is connected, $\varphi'(u) \subseteq \varphi(u) \subseteq \varphi'(u) \cup \{a\} \cup S(T)$, and if u is unlabeled, then $\varphi(u) = \varphi'(u)$. For each $u, u' \in B_t$ with $c(u) = c(u')$, $\varphi(u) \cap \varphi(u') = \emptyset$.
- *Definition of the edge-model function:* For each $e \in E(T) \setminus \{\{a, s\}, \{s, x\}\}$, $\psi(e) = \psi'(e)$, and $\psi(\{a, s\}) = \psi(\{s, x\}) = \emptyset$.
- *Definition of the coloring function:* For each $z \in V(T) \setminus \{a, s\}$, $\gamma(z) = \gamma'(z) \cup \{c(v) \mid z \in \varphi(v)\}$. For each $z \in \{a, s\}$, $\gamma(z) = \{i \mid \exists u \in B_t : c(u) = i \text{ and } z \in \varphi(u)\}$.

(iv) $\rho(v) = a$ **with** $a \notin V(T')$ **and** a **subdivides an edge** $\{x, y\}$ **of** T'. See Fig. 1(iv) for an example. Again, we may assume that $x \in O(T')$ and $y \in S(T')$. Then T is obtained from T' by removing the edge $\{x, y\}$, adding four vertices $a, b \in O(T)$ and $s_1, s_2 \in S(T)$, and five edges $\{x, s_1\}$, $\{s_1, b\}$, $\{b, y\}$, $\{a, s_2\}$, and $\{s_2, b\}$. Let us define φ, ψ, and γ.

- *Definition of the vertex-model function:* $T[\varphi(v)]$ is connected, contains a and, for every $z \in \varphi(v)$, $c(v) \notin \gamma'(z)$. For each $u \in B_{t'}$, $T[\varphi(u)]$ is connected, $\varphi'(u) \subseteq \varphi(u) \subseteq \varphi'(u) \cup \{a, b\} \cup S(T)$, and if u is unlabeled, then $\varphi(u) = \varphi'(u)$. For each $u, u' \in B_t$ with $c(u) = c(u')$, $\varphi(u) \cap \varphi(u') = \emptyset$.

- *Definition of the edge-model function*: For each edge $e \in E(T) \setminus \{\{a, s_2\}, \{s_2, b\}, \{x, s_1\}, \{s_1, b\}, \{b, y\}\}$, $\psi(e) = \psi'(e)$. $\psi(\{x, s_1\}) = \psi(\{s_1, b\}) = \psi(\{b, y\}) = \psi'(\{x, y\})$, and $\psi(\{a, s_2\}) = \psi(\{s_2, b\}) = \emptyset$.
- *Definition of the coloring function*: For every $z \in O(T) \setminus \{a, b\}$, $\gamma(z) = \gamma'(z) \cup \{c(v) \mid z \in \varphi(v)\}$. For every $z \in \{a, s_2\}$, $\gamma(z) = \{i \mid \exists u \in B_t : c(u) = i$ and $z \in \varphi(u)\}$. For every $z \in \{b, s_1\}$, $\gamma(z) = \{i \mid \exists u \in B_t : c(u) = i$ and $z \in \varphi(u)\} \cup \psi'(\{x, y\})$. For every $z \in S(T')$, $\gamma(z) = \gamma'(z) \cup \{i \mid \exists u \in B_t : c(u) = i$ and $z \in \varphi(u)\}$.

3. **t is an introduce-vertex node such that the introduced vertex v is labeled:** This case is very similar to Case 2 but, as vertex v is a leaf, only Case 2(iii) and (iv) can be applied. In both cases, we further impose that $\varphi(v) = \{a\}$ and $\gamma(v) = \{i \in [k] \mid v \in L(T_i), T_i \in \mathcal{T}\}$.

4. **t in an introduce-edge node for an edge $\{v, w\}$ with $c(\{v, w\}) = i$:** Let $(\mathfrak{T}' = (T', \varphi', \psi', \rho'), \gamma')$ be an element of $\mathcal{R}_{t'}$ such that there exist $a \in \varphi'(v)$ and $b \in \varphi'(w)$ such that $\{a, b\} \in E(T)$ and $i \notin \psi'(\{a, b\})$. We construct $(\mathfrak{T} = (T, \varphi, \psi, \rho), \gamma)$ as an element of $\mathcal{R}_t$ as follows: $T = T'$. For every $v \in B_t$, $\varphi(v) = \varphi'(v)$. For every $e \in E(T) \setminus \{\{a, b\}\}$, $\psi(e) = \psi'(e)$. $\psi(\{a, b\}) = \psi'(\{a, b\}) \cup \{i\}$. For every $v \in V(T)$, $\gamma(v) = \gamma'(v)$.

5. **t is a forget-vertex node for a vertex v:** Let $(\mathfrak{T}' = (T', \varphi', \psi', \rho'), \gamma')$ be an element of $\mathcal{R}_{t'}$. We construct $(\mathfrak{T} = (T, \varphi, \psi, \rho), \gamma)$ as an element of $\mathcal{R}_t$ as follows: $\mathfrak{T} = \mathfrak{T}'|_{B_{t'}}^z$. For every $a \in O(T)$, $\gamma(a) = \gamma'(a)$. For every $z \in S(T)$, if x and y are the neighbors of z in T, then $\gamma(z) = \{i \mid \exists a \in V(T')$ on the path between x and y in $T' : (i \in \gamma'(a))$ and $(\forall u \in B_t : a \notin \varphi'(u))\}$.

6. **t is a join node:** Let $(\mathfrak{T}' = (T, \varphi, \psi', \rho), \gamma')$ be an element of $\mathcal{R}_{t'}$ and let $(\mathfrak{T}'' = (T, \varphi, \psi'', \rho), \gamma'')$ be an element of $\mathcal{R}_{t''}$ such that for every $z \in V(T)$, $\gamma'(z) \cap \gamma''(z) = \emptyset$ and for every $e \in E(T)$, $\psi'(z) \cap \psi''(z) = \emptyset$. We construct $(\mathfrak{T} = (T, \varphi, \psi, \rho), \gamma)$ as an element of $\mathcal{R}_t$ as follows: For every $e \in E(T)$, $\psi(e) = \psi'(e) \cup \psi''(e)$, and for every $z \in V(T)$, $\gamma(z) = \gamma'(z) \cup \gamma''(z)$.

4   Further Research

In this paper we give the first "reasonable" FPT algorithms for the COMPATIBILITY and the AGREEMENT problems for unrooted phylogenetic trees. Even though this is, from a theoretical point of view, a big step further toward solving this problem in reasonable time, our running times are still prohibitive to be of any use in real-life phylogenomic studies, where k can go up very quickly [7]. One possibility to design a practical algorithm is to devise reduction rules to keep k small. Another possibility would be to design an FPT algorithm with respect to a parameter that is smaller than the number of gene trees in phylogenomic studies.

From a more theoretical perspective, a natural question is whether the function $2^{O(k^2)}$ in the running times of our algorithms can be improved. It would also be interesting to prove lower bounds for algorithms parameterized by treewidth to solve these problems, assuming the Exponential Time Hypothesis [13].

References

1. Aho, A.V., Sagiv, Y., Szymanski, T.G., Ullman, J.D.: Inferring a tree from lowest common ancestors with an application to the optimization of relational expressions. SIAM J. Comput. **10**(3), 405–421 (1981)
2. Bininda-Emonds, O.R.P. (ed.): Phylogenetic Supertrees: Combining Information to Reveal the Tree of Life. Computational Biology, vol. 4. Springer, Netherlands (2004)
3. Bininda-Emonds, O.R., Gittleman, J.L., Steel, M.A.: The (super) tree of life: procedures, problems, and prospects. Ann. Rev. Ecol. Syst. **33**, 265–289 (2002)
4. Bodlaender, H.L., Drange, P.G., Dregi, M.S., Fomin, F.V., Lokshtanov, D., Pilipczuk, M.: An $O(c^k n)$ 5-approximation algorithm for treewidth. In: Proceedings of the IEEE 54th Annual Symposium on Foundations of Computer Science (FOCS), pp. 499–508 (2013)
5. Bryant, D., Lagergren, J.: Compatibility of unrooted phylogenetic trees is FPT. Theor. Comput. Sci. **351**(3), 296–302 (2006)
6. Cygan, M., Nederlof, J., Pilipczuk, M., Pilipczuk, M., van Rooij, J.M.M., Wojtaszczyk, J.O.: Solving connectivity problems parameterized by treewidth in single exponential time. In: Proceedings of the IEEE 52nd Annual Symposium on Foundations of Computer Science (FOCS), pp. 150–159 (2011)
7. Delsuc, F., Brinkmann, H., Philippe, H.: Phylogenomics and the reconstruction of the tree of life. Nat. Rev. Genet. **6**(5), 361–375 (2005)
8. Diestel, R.: Graph Theory. Graduate Texts in Mathematics, vol. 173, 4th edn. Springer, Heidelberg (2010)
9. Felsenstein, J.: Inferring Phylogenies. Sinauer Associates, Incorporated, Sunderland (2004)
10. Frick, M., Grohe, M.: The complexity of first-order and monadic second-order logic revisited. Ann. Pure Appl. Logic **130**(1–3), 3–31 (2004)
11. Gordon, A.D.: Consensus supertrees: the synthesis of rooted trees containing overlapping sets of labeled leaves. J. Classif. **3**(2), 335–348 (1986)
12. Kloks, T.: Treewidth, Computations and Approximations. LNCS, vol. 842. Springer, Heidelberg (1994)
13. Lokshtanov, D., Marx, D., Saurabh, S.: Lower bounds based on the exponential time hypothesis. Bull. EATCS **105**, 41–72 (2011)
14. Maddison, W.: Reconstructing character evolution on polytomous cladograms. Cladistics **5**(4), 365–377 (1989)
15. Ng, M., Wormald, N.C.: Reconstruction of rooted trees from subtrees. Discrete Appl. Math. **69**(1–2), 19–31 (1996)
16. Scornavacca, C.: Supertree methods for phylogenomics. Ph.D. thesis, Université Montpellier II-Sciences et Techniques du Languedoc (2009)
17. Scornavacca, C., van Iersel, L., Kelk, S., Bryant, D.: The agreement problem for unrooted phylogenetic trees is FPT. J. Graph Algorithms Appl. **18**(3), 385–392 (2014)
18. Steel, M.: The complexity of reconstructing trees from qualitative characters and subtrees. J. Classif. **9**, 91–116 (1992)

A Very Fast String Matching Algorithm Based on Condensed Alphabets

Simone Faro[✉]

Università di Catania, Viale A.Doria n.6, 95125 Catania, Italy
faro@dmi.unict.it

Abstract. String matching is the problem of finding all the substrings of a text which correspond to a given pattern. It's one of the most investigated problem in computer science, mainly due to its various applications in many fields. In recent years most solutions to the problem focused on efficiency and flexibility of the searching procedure and effective techniques appeared to speed-up previous solutions. In this paper we present a simple and very efficient algorithm for string matching. It can be seen as an extension of the Skip-Search algorithm to condensed alphabets with the aim of reducing the number of verifications during the searching phase. From our experimental results it turns out that the new variant obtains in most cases the best running time when compared against the most effective algorithms in literature. This makes the new algorithm one of the most flexible solutions in practical cases.

Keywords: Exact text analysis · String matching · Experimental algorithms · Text processing

1 Introduction

The *exact string matching problem* is one of the most studied problem in computer science. It consists in finding all the (possibly overlapping) occurrences of an input pattern x in a text y, over the same alphabet Σ of size σ. A huge number of solutions have been devised since the 1980s [6,17] and, in spite of such wide literature, much work has been produced in the last few years, indicating that the need for efficient solutions to this problem is high.

Solutions to such problem can be divided in two classes, *counting* solutions simply counts the number of occurrences of the pattern in the text, while *reporting* solutions are also able to report the exact positions in which the pattern occurs. Solutions in the first class are faster in general. In this paper we are interested in the second class of algorithms.

From a theoretical point of view the exact string matching problem has been extensively studied. If we indicate with m and n the lengths of the pattern and of the text, respectively, the problem can be solved in $\mathcal{O}(n)$ worst case time

---

This work has been supported by G.N.C.S., Istituto Nazionale di Alta Matematica "Francesco Severi".

© Springer International Publishing Switzerland 2016
R. Dondi et al. (Eds.): AAIM 2016, LNCS 9778, pp. 65–76, 2016.
DOI: 10.1007/978-3-319-41168-2_6

complexity [20]. However, in many practical cases it is possible to avoid reading all the characters of the text achieving sub-linear performances on the average. The optimal average $\mathcal{O}(\frac{n\log_\sigma m}{m})$ time complexity [23] was reached for the first time by the Backward-DAWG-Matching algorithm [8] (BDM). Interested readers can refer to [6,14,17] for a survey of the most efficient solutions to the problem.

In recent years most solutions to the problem focused on efficiency and flexibility and effective techniques appeared to speed-up previous formerly efficient solutions. Among such techniques *bit-parallelism, string-packing, q-grams, filtering* and *hashing* deserve a special mention since they inspired a lot of work. *Filtering* and *hashing* are two techniques particularly relevant in this paper.

Specifically, instead of checking at each position of the text if the pattern occurs, it seems to be more efficient to *filter* positions of the text by checking only if the corresponding content *looks like* the input pattern. When a resemblance is detected a more detailed check is performed.

The first algorithm to take advantage of such technique was the well known Karp-Rabin algorithm [19] in 1987. It uses *hashing* function for computing a fingerprint value of the pattern. Subsequently a fingerprint value for each text substring of length m is computed. Then a naive check at a given position of the text is performed only if the fingerprint value of the corresponding substring is equal to the fingerprint value of the pattern. The overall worst case time complexity of the algorithm is $\mathcal{O}(nm)$ but a linear behavior can be observed on average. The first solution based on filtering and hashing, showing a sub-linear average behavior, was presented by Lecroq in 2007 [21]. The algorithm, named Hash$q$, is simply a generalization of the Boyer-Moore-Horspool algorithm to condensed alphabets. In this case groups of q characters (or q-grams) are hashed in a single fingerprint value, generating an extended condensed alphabet. Such condensed alphabet correspond to the set of all fingerprint values generated by all possible combinations of q characters drawn from the original alphabet.

The idea of extending efficient solutions by condensed alphabets has been later extensively adopted in string matching [17]. However, although several algorithms have been proposed in the last decade, the Hash$q$ algorithm is still one of the most effective solutions in practical cases [14].

In this paper we present a simple, yet very efficient, algorithm for the exact string matching problem based on a well known filtering solution, the Skip-Search algorithm [7], extended with condensed alphabets. We will observe how the use of a condensed alphabet allows to drastically reduce the number of verifications of such filtering approach. The worst case time complexity of the algorithm is $\mathcal{O}(nm)$. However, despite its quadratic worst case behavior, we will show in our experimental evaluation that such extension leads to one of the most efficient and flexible algorithms for string matching. Specifically it turns out that the new solution obtains the best results, in terms of running times, in most cases and especially for small alphabets and long patterns.

The paper is organized as follows. In Sect. 2 we describe in detail the Skip-Search algorithm and its variants. In Sect. 3 we introduce and analyze the new algorithm based on condensed alphabets. In Sect. 4 we compare the new presented algorithm against the most effective solutions known in literature. We drawn our conclusions in Sect. 5.

2   The Skip Search and the Alpha Skip Search Algorithms

The Skip Search algorithm is an elegant and efficient solution to the exact pattern matching problem, firstly presented in [7] and subsequently adapted to many other problems and variants of exact pattern matching.

Let x and y be a pattern and a text of length m and n, respectively, over a common alphabet Σ of size σ. For each character c of the alphabet, the Skip Search algorithm collects in a bucket $B[c]$ all the positions of that character in the pattern x, so that for each $c \in \Sigma$ we have:

$$B[c] = \{i \ : \ 0 \leq i \leq m - 1 \text{ and } x[i] = c\}.$$

Plainly, the space and time complexity needed for the construction of the array B of buckets is $\mathcal{O}(m + \sigma)$.

Thus if a character occurs k times in the pattern, there are k corresponding positions in the bucket of the character. Notice that when the pattern is shorter than the alphabet size, some buckets are empty. This observation turns out to be particularly suitable for our purpose.

The search phase of the Skip Search algorithm examines all the characters $y[j]$ in the text at positions $j = km - 1$, for $k = 1, 2, \ldots, \lfloor n/m \rfloor$. For each such character $y[j]$, the bucket $B[y[j]]$ allows one to compute the possible positions h of the text in the neighborhood of j at which the pattern could occur.

By performing a character-by-character comparison between x and the substring $y[h \mathbin{..} h + m - 1]$ until either a mismatch is found, or all the characters in the pattern x have been considered, it can be tested whether x actually occurs at position h of the text.

The Skip Search algorithm has a quadratic worst-case time complexity, however, as shown in [7], the expected number of text character inspections is $\mathcal{O}(n)$. In addition it is interesting to observe that the Skip Search algorithms performs better in the case of large alphabets since most of the buckets in the array B are empty.

Among the variants of the Skip Search algorithm, the most relevant one for our purposes is the Alpha Skip Search algorithm [7], which collects buckets for substrings of the pattern rather than for its single characters.

During the preprocessing phase of the Alpha Skip Search algorithm, all the factors of length $\ell = \lfloor \log_\sigma m \rfloor$ occurring in the pattern x are arranged in a trie T_x, for fast retrieval. In addition, for each leaf ν of T_x a bucket is maintained which stores the positions in x of the factor corresponding to ν. Provided that the alphabet size is considered as a constant, the worst-case running time of the preprocessing phase is linear.

The searching phase consists in looking into the buckets of the text factors $y[j \mathbin{..} j + \ell - 1]$, for all $j = k(m - \ell + 1) - 1$ such that $1 \leq k \leq \lfloor (n - \ell)/m \rfloor$, and then test, as in the previous case, whether there is an occurrence of the pattern at the indicated positions of the text.

The worst-case time complexity of the searching phase is quadratic, though the expected number of text character comparisons is $\mathcal{O}(n \log_\sigma m/(m - \log_\sigma m))$.

3   A New Fast Variant of the Skip-Search Algorithm

In this section we present an efficient extension of the Skip-Search algorithm using condensed alphabets. The resulting algorithm has a quadratic worst case time complexity while on average it shows a sublinear behavior.

Let x be a pattern of length m and let y be a text of length n. Moreover suppose both strings are drawn from a common alphabet Σ of size σ and suppose q is a constant value, with $1 \leq q \leq m$. The algorithm can be divided in a preprocessing and a searching phase.

The preprocessing phase of the algorithm indexes all subsequences of the pattern (of length q) in order to be able to locate them during the searching phase. For efficiency reasons, each substring of length q is converted into a numeric value, called *fingerprint*, which is used to index the substring. A fingerprint value ranges in the interval $\{0 .. 2^\alpha - 1\}$, for a given bound α. In our setting the value α is set to 16, so that a fingerprint can fit into a single 16-bit register.

The procedure FNG for computing the fingerprints is shown in Fig. 1 (on the left). Given a sequence x of length m, an index i such that $0 \leq i < m - q$, and two integers k and q such that $kq \leq \alpha$, the procedure FNG computes the fingerprint v of the substring $x[i .. i + q - 1]$. Specifically the fingerprint v is computed as

$$v = \sum_{j=0}^{q-1} (x[i+j] \ll kj).$$

Plainly, the time complexity of the procedure FNG is $\mathcal{O}(q)$. Observe that the the fingerprint value is not unique for each substring of length q, i.e. two different strings can be associated with the same fingerprint value. The preprocessing phase of the algorithm, which is reported in Fig. 1 (on the left), consists in compiling the fingerprints of all possible substrings of length q contained in the pattern x. Thus a fingerprint value v, with $0 \leq v < 2^\alpha$, is computed for each subsequence $x[i .. i + q - 1]$, for $0 \leq i < m - q$. To this purpose a table F of size 2^α is maintained for storing, for any possible fingerprint value v, the set of positions i such that $\text{FNG}(x, i, q, k) = v$. More precisely, for $0 \leq v < 2^\alpha$, we have

$$F[v] = \Big\{ i \mid 0 \leq i < m - q \text{ and } \text{FNG}(x, i, q, k) = v \Big\}.$$

The preprocessing phase of the algorithm requires some additional space to store the $(m - q)$ possible alignments in the 2^α locations of the table F. Thus, the space requirement of the algorithm is $\mathcal{O}(m - q + 2^\alpha)$ that approximates to $\mathcal{O}(m)$, since α is constant. The first loop of the preprocessing phase just initializes the

```
FNG(x, i, q, k)                          SKIPq(x, r, y, n, q, k)
2.  v ← 0                                1.  F ←Preprocessing(x, q, m, k)
3.  for j ← q − 1 downto 0 do            2.  for j ← m − 1 to n step m − q + 1 do
4.     v ← (v ≪ k) + x[i + j]            3.     v ← FNG(y, j, q, k)
5.  return v                             4.     for each i ∈ F[v] do
                                         5.        if x = y[j − i . . j − i + m − 1]
                                         6.        then output (j − i)
PREPROCESSING(x, q, m, k)
1.  for v ← 0 to 2^α − 1 do
2.     F[v] ← ∅
3.  for i ← 0 to m − q do
4.     v ← FNG(x, i, q, k)
5.     F[v] ← F[v] ∪ {(i + q − 1)}
6.  return F
```

Fig. 1. The pseudo-code of the Skip$q$ algorithm for the exact string matching problem.

table F, while the second loop is run $(m − q)$ times, which makes the overall time complexity of such phase $\mathcal{O}(m + 2^\alpha)$ that, again, approximates to $\mathcal{O}(m)$.

Along the same line of the Skip Search algorithm, the basic idea of the searching phase is to compute a fingerprint value every $(m − q + 1)$ positions of the text y and to check whether the pattern appears in y, involving the block $y[j . . j+q−1]$. If the fingerprint value indicates that some of the alignments are possible, then the candidate positions are checked naively for matching.

The pseudo-code provided in Fig. 1 (on the right) reports the skeleton of the algorithm. The main loop investigates the blocks of the text y in steps of $(m − q + 1)$ blocks. If the fingerprint v computed on $y[j . . j + q − 1]$ points to a nonempty bucket of the table F, then the positions listed in $F[v]$ are verified accordingly.

In particular $F[v]$ contains a linked list of the values i marking the pattern x and the beginning position of the pattern in the text. While looking for occurrences on $y[j . . j + q − 1]$, if $F[v]$ contains the value i, this indicates the pattern x may potentially begin at position $(j − i)$ of the text. In that case, a matching test is to be performed between x and $y[j − i . . j − i + m − 1]$ via a character-by-character inspection.

The total number of filtering operations is exactly $n/(m−q)$. At each attempt, the maximum number of verification requests is $(m − q)$, since the filter provides information about that number of appropriate alignments of the pattern. On the other hand, if the computed fingerprint points to an empty location in F, then there is obviously no need for verification. The verification cost for a pattern x of length m is assumed to be $\mathcal{O}(m)$, with the brute-force checking approach. Hence, in the worst case the time complexity of the verification is $\mathcal{O}(m(m − q))$, which happens when all alignments in x must be verified at any possible beginning

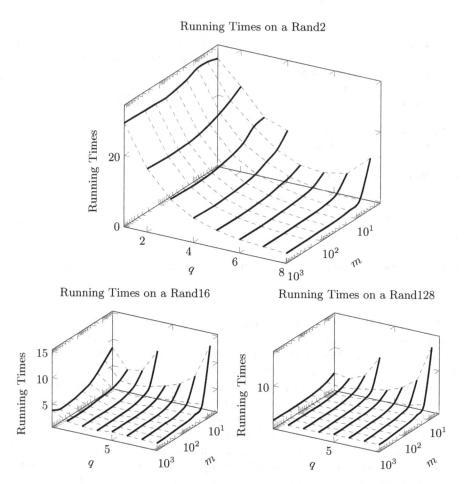

Fig. 2. Running times of the Skip-Search extended with condensed alphabets using groups of q characters. We report running times of the algorithms for different values of q. Experimental test have been conducted on random text over alphabets of size 2, 16 and 128, respectively, with a uniform distribution of characters.

position. Hence, the best case complexity is $\mathcal{O}(n/(m-q))$, while the worst case complexity is $\mathcal{O}(nm)$.

Figure 2 shows experimental evaluations to campare the performances of the algorithm under various conditions and for different values of the parameter q. Experimental evaluations have been conducted on random text of 4Mb over alphabets of size 2, 16 and 128, respectively, with a uniform distribution of characters (a detailed description of the experimental settings can be found in Sect. 4).

It turns out from experimental evaluations shown in Fig. 2 that the performances of the algorithm strongly depend on the values of m, q and σ. When the size of the alphabet is small then larger values of the parameter q are more effective. Such difference is less sensible when the size of the alphabet gets larger. However it turns out that the smaller is the length of the pattern the lower is the performance of the algorithm. This behavior is more evident for larger values of the parameter q. Thus, the choice of the parameter q should be directed to larger values when the size of alphabet decreases or when the length of the pattern increases. Conversely the values of q should get smaller.

4   Experimental Results

In this section we evaluate the performance of the new presented algorithms against the most efficient solution known in literature for the online exact string matching problem. Specifically we compare the following 15 algorithms implemented in 79 variants, depending on the values of their parameters:

- AOSOq: the Average-Optimal variant [18] of the Shift-Or algorithm [2] using q.grams, with $1 \leq q \leq 6$;
- BNDMq: the Backward-Nondeterministic-DAWG-Matching algorithm [22] implemented using q-grams with $1 \leq q \leq 8$;
- BSDMq: the Backward-SNR-DAWG-Matching algorithm [15] using condensed alphabets with groups of q characters, with $1 \leq q \leq 8$;
- BXSq: the Backward-Nondeterministic-DAWG-Matching algorithm [22] with Extended Shift [10] implemented using q-grams and $1 \leq q \leq 8$;
- EBOM: the extended version [13] of the BOM algorithm [1];
- FSBNDMqs: the Forward Simplified version [13] of the BNDM algorithm [22] implemented using q-grams and $1 \leq q \leq 8$;
- KBNDM: the Factorized variant [5] BNDM algorithm [22];
- SBNDMq: the Simplified version of the Backward-Nondeterministic-DAWG-Matching algorithm [1] implemented using q-grams and $1 \leq q \leq 8$;
- FS-w: the Multiple Windows version [16] of the Fast Search algorithm [3] implemented using w sliding windows, with $2 \leq w \leq 6$;
- HASHq: the Hashing algorithm [21] using q-grams, with $3 \leq q \leq 5$;
- IOM: the Improved Occurrence Matcher [4];
- WOM: the Worst Occurrence Matcher [4];
- JOM: the Jumping Occurrence Matcher [4];
- ASKIP the Alpha variant of the SKip-Search algorithm [7];
- SKIPq: the new Skip Search variants using q-grams, with $1 \leq q \leq 8$ (observe that when $q = 1$ we have the original Skip-Search algorithm [7]);

For the sake of completeness we evaluate also the following two string matching algorithms for *counting* occurrences.

- EPSM: the Exact Packed String Matching algorithm [12];
- TSOq: the Two-Way variant of [9] the Shift-Or algorithm [2] implemented with a loop unrolling of q characters, with $q = 5$;

All algorithms have been implemented in the C programming language and have been tested using the SMART tool[1]. The experiments were executed locally on an MacBook Pro with 4 Cores, a 2 GHz Intel Core i7 processor, 16 GB RAM 1600 MHz DDR3, 256 KB of L2 Cache and 6 MB of Cache L3. Algorithms have been compared in terms of running times, including any preprocessing time.

We report experimental evaluations on a three random sequences (see Tables 1 and 2) and on three real data (see Tables 3, 4 and 5). Specifically random sequences are over alphabets of 2 and 16 characters, with a uniform distribution. For the case of real data evaluations we used a genome sequence, a protein sequence and an english text. All sequences have a length of 5 MB, are provided by theSMART research tool and are available online for download.

During the experimental evaluations patterns of length m were randomly extracted from the sequences, with m ranging over the set of values $\{2^i \mid 2 \leq i \leq 10\}$. For each case, the mean over the running times, expressed in hundredths of seconds, of 500 runs has been reported.

In the following tables we report the running times of our evaluations. Each table is divided in four blocks. The first and the second block present the most effective algorithms known in literature based on automata and comparison of characters, respectively. Best results among this two sets of algorithms have been bold-faced in order to easily locate the best solutions among previous known algorithms. The third block contains the running times of the new algorithm, including the speed up (in percentage) obtained against the best running time in the first two blocks. Positive values indicate a braking of the running time while

Table 1. Experimental results on a random sequence over an alphabet of 2 characters.

m	4	8	16	32	64	128	256	512	1024
AOSOq	$41.70^{(2)}$	$35.62^{(4)}$	$14.37^{(4)}$	$4.54^{(4)}$	$4.64^{(4)}$	$4.59^{(4)}$	$4.58^{(4)}$	$4.62^{(4)}$	$4.57^{(4)}$
BNDMq	$16.55^{(4)}$	$8.67^{(6)}$	$5.10^{(6)}$	$4.06^{(6)}$	$4.92^{(4)}$	$4.92^{(4)}$	$4.92^{(4)}$	$4.89^{(4)}$	$4.90^{(4)}$
BSDMq	$\mathbf{15.35^{(4)}}$	$\mathbf{7.52^{(6)}}$	$\mathbf{4.30^{(8)}}$	$\mathbf{3.30^{(8)}}$	$\mathbf{2.99^{(8)}}$	$\mathbf{2.79^{(8)}}$	$\mathbf{2.78^{(8)}}$	$2.74^{(8)}$	$2.73^{(8)}$
BXSq	$22.33^{(4)}$	$10.97^{(6)}$	$5.47^{(8)}$	$3.74^{(8)}$	$3.75^{(8)}$	$3.75^{(8)}$	$3.76^{(8)}$	$3.74^{(8)}$	$3.75^{(8)}$
EBOM	25.12	20.32	13.65	8.72	6.27	4.62	3.77	3.28	2.98
FSBNDMqs	$16.66^{(4,1)}$	$7.80^{(6,1)}$	$5.08^{(6,1)}$	$4.05^{(6,1)}$	$4.00^{(6,1)}$	$4.07^{(6,1)}$	$4.05^{(6,1)}$	$4.05^{(6,1)}$	$4.05^{(6,1)}$
KBNDM	27.67	19.08	11.31	7.07	5.82	5.71	5.75	5.78	5.79
SBNDMq	$15.36^{(4)}$	$8.33^{(6)}$	$5.05^{(6)}$	$4.01^{(6)}$	$4.05^{(6)}$	$4.08^{(6)}$	$4.07^{(6)}$	$4.08^{(6)}$	$4.09^{(6)}$
FS-w	$26.30^{(2)}$	$20.20^{(2)}$	$15.16^{(2)}$	$11.63^{(2)}$	$9.42^{(2)}$	$8.00^{(2)}$	$6.90^{(2)}$	$6.11^{(2)}$	$5.52^{(2)}$
FJS	28.78	33.80	36.59	34.94	36.25	36.02	36.51	36.36	36.55
HASHq	$25.44^{(3)}$	$11.94^{(5)}$	$6.07^{(5)}$	$3.99^{(8)}$	$3.20^{(8)}$	$2.97^{(8)}$	$2.95^{(8)}$	$\mathbf{2.71^{(8)}}$	$\mathbf{2.62^{(8)}}$
ASKIP	34.24	22.43	13.14	6.50	4.62	3.62	3.27	3.16	3.52
IOM	23.90	24.64	26.64	26.58	26.76	26.60	26.69	26.51	26.67
WOM	30.32	26.31	23.00	20.37	17.94	16.47	14.94	13.78	12.76
SKIPq	$\mathbf{14.37^{(4)}}$	$\mathbf{7.24^{(6)}}$	$\mathbf{4.30^{(8)}}$	$\mathbf{3.20^{(8)}}$	$\mathbf{2.77^{(8)}}$	$\mathbf{2.65^{(8)}}$	$\mathbf{2.65^{(8)}}$	$\mathbf{2.55^{(8)}}$	$\mathbf{2.51^{(8)}}$
speed-up	-6.4%	-3.7%	0.0%	-3.0%	-7.3%	-5.0%	-4.7%	-5.9%	-4.2%
EPSM	6.51	8.46	3.20	2.45	2.25	2.22	2.30	2.31	2.30
TSOq	$12.67^{(5)}$	$10.01^{(5)}$	$6.67^{(5)}$	$4.39^{(5)}$	$3.17^{(5)}$	-	-	-	-

---

[1] The SMART tool is available online at http://www.dmi.unict.it/~faro/smart/.

Table 2. Experimental results on a random sequence over an alphabet of 16 characters.

m	4	8	16	32	64	128	256	512	1024
AOSOq	$10.70^{(2)}$	$4.29^{(4)}$	$3.75^{(4)}$	$3.75^{(4)}$	$3.07^{(6)}$	$3.11^{(6)}$	$3.13^{(6)}$	$3.09^{(6)}$	$3.14^{(6)}$
BNDMq	$11.99^{(4)}$	$4.20^{(4)}$	$2.96^{(4)}$	$2.38^{(4)}$	$2.39^{(4)}$	$2.41^{(4)}$	$2.38^{(4)}$	$2.35^{(4)}$	$2.39^{(4)}$
BSDMq	$4.82^{(2)}$	$3.79^{(4)}$	$2.67^{(4)}$	$2.27^{(4)}$	**$2.10^{(4)}$**	**$2.02^{(4)}$**	**$1.98^{(4)}$**	**$1.96^{(4)}$**	**$1.99^{(4)}$**
BXSq	$6.86^{(2)}$	$4.29^{(2)}$	$3.08^{(2)}$	$2.46^{(2)}$	$2.47^{(4)}$	$2.50^{(4)}$	$2.51^{(4)}$	$2.52^{(4)}$	$2.48^{(4)}$
EBOM	**3.75**	**2.92**	**2.51**	**2.25**	2.16	2.19	2.16	2.16	2.30
FSBNDMqs	$4.28^{(2,0)}$	$3.21^{(2,0)}$	$2.55^{(3,1)}$	$2.18^{(3,1)}$	$2.20^{(3,1)}$	$2.18^{(3,1)}$	$2.19^{(3,1)}$	$2.19^{(3,1)}$	$2.21^{(3,1)}$
KBNDM	7.38	4.94	3.76	3.17	2.95	3.04	2.90	3.00	2.98
SBNDMq	$5.26^{(2)}$	$3.62^{(2)}$	$2.73^{(2)}$	$2.31^{(2)}$	$2.36^{(4)}$	$2.37^{(4)}$	$2.37^{(4)}$	$2.39^{(4)}$	$2.42^{(4)}$
FS-w	$5.15^{(6)}$	$3.69^{(6)}$	$3.05^{(6)}$	$2.72^{(6)}$	$2.72^{(6)}$	$2.68^{(6)}$	$2.64^{(6)}$	$2.64^{(6)}$	$2.66^{(6)}$
FJS	9.38	7.08	5.27	4.94	4.65	4.54	4.18	4.65	4.49
HASHq	$19.81^{(3)}$	$8.35^{(3)}$	$5.07^{(3)}$	$3.61^{(5)}$	$3.10^{(5)}$	$2.97^{(5)}$	$2.89^{(5)}$	$2.73^{(5)}$	$2.62^{(5)}$
ASKIP	9.04	7.48	4.98	3.35	2.84	2.82	3.07	3.76	6.23
IOM	8.96	6.40	5.11	4.49	4.35	4.28	4.28	4.31	4.35
WOM	9.39	6.66	5.16	4.43	4.17	3.97	3.85	3.80	3.72
SKIPq	$4.85^{(2)}$	$3.70^{(3)}$	$3.13^{(3)}$	**$2.24^{(4)}$**	**$2.06^{(4)}$**	**$1.99^{(4)}$**	**$1.93^{(4)}$**	**$1.86^{(4)}$**	**$1.84^{(4)}$**
speed-up	$+29\%$	$+26\%$	$+24\%$	-0.4%	-1.9%	-1.5%	-2.5%	-5.1%	-7.5%
EPSM	6.62	25.55	2.72	2.10	1.94	1.89	1.81	1.74	1.78
TSOq	$5.45^{(5)}$	$3.88^{(5)}$	$3.11^{(5)}$	$2.51^{(5)}$	$2.20^{(5)}$	-	-	-	-

Table 3. Experimental results on a genome sequence.

m	4	8	16	32	64	128	256	512	1024
AOSOq	$16.98^{(2)}$	$9.63^{(2)}$	$3.93^{(4)}$	$3.39^{(4)}$	$2.98^{(6)}$	$2.97^{(6)}$	$2.99^{(6)}$	$3.00^{(6)}$	$3.03^{(6)}$
BNDMq	$11.13^{(4)}$	$4.10^{(4)}$	$2.99^{(4)}$	$2.47^{(4)}$	$2.38^{(4)}$	$2.39^{(4)}$	$2.41^{(4)}$	$2.47^{(4)}$	$2.45^{(4)}$
BSDMq	$8.37^{(4)}$	**$3.71^{(4)}$**	**$2.78^{(4)}$**	**$2.46^{(4)}$**	**$2.25^{(8)}$**	**$2.15^{(8)}$**	**$2.11^{(8)}$**	**$2.16^{(6)}$**	**$2.11^{(6)}$**
BXSq	$11.86^{(2)}$	$4.78^{(4)}$	$3.25^{(4)}$	$2.53^{(6)}$	$2.50^{(6)}$	$2.52^{(4)}$	$2.49^{(4)}$	$2.55^{(4)}$	$2.54^{(4)}$
EBOM	7.72	7.15	5.66	4.10	3.17	2.67	2.40	2.32	2.41
FSBNDMqs	**$6.46^{(3,1)}$**	$3.87^{(4,1)}$	$2.94^{(4,1)}$	$2.38^{(4,1)}$	$2.35^{(6,2)}$	$2.31^{(6,1)}$	$2.33^{(6,1)}$	$2.38^{(3,1)}$	$2.37^{(6,1)}$
KBNDM	10.88	8.21	6.15	4.17	3.27	3.09	3.10	3.13	3.14
SBNDMq	$8.75^{(2)}$	$3.95^{(4)}$	$2.97^{(4)}$	$2.47^{(4)}$	$2.39^{(4)}$	$2.39^{(4)}$	$2.36^{(4)}$	$2.38^{(4)}$	$2.38^{(4)}$
FS-w	$12.33^{(2)}$	$9.39^{(2)}$	$7.76^{(2)}$	$6.89^{(2)}$	$6.16^{(2)}$	$5.63^{(2)}$	$5.06^{(2)}$	$4.73^{(2)}$	$4.42^{(2)}$
FJS	18.60	16.69	16.96	15.96	16.09	16.80	16.71	16.61	16.59
HASHq	$18.09^{(3)}$	$7.68^{(3)}$	$4.67^{(5)}$	$3.31^{(5)}$	$2.78^{(5)}$	$2.60^{(5)}$	$2.63^{(5)}$	$2.51^{(5)}$	$2.40^{(5)}$
ASKIP	17.19	7.45	4.32	3.17	2.66	2.60	2.79	3.24	5.18
IOM	14.41	11.88	11.08	11.17	11.17	11.13	11.03	11.03	10.98
WOM	16.69	12.48	9.88	8.61	7.75	7.16	6.72	6.29	6.11
SKIPq	**$6.02^{(3)}$**	**$3.71^{(4)}$**	**$2.76^{(4)}$**	**$2.34^{(4)}$**	**$2.15^{(4)}$**	**$2.08^{(6)}$**	**$2.01^{(8)}$**	**$1.91^{(8)}$**	**$1.88^{(8)}$**
speed-up	-6.8%	0.0%	-0.7%	-4.9%	-4.4%	-3.3%	-4.7%	-12%	-11%
EPSM	5.87	3.72	2.50	1.93	1.75	1.72	1.66	1.62	1.65
TSOq	$5.54^{(5)}$	$3.85^{(5)}$	$3.08^{(5)}$	$2.42^{(5)}$	$2.05^{(5)}$	-	-	-	-

a negative percentage represent and improvements of the performance. Running times with an improvement of the performance have been bold-faced.

The last block reports the running times obtained by the best two algorithms for *counting* occurrences (we do not compare them against the other algorithms).

Among the previous solutions it turns out that the BSDMq algorithm is fastest in the case of small alphabets (2 and 16 characters), however it is second to the Hashq algorithm for $\sigma = 2$ and very long patterns, and to the EBOM algorithm for $\sigma = 16$ and short patterns. Regarding the performance of the new

Table 4. Experimental results on a protein sequence.

m	4	8	16	32	64	128	256	512	1024
AOSOq	$10.80^{(2)}$	$4.27^{(4)}$	$3.84^{(4)}$	$3.81^{(4)}$	$3.18^{(4)}$	$3.17^{(4)}$	$3.16^{(4)}$	$3.16^{(4)}$	$3.16^{(4)}$
BNDMq	$12.20^{(4)}$	$4.29^{(4)}$	$3.06^{(4)}$	$2.46^{(4)}$	$2.45^{(4)}$	$2.43^{(4)}$	$2.42^{(4)}$	$2.40^{(4)}$	$2.40^{(4)}$
BSDMq	$4.68^{(2)}$	$3.71^{(2)}$	$2.75^{(4)}$	$2.35^{(4)}$	$\mathbf{2.06}^{(4)}$	$\mathbf{1.98}^{(4)}$	$\mathbf{1.97}^{(4)}$	$\mathbf{1.97}^{(4)}$	$\mathbf{1.94}^{(4)}$
BXSq	$6.91^{(2)}$	$4.29^{(2)}$	$3.12^{(2)}$	$2.52^{(2)}$	$2.48^{(2)}$	$2.52^{(2)}$	$2.50^{(2)}$	$2.51^{(2)}$	$2.52^{(2)}$
EBOM	$\mathbf{3.87}$	$\mathbf{2.94}$	$\mathbf{2.57}$	$\mathbf{2.29}$	2.11	2.18	2.20	2.24	2.42
FSBNDMqs	$4.32^{(2,0)}$	$3.28^{(2,0)}$	$2.59^{(3,1)}$	$2.26^{(3,1)}$	$2.22^{(3,1)}$	$2.25^{(3,1)}$	$2.25^{(3,1)}$	$2.20^{(3,1)}$	$2.26^{(3,1)}$
KBNDM	7.46	4.97	3.81	3.24	3.04	3.01	2.95	2.96	2.95
SBNDMq	$5.25^{(2)}$	$3.67^{(2)}$	$2.79^{(2)}$	$2.34^{(2)}$	$2.45^{(4)}$	$2.41^{(4)}$	$2.42^{(4)}$	$2.41^{(4)}$	$2.40^{(4)}$
FS-w	$6.18^{(2)}$	$4.33^{(2)}$	$3.55^{(2)}$	$3.20^{(2)}$	$3.05^{(2)}$	$2.94^{(2)}$	$2.90^{(2)}$	$2.87^{(2)}$	$2.86^{(2)}$
FJS	9.68	18.54	4.18	3.02	2.92	2.89	2.82	3.16	4.11
HASHq	$19.92^{(3)}$	$8.36^{(3)}$	$5.05^{(3)}$	$3.75^{(5)}$	$3.19^{(5)}$	$2.99^{(5)}$	$2.92^{(5)}$	$2.76^{(5)}$	$2.66^{(5)}$
ASKIP	8.63	7.06	5.07	3.52	2.87	2.87	3.13	3.92	6.56
IOM	8.87	6.36	5.02	4.41	4.04	3.92	3.86	3.86	3.79
WOM	9.31	6.61	5.13	4.32	4.03	3.72	3.56	3.43	3.33
SKIPq	$4.56^{(2)}$	$3.41^{(3)}$	$2.64^{(3)}$	$\mathbf{2.26}^{(3)}$	$\mathbf{2.00}^{(3)}$	$\mathbf{1.94}^{(4)}$	$\mathbf{1.92}^{(4)}$	$\mathbf{1.90}^{(4)}$	$\mathbf{1.92}^{(4)}$
speed-up	$+17\%$	$+16\%$	$+2.7\%$	-1.3%	-2.2%	-2.0%	-2.5%	-2.5%	-1.0%
EPSM	6.67	25.55	2.77	2.16	1.91	1.91	1.90	1.83	1.86
TSOq	$5.41^{(5)}$	$3.90^{(5)}$	$3.29^{(5)}$	$2.59^{(5)}$	$2.17^{(5)}$	-	-	-	-

Table 5. Experimental results on a natural language sequence.

m	4	8	16	32	64	128	256	512	1024
AOSOq	$11.14^{(2)}$	$4.58^{(4)}$	$3.89^{(4)}$	$3.76^{(4)}$	$3.16^{(6)}$	$3.16^{(6)}$	$3.18^{(6)}$	$3.21^{(6)}$	$3.16^{(6)}$
BNDMq	$12.30^{(4)}$	$4.35^{(4)}$	$3.17^{(4)}$	$2.49^{(4)}$	$2.53^{(4)}$	$2.52^{(4)}$	$2.51^{(4)}$	$2.54^{(4)}$	$2.51^{(4)}$
BSDMq	$4.73^{(2)}$	$3.85^{(2)}$	$\mathbf{2.86}^{(4)}$	$\mathbf{2.35}^{(4)}$	$\mathbf{2.20}^{(4)}$	$\mathbf{2.09}^{(4)}$	$\mathbf{2.07}^{(4)}$	$\mathbf{2.02}^{(4)}$	$\mathbf{2.00}^{(4)}$
BXSq	$7.38^{(2)}$	$4.85^{(2)}$	$3.43^{(4)}$	$2.59^{(4)}$	$2.59^{(4)}$	$2.64^{(4)}$	$2.62^{(4)}$	$2.62^{(4)}$	$2.63^{(4)}$
EBOM	$\mathbf{4.33}$	$\mathbf{3.47}$	3.05	2.74	2.54	2.51	2.40	2.40	2.57
FSBNDMqs	$4.66^{(2,0)}$	$3.55^{(3,1)}$	$2.77^{(3,1)}$	$2.39^{(3,1)}$	$2.39^{(3,1)}$	$2.38^{(3,1)}$	$2.41^{(3,1)}$	$2.42^{(3,1)}$	$2.43^{(3,1)}$
KBNDM	7.84	5.49	4.22	3.59	3.28	3.08	3.04	3.03	3.03
SBNDMq	$5.75^{(2)}$	$4.18^{(2)}$	$3.13^{(4)}$	$2.43^{(4)}$	$2.52^{(4)}$	$2.50^{(4)}$	$2.52^{(4)}$	$2.51^{(4)}$	$2.52^{(4)}$
FS-w	$6.05^{(6)}$	$4.25^{(6)}$	$3.39^{(6)}$	$2.89^{(6)}$	$2.73^{(6)}$	$2.54^{(6)}$	$2.43^{(6)}$	$2.40^{(6)}$	$2.39^{(6)}$
FJS	7.06	25.33	3.68	2.95	2.96	2.81	3.18	3.42	3.83
HASHq	$19.96^{(3)}$	$8.34^{(3)}$	$5.02^{(3)}$	$3.68^{(5)}$	$3.17^{(5)}$	$2.95^{(5)}$	$2.96^{(5)}$	$2.76^{(5)}$	$2.65^{(5)}$
ASKIP	10.17	7.78	5.25	3.65	2.97	2.89	3.14	3.77	5.90
IOM	9.37	6.67	5.26	4.38	3.96	3.73	3.47	3.30	3.20
WOM	9.98	7.01	5.28	4.32	3.91	3.53	3.25	3.11	3.02
SKIPq	$4.62^{(2)}$	$3.58^{(3)}$	$\mathbf{2.78}^{(3)}$	$\mathbf{2.30}^{(3)}$	$\mathbf{2.13}^{(4)}$	$\mathbf{2.07}^{(4)}$	$\mathbf{2.06}^{(4)}$	$\mathbf{1.97}^{(4)}$	$\mathbf{1.96}^{(4)}$
speed-up	$+6.7\%$	-3.2%	-2.8%	-2.1%	-3.2%	-1.0%	-0.5%	-2.5%	-2.0%
EPSM	6.72	26.36	2.86	2.13	1.94	1.94	1.92	1.86	1.87
TSOq	$5.54^{(5)}$	$4.05^{(5)}$	$3.26^{(5)}$	$2.61^{(5)}$	$2.23^{(5)}$	-	-	-	-

algorithm, it obtains always the best results in the case of small alphabets. In such cases the gain in performance is up to 7 %. When the size of the alphabet increases the new algorithm maintain the best results only in the case of medium and long patterns (i.e. for $m \geq 32$).

The same behavior can be observed in the case of real data experimental results, where the new solution obtains the best running time in most cases. In the case of a genome sequence it is always the best choice, with a gain in performance up to 12 %. Such gain is less evident when the size of the alphabet increases. It is up to 2.5 % when searching protein sequences and natural language texts.

5   Conclusions

In this paper we presented a simple, yet efficient, variant of the Skip-Search algorithm, based on condensed alphabets. Although such extension has been applied to many algorithms in recent years, it turns out that when applied to the Skip-Search algorithm, it produces a very fast searching procedure. It will be interesting to investigate whether the use of multiple hash function can reduce the number of false positives detected during the filtering phase.

References

1. Allauzen, C., Crochemore, M., Raffinot, M.: Factor oracle: A new structure for pattern matching. In: Bartosek, M., Tel, G., Pavelka, J. (eds.) SOFSEM 1999. LNCS, vol. 1725, pp. 295–310. Springer, Heidelberg (1999). http://dx.doi.org/10.1007/3-540-47849-3_18
2. Baeza-Yates, R., Gonnet, G.H.: A new approach to text searching. Commun. ACM **35**(10), 74–82 (1992). http://doi.acm.org/10.1145/135239.135243
3. Cantone, D., Faro, S.: Fast-search algorithms: New efficient variants of the boyer-moore pattern-matching algorithm. J. Automata Lang. Comb. **10**(5/6), 589–608 (2005)
4. Cantone, D., Faro, S.: Improved and self-tuned occurrence heuristics. In: Holub, J., Zdárek, J. (eds.) Proceedings of the Prague Stringology Conference 2013, Prague, Czech Republic, 2–4 September 2013, pp. 92–106. Department of Theoretical Computer Science, Faculty of Information Technology, Czech Technical University in Prague (2013). http://www.stringology.org/event/2013/p09.html
5. Cantone, D., Faro, S., Giaquinta, E.: A compact representation of nondeterministic (suffix) automata for the bit-parallel approach. Inf. Comput. **213**, 3–12 (2012). http://dx.doi.org/10.1016/j.ic.2011.03.006
6. Charras, C., Lecroq, T.: Handbook of Exact String Matching Algorithms. College Publications (2004)
7. Charras, C., Lecroq, T., Pehoushek, J.D.: A very fast string matching algorithm for small alphabeths and long patterns (extended abstract). In: Farach-Colton [11], pp. 55–64. http://dx.doi.org/10.1007/BFb0030780
8. Crochemore, M., Czumaj, A., Gasieniec, L., Jarominek, S., Lecroq, T., Plandowski, W., Rytter, W.: Speeding up two string-matching algorithms. Algorithmica **12**(4/5), 247–267 (1994). http://dx.doi.org/10.1007/BF01185427
9. Durian, B., Chhabra, T., Ghuman, S.S., Hirvola, T., Peltola, H., Tarhio, J.: Improved two-way bit-parallel search. In: Holub, J., Zdárek, J. (eds.) Proceedings of the Prague Stringology Conference 2014, Prague, Czech Republic, 1–3 September 2014, pp. 71–83. Department of Theoretical Computer Science, Faculty of Information Technology, Czech Technical University in Prague (2014)
10. Ďurian, B., Peltola, H., Salmela, L., Tarhio, J.: Bit-parallel search algorithms for long patterns. In: Festa, P. (ed.) SEA 2010. LNCS, vol. 6049, pp. 129–140. Springer, Heidelberg (2010)

11. Farach-Colton, M. (ed.): CPM 1998. LNCS, vol. 1448. Springer, Heidelberg (1998)
12. Faro, S., Külekci, M.O.: Fast and flexible packed string matching. J. Discrete Algorithms **28**, 61–72 (2014). http://dx.doi.org/10.1016/j.jda.2014.07.003
13. Faro, S., Lecroq, T.: Efficient variants of the backward-oracle-matching algorithm. In: Holub, J., Žďárek, J. (eds.) Proceedings of the Prague Stringology Conference 2008, pp. 146–160. Czech Technical University in Prague, Czech Republic (2008)
14. Faro, S., Lecroq, T.: The exact string matching problem: a comprehensive experimental evaluation. CoRR abs/1012.2547 (2010)
15. Faro, S., Lecroq, T.: A fast suffix automata based algorithm for exact online string matching. In: Moreira, N., Reis, R. (eds.) CIAA 2012. LNCS, vol. 7381, pp. 149–158. Springer, Heidelberg (2012). http://dx.doi.org/10.1007/978-3-642-31606-7_13
16. Faro, S., Lecroq, T.: A multiple sliding windows approach to speed up string matching algorithms. In: Klasing, R. (ed.) SEA 2012. LNCS, vol. 7276, pp. 172–183. Springer, Heidelberg (2012)
17. Faro, S., Lecroq, T.: The exact online string matching problem: A review of the most recent results. ACM Comput. Surv. **45**(2), 13 (2013). http://doi.acm.org/10.1145/2431211.2431212
18. Fredriksson, K., Grabowski, S.: Practical and optimal string matching. In: Consens, M.P., Navarro, G. (eds.) SPIRE 2005. LNCS, vol. 3772, pp. 376–387. Springer, Heidelberg (2005). http://dx.doi.org/10.1007/11575832_42
19. Karp, R.M., Rabin, M.O.: Efficient randomized pattern-matching algorithms. IBM J. Res. Dev. **31**(2), 249–260 (1987)
20. Knuth, D.E., Morris, J.H., Pratt, V.R.: Fast pattern matching in strings. SIAM J. Comput. **6**(1), 323–350 (1977)
21. Lecroq, T.: Fast exact string matching algorithms. Inf. Process. Lett. **102**(6), 229–235 (2007). http://dx.doi.org/10.1016/j.ipl.2007.01.002
22. Navarro, G., Raffinot, M.: A bit-parallel approach to suffix automata: Fast extended string matching. In: Farach-Colton [11], pp. 14–33. http://dx.doi.org/10.1007/BFb0030778
23. Yao, A.C.: The complexity of pattern matching for a random string. SIAM J. Comput. **8**(3), 368–387 (1979). http://dx.doi.org/10.1137/0208029

Minimum-Density Identifying Codes in Square Grids

Marwane Bouznif[1], Frédéric Havet[2(✉)], and Myriam Preissmann[3]

[1] A-SIS, Saint Étienne, France
marwane.bouznif@a-sis.com
[2] Projet COATI, I3S (CNRS, UNSA) and INRIA, Sophia Antipolis, France
Frederic.Havet@sophia.inria.fr
[3] Univ. Grenoble Alpes and CNRS, G-SCOP, Grenoble, France
myriam.preissmann@grenoble-inp.fr

Abstract. An identifying code in a graph G is a subset of vertices with the property that for each vertex $v \in V(G)$, the collection of elements of C at distance at most 1 from v is non-empty and distinct from the collection of any other vertex. We consider the minimum density $d^*(\mathcal{S}_k)$ of an identifying code in the square grid $\mathcal{S}_k$ of height k (i.e. with vertex set $\mathbb{Z} \times \{1, \ldots, k\}$). Using the Discharging Method, we prove $\frac{7}{20} + \frac{1}{20k} \leq d^*(\mathcal{S}_k) \leq \min\left\{\frac{2}{5}, \frac{7}{20} + \frac{3}{10k}\right\}$, and $d^*(\mathcal{S}_3) = \frac{7}{18}$.

Keywords: Identifying code · Square grid · Discharging method

1 Introduction

The *two-way infinite path*, denoted $P_{\mathbb{Z}}$, is the graph with vertex set $\mathbb{Z}$ and edge set $\{\{i, i+1\} : i \in \mathbb{Z}\}$. For every positive integer k, the *finite path of length* $k - 1$, denoted P_k, is the subgraph of $P_{\mathbb{Z}}$ induced by $\{1, 2, \ldots, k\}$.

The *cartesian product* of two graphs G and H, denoted $G \square H$, is the graph with vertex set $V(G) \times V(H)$ and edge set $\{(a, x)(b, y) \mid \text{either } (a = b \text{ and } xy \in E(H)) \text{ or } (ab \in E(G) \text{ and } x = y)\}$. A *square grid* is the cartesian product of two paths, which can be finite or infinite. The *square lattice* is the cartesian product $P_{\mathbb{Z}} \square P_{\mathbb{Z}}$ of two two-way infinite paths and is denoted by $\mathcal{G}$. For every positive integer k, we denote by $\mathcal{S}_k$ the square grid $P_{\mathbb{Z}} \square P_k$.

Let G be a graph. The *closed neighbourhood* of v, denoted $N[v]$, is the set of vertices that are either v or adjacent to v in G. A set $C \subseteq V(G)$ is an *identifying code* in G if for every vertex $v \in V(G)$, $N[v] \cap C \neq \emptyset$, and for any two distinct vertices $u, v \in V(G)$, $N[u] \cap C \neq N[v] \cap C$.

Let G be a (finite or infinite) graph. For any non-negative integer r and vertex v, we denote by $B_r(v)$ the ball of radius r in G, that is $B_r(v) = \{x \mid$

F. Havet—Partly supported by ANR Blanc STINT.

© Springer International Publishing Switzerland 2016
R. Dondi et al. (Eds.): AAIM 2016, LNCS 9778, pp. 77–88, 2016.
DOI: 10.1007/978-3-319-41168-2_7

dist$(v, x) \leq r$}. For any set of vertices $C \subseteq V(G)$, the *density* of C in G, denoted by $d(C, G)$, is defined by

$$d(C, G) = \limsup_{r \to +\infty} \frac{|C \cap B_r(v_0)|}{|B_r(v_0)|},$$

where v_0 is an arbitrary vertex in G. The infimum of the density of an identifying code in G is denoted by $d^*(G)$. Observe that if G is finite, then $d^*(G) = |C^*|/|V(G)|$, where C^* is a minimum-size identifying code in G.

The problem of finding identifying codes of small density was introduced in [10] in relation to fault diagnosis in arrays of processors. Identifying codes are also used in [11] to model a location detection problem with sensor networks. Identifying codes of the grids have been studied [2,4,5,7,9,10] as well as variations where instead of considering the closed neighbourhood to identify a vertex, the ball of radius r (for some fixed r) is considered [2,8]. The closely related problem of finding a locating-dominating set with minimum density has also been studied [12].

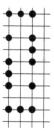

Fig. 1. Tile generating an optimal identifying code of the grid.

In this paper, we are interested in identifying codes of square grids, and more specifically the $\mathcal{S}_k$. The tile depicted in Fig. 1 was given in [3]. It generates a periodic tiling of the plane with periods $(0, 10)$ and $(4, 1)$, yielding an identifying code $C_{\mathcal{G}}^*$ of the square lattice with density $\frac{7}{20}$. Ben-Haim and Litsyn [1] proved that this density is optimal, that is $d^*(\mathcal{G}) = \frac{7}{20}$.

Daniel et al. [6] showed that $d^*(\mathcal{S}_1) = \frac{1}{2}$ and $d^*(\mathcal{S}_2) = \frac{3}{7}$. For larger value of k, they proved the following lower and upper bound on $d^*(\mathcal{S}_k)$: $\frac{7}{20} - \frac{1}{2k} \leq d^*(\mathcal{S}_k) \leq \min\left\{\frac{2}{5}, \frac{7}{20} + \frac{2}{k}\right\}$. In this paper, we improve on both the lower and upper bounds of $d^*(\mathcal{S}_k)$. We prove

$$\frac{7}{20} + \frac{1}{20k} \leq d^*(\mathcal{S}_k) \leq \min\left\{\frac{2}{5}, \frac{7}{20} + \frac{3}{10k}\right\}.$$

The upper bound is obtain by deriving an identifying code of $\mathcal{S}_k$ with density $\frac{7}{20} + \frac{3}{10k}$ from the optimal identifying code $C_{\mathcal{G}}^*$ of the square lattice.

The lower bound is obtained using the Discharging Method and proceeds in two phases. The first one is a rewriting of the proof of Ben-Haim and Litsyn [1] as a Discharging Method proof. Doing so, it becomes clear that it extends to any square grid, and so that $d^*(G) \geq \frac{7}{20}$ for any square grid. It makes it also possible to improve on this bound when $G = S_k$ with $k \geq 3$ in a second phase.

We strongly believe that both our upper and lower bounds may be improved using the same general techniques. In fact, to obtain the upper bound, we only alter the code $C_{\mathcal{G}}^*$ on the top two rows and the bottom two rows of S_k. Looking for alterations on more rows, possibly with the help of a computer, will certainly yield codes with smaller density. We made no attempt to optimize the second phase in the lower bound proof. Doing more complicated discharging rules, based on more complicated properties of identifying codes will surely give better bounds. However, we do not see any way to make the two bounds meet for all k. Nevertheless, we are able to do it for $k = 3$: we show that $d^*(S_3) = \frac{7}{18}$.

2   General Upper Bounds

Theorem 1. *For all $k \geq 7$, we have $d^*(S_k) \leq \dfrac{7}{20} + \dfrac{3}{10k}$.*

Proof. Let K_k be the code of S_k obtained from $C_{\mathcal{G}}^*$ on S_k by replacing the rows $\mathbb{Z} \times \{1\}$ and $\mathbb{Z} \times \{2\}$ by the rows depicted in Fig. 2 and the rows $\mathbb{Z} \times \{k-1\}$ and $\mathbb{Z} \times \{k\}$ by the ones obtained symmetrically.

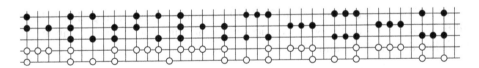

Fig. 2. The bottom rows (white disks) of K_k.

We claim the code K_k is identifying. Indeed since $C_{\mathcal{G}}^*$ is an identifying code of the square lattice, it suffices to check that for every vertex $v \in \mathbb{Z} \times \{1, 2, 3, k-2, k-1, k\}$, there is no vertex w such that $N[v] \cap K_k = N[w] \cap K_k$. This can be easily done.

The density of K_k on $\mathbb{Z} \times \{1, 2, k, k-1\}$ is $\frac{34}{80}$. So, the density of K_k is $\frac{7}{20}(1 - \frac{4}{k}) + \frac{34}{80} \times \frac{4}{k} = \frac{7}{20} + \frac{3}{10k}$.

3   Lower Bounds on $d^*(S_k)$

The aim of this section is to show that $d^*(S_k) \geq \frac{7}{20} + \frac{1}{20k}$.

The general idea is to consider an identifying code C in a square grid G. We assign an initial weight $w(v)$ to each vertex where $w(v) = 1$ if $v \in C$ and $w(v) = 0$ otherwise. We then apply some local discharging rules. In such rules, some vertices send part of their weight to some other vertices at distance at most s, for some fixed integer s. We then prove that the final weight $w^*(v)$ of every vertex v is at least d^*. We claim that it implies $d(C, G) \geq d^*$. This is trivial if G is bounded. Suppose now that $G = \mathcal{S}_k$. Since a vertex sends at most 1 to vertices at distance at most s, a charge of at most $|B_{r+s}(v_0) \setminus B_r(v_0)| \leq 2sk$ enters $B_r(v_0)$ during the discharging phase. Thus

$$|C \cap B_r(v_0)| = \sum_{v \in B_r(v_0)} w(v) \geq \sum_{v \in B_r(v_0)} w^*(v) - |B_{r+s}(v_0) \setminus B_r(v_0)| \geq d^* \cdot |B_r(v_0)| - 2sk.$$

But $|B_r(v_0)| \geq (2r+1)k - k^2$, thus $d(C, \mathcal{S}_k) \geq \limsup_{r \to +\infty} \left(d^* - \dfrac{2sk}{(2r+1)k - k^2} \right) = d^*$. This proves our claim. We then deduce $d^*(\mathcal{S}_k) \geq d^*$.

Let C be an identifying code in a square grid G. We denote by U the set of vertices not in C. For $1 \leq i \leq 5$, we define $L_i = \{v \in G \mid |N[v] \cap C| = i\}$, and we set $C_i = L_i \cap C$ and $U_i = L_i \cap U$. Observe that U_5 is empty. For $X \in \{C, L, U\}$ we set $X_{\geq i} = \bigcup_{j=i}^{5} X_j$ and $X_{\leq i} = \bigcup_{j=1}^{i} X_j$.

For every set $S \subseteq V(G)$, a vertex in S is called an S-vertex.

The following proposition is a direct consequence of the definition of identifying code.

Proposition 1. *Let C be an identifying code in a square grid G.*

(i) Every vertex in C has at most one neighbour in U_1.
(ii) Every vertex in C_1 has no neighbour in U_1.
(iii) Two vertices in C_2 are not adjacent.

Let C' be the set of vertices in C_1 that have four neighbours in G that belong all to $U_{\leq 2}$. Let $\tilde{L}_3$ be the set of vertices in C_3 having at least one neighbour in $C_{\geq 3}$. Set $\overline{L}_3 = L_3 \setminus \tilde{L}_3$ and $\overline{C}_3 = C_3 \setminus \tilde{L}_3$

Proposition 2. (Ben-Haim and Litsyn [1]). *Let C be an identifying code in a square grid G. There is a bipartite graph H with bipartition $(C', L_{\geq 3})$ such that*

(i) the degree of every element of C' is at least 4,
(ii) the degree of every element of $\overline{L}_3$ is at most 2,
(iii) the degree of every element of $\tilde{L}_3$ is at most 6, and
(iii) the degree of every element of $L_{\geq 4}$ is at most 4.

Proof. Ben-Haim and Litsyn [1] proved Proposition 2 for another definition of the set $\tilde{L}_3$: their set is larger than ours. However it happens that using word for word the same rules as Ben-Haim and Litsyn (Steps 1 to 10 in [1]) for building the bipartite graph from an identifying code of the square lattice, regardless to the fact that $\tilde{L}_3$ is not the same set, we get a bipartite graph that may be different but has exactly the same degree properties; the proof of this fact is exactly the same as the one of [1].

Furthermore, in the construction of Ben-Haim and Litsyn, the neighbours in H of an element c' of C' are always in the rectangle with corners c' and another vertex $c \in C$. Therefore it is in any square grid containing those two vertices, and so their proof works for any square grid. However, it might be possible that this vertex is not in G if this graph is not a square grid and the proof of Proposition 2 does not work for any induced subgraph G of $\mathcal{G}$.

Remark 1. The graph H in Proposition 2 may have some double edges.

Theorem 2. *Let G be a square grid. Then $d^*(G) \geq \frac{7}{20}$.*

Proof. Let C be an indentifying code in G and H be a bipartite graph associated to C as described in Proposition 2. We give an initial weight 1 to the vertices of C and 0 to the vertices in U. We then apply the following discharging rules, one after another. So if several rules must be applied to a same vertex, then it will send charge several times.

(R1) Every vertex of C sends $\frac{7}{20}$ to each neighbour in U_1 and $\frac{7}{40}$ to each neighbour in $U_{\geq 2}$.

(R2) Every vertex of $L_{\geq 3}$ sends $\frac{1}{20}$ to its neighbours in $C_{\leq 2}$.

(R3) Every vertex of $L_{\geq 3}$ sends $\frac{1}{80}$ to each C'-vertex to which it is adjacent in H by one edge and $\frac{2}{80}$ to each C'-vertex to which it is adjacent in H by two edges.

Let us prove that the final weight $w'(v)$ of each vertex v is at least $7/20$.

If $v \in C'$, then its original weight is 1. By Proposition 1-(ii), it has no U_1 neighbour. Hence it sends $\frac{7}{40}$ to each of its four neighbours in U by (R1), and receives $\frac{1}{80}$ from each of its at least four edges in H by (R3). Hence $w'(v) \geq 1 - 4 \cdot \frac{7}{40} + 4 \cdot \frac{1}{80} = \frac{7}{20}$.

If $v \in C_1 \setminus C'$, then its original weight is 1. By Proposition 1-(ii), it has no U_1 neighbour. By definition of C', it has a neighbour in $L_{\geq 3}$ from which it receives $\frac{1}{20}$ by (R2). Hence $w'(v) \geq 1 - 4 \cdot \frac{7}{40} + \frac{1}{20} = \frac{7}{20}$.

If $v \in U_1 \cup U_2$, then its original weight is 0. It receives $\frac{7}{20}$ by (R1), and it does not send anything. Hence $w'(v) = \frac{7}{20}$.

If $v \in C_2$, then its original weight is 1. By Proposition 1-(i), it has at most one U_1 neighbour, so its sends at most $\frac{7}{20} + 2 \cdot \frac{7}{40}$ by (R1). Moreover, by Proposition 1-(iii), it has a neighbour in $C_{\geq 3}$ from which it receives $\frac{1}{20}$ by (R2). Hence $w'(v) \geq 1 - \frac{7}{20} - 2 \cdot \frac{7}{40} + \frac{1}{20} = \frac{7}{20}$.

If $v \in \tilde{L}_3$, then it is in C_3, so its original weight is 1. By Proposition 1-(i), it has at most one U_1 neighbour, so it sends at most $\frac{7}{20} + \frac{7}{40}$ by (R1). By definition of $\tilde{L}_3$, v has at most one neighbour in $C_{\leq 2}$, so its sends at most $\frac{1}{20}$ by (R2). Finally, it has degree at most 6 in H, so it sends at most $6 \cdot \frac{1}{80}$ by (R3). Hence $w'(v) \geq 1 - \frac{7}{20} - \frac{7}{40} - \frac{1}{20} - 6 \cdot \frac{1}{80} = \frac{7}{20}$.

If $v \in \overline{C}_3$, then it is in C_3, so its original weight is 1. By Proposition 1-(i), it has at most one U_1 neighbour, so it sends at most $\frac{7}{20} + \frac{7}{40}$ by (R1). It has at most two neighbours in $C_{\leq 2}$, so its sends at most $2 \cdot \frac{1}{20}$ by (R2). And it has degree at most 2 in H, so it sends at most $2 \cdot \frac{1}{80}$ by (R3). Hence $w'(v) \geq 1 - \frac{7}{20} - \frac{7}{40} - 2 \cdot \frac{1}{20} - 2 \cdot \frac{1}{80} = \frac{7}{20}$.

If $v \in U_3$, then its original weight is 0. It receives $3 \cdot \frac{7}{40}$ by (R1). It has at most three neighbours in $C_{\leq 2}$, so its sends at most $3 \cdot \frac{1}{20}$ by (R2). And it has degree at most 2 in H, so it sends at most $2 \cdot \frac{1}{80}$ by (R3). Hence $w'(v) \geq 3 \cdot \frac{7}{40} - 3 \cdot \frac{1}{20} - 2 \cdot \frac{1}{80} = \frac{7}{20}$.

If $v \in C_4$, then its original weight is 1. It send at most $\frac{7}{20}$ to its unique U-neighbour by (R1). It has at most three neighbours in $C_{\leq 2}$, so its sends at most $3 \cdot \frac{1}{20}$ by (R2). And it has degree at most 4 in H, so it sends at most $4 \cdot \frac{1}{80}$ by (R3). Hence $w'(v) \geq 1 - \frac{7}{20} - 3 \cdot \frac{1}{20} - 4 \cdot \frac{1}{80} = \frac{9}{20}$.

If $v \in U_4$, then its original weight is 0. It receives $4 \cdot \frac{7}{40}$ by (R1). It has at most four neighbours in $C_{\leq 2}$, so its sends at most $4 \cdot \frac{1}{20}$ by (R2). And it has degree at most 4 in H, so it sends at most $4 \cdot \frac{1}{80}$ by (R3). Hence $w'(v) \geq 4 \cdot \frac{7}{40} - 4 \cdot \frac{1}{20} - 4 \cdot \frac{1}{80} = \frac{9}{20}$.

If $v \in C_5$, then its original weight is 1. It has no U-neighbour. It has at most four neighbours in $C_{\leq 2}$, so its sends at most $4 \cdot \frac{1}{20}$ by (R2). And it has degree at most 4 in H, so it sends at most $4 \cdot \frac{1}{80}$ by (R3). Hence $w'(v) \geq 1 - 4 \cdot \frac{1}{20} - 4 \cdot \frac{1}{80} = \frac{15}{20}$.

Thus at the end, $w'(v) \geq \frac{7}{20}$ for all vertex v, so $d(C, G) \geq \frac{7}{20}$.

Theorem 2 is tight because $d^*(\mathcal{G}) = \frac{7}{20}$. However, for $\mathcal{S}_k$, we can improve on 7/20.

Theorem 3. *For any $k \geq 3$, $d^*(\mathcal{S}_k) = \frac{7}{20} + \frac{1}{20k}$.*

Proof. Let us first give some definition. In $\mathcal{S}_k$, the *row* of index i, denoted R_i, is the set of vertices $\mathbb{Z} \times \{i\}$, the *column* of index j, denoted Q_j, is the set of vertices $\{j\} \times \{1, \ldots, k\}$. The *border vertices* are those of $R_1 \cup R_k$.

Let C be an identifying code in $\mathcal{S}_k$.

We first apply the discharging phase as in the proof of Theorem 2. At the end of this phase every vertex has weight at least 7/20. But some of them may have a larger weight.

It is for example the case of C_4-vertices which have weight at least 9/20. Let D_3 be the set of vertices of C_3 having no neighbour in $C_{\leq 2}$. Observe that $D_3 \subseteq \tilde{L}_3$. A vertex of D_3 do not send anything by (R2), hence its weight is at least $\frac{8}{20}$. Set $D = D_3 \cup C_4$.

Consider also border C-vertices. Such vertices are missing one neighbour, so for any $1 \leq i \leq 4$, border C_i-vertices gives to one U-neighbour less than non-border C_i-vertices by (R1). It follows that if v is a border C-vertex, then $w(v) \geq \frac{7}{20} + \frac{7}{40}$.

The following claim shows that there are many vertices in $R_1 \cup R_2$ with a weight larger than 7/20.

Claim. Let C be a code of $\mathcal{S}_k$. If $\{(a-3,1), (a-2,1), (a-1,1), (a,1), (a+1,1), (a+2,1), (a+3,1)\} \cap C = \emptyset$, then $(a,2)$ is in D.

Proof. If $\{(a-3,1), (a-2,1), (a-1,1), (a,1), (a+1,1), (a+2,1), (a+3,1)\} \cap C = \emptyset$, then necessarily $(a-2,2)$, $(a-1,2)$, $(a,2)$, $(a+1,2)$, and $(a+2,2)$ are in C, because each vertex has a neighbour in C. Therefore $(a-1,2)$, $(a,2)$, and $(a+1,2)$ are in $C_{\geq 3}$ and so $(a,2)$ is in D.

We then proceed to a second discharging phase. Set $S_j = Q_{j-3} \cup Q_{j-2} \cup Q_{j-1} \cup Q_j \cup Q_{j+1} \cup Q_{j+2} \cup Q_{j+3}$.

(R4) Every vertex in D gives $\frac{1}{20k}$ to every vertex in its column.

(R5) Every border C-vertex in column Q_j gives $\frac{1}{40k}$ to every vertex in S_j.

Let us examine the weight $w^*(v)$ of a vertex v after this phase.

Observe first that every vertex receives at least $\frac{1}{20k}$ during this second phase. Indeed, if $v = (a, b)$ has a D-vertex in its column, then it receives $\frac{1}{20k}$ from it by (R4). If it has no D-vertex in its column, then by Claim 3, a vertex in $\{(a-3,1), (a-2,1), (a-1,1), (a,1), (a+1,1), (a+2,1), (a+3,1)\}$ is a border C-vertex, and symmetrically, a vertex in $\{(a_3, k), (a-2, k), (a-1, k), (a, k), (a+1, k), (a+2, k), (a+3, k)\}$ is a border C-vertex. And these two vertices send $\frac{1}{40k}$ each to v by (R5), so v receives at least $\frac{1}{20k}$ in total.

If $v \in D$, then $w'(v) \geq \frac{8}{20}$. By (R4), it sends $\frac{1}{20k}$ to the k vertices of its column. Hence it sends $\frac{1}{20}$. Since it received at least $\frac{1}{20k}$, $w^*(v) \geq \frac{7}{20} + \frac{1}{20k}$.

If v is a border C-vertex, then $w'(v) \geq \frac{7}{20} + \frac{7}{40}$. By (R4), it sends $\frac{1}{40k}$ to the $7k$ vertices of S_j. Hence it sends $\frac{7}{40}$. It also receives at least $\frac{1}{20k}$. So $w^*(v) \geq \frac{7}{20} + \frac{1}{20k}$.

If v is neither a border C-vertex, nor a D-vertex, then it does not send anything. So $w^*(v) \geq w'(v) + \frac{1}{20k} \geq \frac{7}{20} + \frac{1}{20k}$.

To conclude, after the second phase, each vertex has weight at least $\frac{7}{20} + \frac{1}{20k}$. Thus $d(C, S_k) \geq \frac{7}{20} + \frac{1}{20k}$.

4 Optimal Identifying Code in S_3

Theorem 4. $d^*(S_3) = \frac{7}{18}$.

It is straightforward to check that repeating the tile of Fig. 3 with period $(12, 0)$, we obtain an identifying code of density $\frac{7}{18}$.

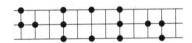

Fig. 3. Tile of a minimum-density identifying code in S_3.

It remains to show that every identifying code in S_3 has density at least $\frac{7}{18}$. We again use the Discharging Method: the technical details are more complicated than in the previous section, but the general framework is the same.

4.1 Properties of Codes in S_3

The *lower row*, (resp. *central row, upper row*) of S_3, is the set of vertices in $\mathbb{Z} \times \{1\}$, (resp. $\mathbb{Z} \times \{2\}$, $\mathbb{Z} \times \{3\}$). A *border vertex* is a vertex on the upper or lower row. A *central vertex* is a vertex on the central row. The *column* of index a is the set $\{(a, 1), (a, 2), (a, 3)\}$.

For convenience, instead of using the set C_i we use the set B_i, which is defined as follows. A vertex is in B_i if it is in C and adjacent to i vertices in U. Hence, a border vertex in B_i is in C_{4-i} and a central vertex in B_i is in C_{5-i}.

Similarly to Proposition 1, we get the following proposition.

Proposition 3. *Let C be an identifying code in S_3. Every border B_3-vertex has no neighbour in U_1.*

Proposition 4. *Let C be an identifying code in S_3. Every vertex in B_4 has a neighbour in $U_{\geq 3}$.*

Proof. Let x be a vertex in B_4. Necessarily x must be a central vertex, that is, $x = (a, 2)$ for some a. Assume for a contradiction that x has no neighbour in $U_{\geq 3}$. Then by Proposition 1, its four neighbours are in U_2. Consider $u = (a, 3)$: one of its neighbours y is in C. By symmetry, we may assume $y = (a - 1, 3)$. Now the two vertices u and $v = (a - 1, 2)$ are both adjacent to x and y. Hence, since u and v are in U_2, we obtain $N[u] \cap C = \{x, y\} = N[v] \cap C$, a contradiction.

Proposition 5. *Let C be an identifying code in S_3. Every border C-vertex adjacent to a central B_3 is in $B_0 \cup B_1$.*

Proof. Assume for a contradiction that a border B_2-vertex y is adjacent to a central B_3-vertex x. Then $N[x] \cap C = N[y] \cap C = \{x, y\}$, a contradiction.

4.2   Establishing the Lower Bound

We use the Discharging Method. Let C be an indentifying code in S_3. We give an initial weight 1 to the vertices of C and 0 to the vertices in U. We will then apply some discharging rules. Our aim is to prove that at the end the final weight of each vertex will be at least $\frac{7}{18}$.

For sake of clarity and to simplify the proof, we will perform these discharging rules in two stages.

A *generous* vertex is either a B_0-vertex or a border vertex in B_1 having its central neighbour in C. We first apply the following rules.

(R0) For $1 \leq i \leq 4$, every vertex of C gives $\frac{7}{18 \times i}$ to each of its neighbours in U_i.
(R1) Every generous vertex gives $\frac{3}{18}$ to its central neighbour(s).

Let us denote by $w_1(v)$ the weight of the vertex v after applying (R0–R1).

Observe that after (R0–R1) all the vertices of U have weight exactly $\frac{7}{18}$. Indeed for all $1 \leq i \leq 4$, every vertex u in U_i receives $\frac{7}{18 \times i}$ from each of its i neighbours. Hence in total it receives $\frac{7}{18}$ and so $w_1(u) = \frac{7}{18}$.

The weight of the vertices of U will not change anymore and the charge will now only move from C-vertices to other C-vertices.

We define the *excess* of a vertex v of C as $\epsilon(v) = w_1(v) - \frac{7}{18}$. Informally, if it is positive, the excess of v measures how much weight v has above $\frac{7}{18}$ and thus can give to other vertices. If it is negative, the excess of v measures the quantity of weight v must receives from others to get weight $\frac{7}{18}$.

Observe that B_0-, B_1-, B_2- and border B_3-vertices have positive excess:

- If $v \in B_0$, then it gives nothing to U-vertices, and it gives $\frac{3}{18}$ to each (at most two) central neighbours. Thus $\epsilon(v) \geq 1 - \frac{7}{18} - \frac{6}{18} = \frac{5}{18}$.
- If $v \in B_1$, then it gives at most $\frac{7}{18}$ to its U-neighbour. So if it is not generous, $\epsilon(v) \geq 1 - \frac{7}{18} - \frac{7}{18} = \frac{2}{9}$ and if v is generous, $\epsilon(v) \geq 1 - \frac{7}{18} - \frac{7}{18} - \frac{3}{18} = \frac{1}{18}$.
- If $v \in B_2$, then it is adjacent to at most one U_1-vertex by Proposition 1. Hence $\epsilon(v) \geq 1 - \frac{7}{18} - \frac{7}{18} - \frac{7}{36} = \frac{1}{36}$.
- If v is a border B_3-vertex, then by Proposition 3, it is adjacent to no U_1-vertex. Hence it gives at most $\frac{7}{36}$ to each of its U-neighbours. So $\epsilon(v) \geq 1 - \frac{7}{18} - 3 \times \frac{7}{36} = \frac{1}{36}$.

On the opposite, some vertices of $B_3 \cup B_4$ may have negative excess. Such vertices of $B_3 \cup B_4$ will be called *defective*. Observe that defective vertices are on the central line. Moreover it is easy to check that a defective vertex has no generous neighbour. Indeed if a defective vertex x has a generous neighbour y, then it is in B_3. Since x has at most one U_1-neighbour, it sends at most $\frac{7}{18} + 2 \times \frac{7}{36} = \frac{14}{18}$ to its U-neighbours. But it also receives $\frac{3}{18}$ from its generous neighbour. Hence $\epsilon(x) \geq 1 - \frac{7}{18} - \frac{14}{18} + \frac{3}{18} = 0$.

Simple calculations and Propositions 1, 3 and 4 show that a defective vertex is of one of the following kinds:

- a B_4-vertex with at least two U_2-neighbours;
- a central B_3-vertex with one U_1-neighbour and no generous neighbour.

We will now apply some new discharging rules in order to give charge to the defective vertices so that the final excess $\epsilon^*(v)$ of every vertex v is non-negative. The rules are applied one after another, so if several rules must be applied to a same vertex then it will send charge several times.

For $S \in \{C, U\}$, an S-*column* is a column all vertices of which are in S. A *right barrier* (resp. *left barrier*) is a C-column such that the right (resp. left) neighbours of its two border vertices are in U_1. A *lonely barrier* is a barrier such that the columns to its right and its left are U-columns. Let x be a C-vertex. Its *right pal* (resp. *left pal*) is the closest central C-vertex to its right (resp. left). A pal is *good* if it is defective or in a lonely barrier.

(R2) Every border C-vertex x whose central neighbour is not in C sends $\epsilon(x)/2$ to each of its good pals, if it has two of them and $\epsilon(x)$ to its good pal, if it has exactly one.

(R3) Every border vertex x in a right (resp. left) barrier sends $\epsilon(x)$ to its right (resp. left) pal. Every central vertex x of a right (resp. left) barrier sends to its right (resp. left) pal $\epsilon(x)$ if it is in B_2 and $\frac{1}{18}$ if it is in B_1.

(R4) Every generous B_1-vertex not in a barrier sends $\frac{1}{36}$ to each of its pals.

(R5) Every central B_1-vertex whose left (resp. right) neighbour is not in C sends $\frac{3}{18}$ to its right (resp. left) neighbour.

(R6) Every border B_2-vertex whose central neighbour is in B_2 and adjacent to a central B_3 sends $\frac{1}{36}$ to this later vertex.

(R7) Every central B_2-vertex with a border C-neighbour and a central C-neighbour sends $\frac{1}{36}$ to its central C-neighbour.

(R8) Every central B_0-vertex or central B_1-vertex with its two central neighbours in C sends $\frac{1}{18}$ to each of its central neighbours.

(R9) Every central vertex in a right (resp. left) barrier resend to its right (resp. left) pal all the charge its receives from border vertices to its left (rep. right) by (R2).

(R10) A central B_2-vertex with a B_3-neighbour to its right (resp. left) and a B_2-neighbour to its left (resp. right) sends $\frac{1}{36}$ and everything it gets from the left (resp. right) to its right (resp. left) neighbour.

It is routine to check that every non-defective vertex sends at most its excess and that its final excess is non-negative. We now consider defective vertices. Let $v = (a, 2)$ be a defective vertex. Let us show that its final excess $\epsilon^*(v)$ is non-negative.

We first consider the case when v is in B_4.

- Assume first that v has three U_2-neighbours and one neighbour in $U_3 \cup U_4$. Then its original excess $\epsilon(v)$ is at least $1 - \frac{7}{18} - 3 \times \frac{7}{36} - \frac{7}{54} = -\frac{11}{108}$. By symmetry, we may assume that $(a, 3) \in U_2$, $(a-1, 3) \in C$ and $(a+1, 3) \in U$. Hence $(a - 1, 2)$ is in $U_3 \cup U_4$ because $N[(a - 1, 2)] \cap C \neq N[(a, 3)] \cap C$. Thus $(a, 1)$ and $(a + 1, 2)$ are in U_2. Since $N[(a, 1)] \cap C \neq N[(a+1, 2] \cap C$, $(a - 1, 1)$ and $(a + 2, 2)$ are in C and $(a + 1, 1) \in U$. But $(a + 1, 1)$ and $(a + 1, 3)$ must have a neighbour in C, so $(a + 2, 1)$ and $(a + 2, 3)$ are in C. Hence the column of index $a + 2$ is a left barrier. So v receives at least $3 \times \frac{1}{36}$ by (R3) from the vertices of this barrier and $\frac{1}{72}$ from each of $(a - 1, 1)$ and $(a - 1, 3)$ by (R2). Hence $\epsilon^*(v) \geq -\frac{11}{108} + 3 \times \frac{1}{36} + 2 \times \frac{1}{72} > 0$.

- Assume now that v has two U_2-neighbours and two neighbours in $U_3 \cup U_4$. Then its original excess $\epsilon(v)$ is at least $1 - \frac{7}{18} - 2 \times \frac{7}{36} - 2 \times \frac{7}{54} = -\frac{1}{27}$. Observe that $(a, 1)$ and $(a, 3)$ may not both be in U_3 for otherwise $(a - 1, 2)$ and $(a + 1, 2)$ would also be in U_3. Hence without loss of generality, we are in one of the following two subcases:

  • $\{(a - 1, 1), (a - 1, 3), (a + 1, 3)\} \subseteq C$ and $(a + 1, 1)$ is in U. Then $(a + 2, 1)$ must be in C. Hence the three vertices $(a - 1, 1)$, $(a - 1, 3)$, $(a + 1, 3)$ send each $\frac{1}{72}$ to v by (R2). Hence $\epsilon^*(v) \geq -\frac{1}{27} + 3 \times \frac{1}{72} > 0$.

  • $\{(a-2, 2), (a-1, 3), (a+1, 1), (a+2, 2)\} \subseteq C$ and $(a-1, 1)$ and $(a+1, 3)$ are in U. Then $(a - 2, 1)$ and $(a + 2, 3)$ must be in C. Observe that the columns of index $a - 2$ and $a + 2$ are not barriers since $(a - 1, 1)$ and $(a + 1, 3)$ are in U_1 and $(a - 1, 3)$ and $(a + 1, 1)$ are not in U_1. If $(a - 2, 2)$ is not defective, then $(a - 1, 3)$ sends at least $\frac{1}{36}$ to v by (R2) and $(a + 1, 1)$ sends at least $\frac{1}{72}$ to v by (R2). Hence $\epsilon^*(v) \geq -\frac{1}{27} + \frac{1}{36} + \frac{1}{72} > 0$. If $(a-2, 2)$ is defective, then it is in B_3. Since the code is identifying $(a - 3, 1)$ is in C, so $(a - 2, 1)$ is a border B_1-vertex. So it sends $\frac{1}{36}$ to v by (R4). As v receives at least $\frac{1}{72}$ from each of $(a-1, 3)$ and $(a+1, 1)$, we have $\epsilon^*(v) \geq -\frac{1}{27} + \frac{1}{36} + 2 \times \frac{1}{72} > 0$.

We now consider the case when $v \in B_3$. Let w be its C-neighbour. w is a central vertex, for otherwise w would be a generous vertex by Proposition 5 and thus v would not be defective. By symmetry, we may assume that $w = (a - 1, 2)$.

– Assume first that v has one U_1-neighbour and two U_2-neighbours. Up to symmetry, the U_1-neighbour z is either $(a+1, 2)$ or $(a, 3)$.
  - If $z = (a+1, 2)$, then $(a+1, 1)$ and $(a+1, 3)$ are in U and so $(a-1, 1)$ and $(a-1, 3)$ are in C. Hence w is in B_1 because a defective vertex has no generous neighbour. Thus w sends $\frac{3}{18}$ to v by (R5). So $\epsilon^*(v) \geq 0$.
  - Assume that $z = (a, 3)$. Then $(a-1, 3)$ and $(a+1, 3)$ are in U. Moreover $(a+1, 2)$ and $(a, 1)$ are in U_2 and so $(a+1, 1)$ is not in C. It follows that $(a-1, 1)$, $(a+2, 2)$, $(a+2, 1)$ and $(a+2, 3)$ are in C. The column of index $a+2$ is a left barrier.
    We claim that v receives at least $\frac{2}{18}$ from its right and at least $\frac{1}{18}$ from its left. This yields $\epsilon^*(v) \geq 0$.
    Let us show that v receives $\frac{2}{18}$ from its right. If one vertex of the column of index $a+2$ is in B_1, the vertices of the barrier send at least $\frac{1}{18} + 2 \times \frac{1}{36} = \frac{2}{18}$ to v. Hence we may assume that $(a+3, 1)$, $(a+3, 2)$ and $(a+3, 3)$ are in U. Furthermore by Proposition 1, $(a+3, 1)$ and $(a+3, 3)$ are in U_2 so $(a+4, 1)$ and $(a+4, 3)$ are in C. If $(a+4, 2)$ is in C, then $(a+3, 2) \in U_2$ and so $\epsilon((a+2, 2)) = \frac{2}{9}$. Hence v receives at least $\frac{2}{9} + 2 \times \frac{1}{36} > \frac{2}{18}$ from its right. If $(a+4, 2)$ is not in C, then by (R2) $(a+4, 1)$ and $(a+4, 3)$ send in total $\frac{1}{36}$ to $(a+2, 2)$ which redirect it to v by (R9). In addition, the barrier send at least $3 \times \frac{1}{36}$ to v by (R3). Hence v receives at least $\frac{2}{18}$ from its right.
    Now, either $(a-1, 2)$ is in B_1 in which case it sends $\frac{1}{18}$ to v by (R8), or $(a-1, 2)$ is in B_2 and sends $\frac{1}{36}$ to v by (R7) and $(a-1, 1)$ is in B_2 or B_1 and sends $\frac{1}{36}$ to v by (R6) or (R4). In both cases, v receives $\frac{1}{18}$ from its left.

– Assume that v has one U_1-neighbour, one U_2-neighbour and one neighbour in $U_3 \cup U_4$. Then $\epsilon(v) \geq 1 - \frac{7}{18} - \frac{7}{18} - \frac{7}{36} - \frac{7}{54} = -\frac{11}{108}$. Let t be the neighbour of v in $U_3 \cup U_4$. By symmetry, we may assume that $t = (a, 3)$ or $t = (a+1, 2)$.
  - Assume that $t = (a, 3)$. Then $(a-1, 3)$, $(a+1, 3)$ are in C and $(a-1, 1)$, $(a+1, 1)$ and $(a+2, 2)$ are in U. Thus $(a+2, 1)$ is in C.
    If $(a-1, 2)$ is in B_1, then it sends $\frac{1}{18}$ to v by (R8). If not $(a-1, 2)$ is in B_2 and thus sends $\frac{1}{36}$ to v by (R7). Moreover $(a-1, 3)$ is in $B_1 \cup B_2$ and thus sends $\frac{1}{36}$ by (R4) or (R6). Hence v receives at least $\frac{1}{18}$ from its left. Let us now show that v receives at least $\frac{2}{27}$ from its right. Since $N[(a+1, 1)] \cap C \neq N[(a+2, 1)] \cap C$, we have $(a+3, 1) \in C$. Since $N[(a+1, 3)] \cap C \neq N[(a+2, 3)] \cap C$, we have $(a+3, 3) \in C$. If $(a+3, 2)$ is in C, then this vertex is not good. So $(a+1, 3)$ sends all its excess to v by (R2). This excess is at least $\frac{5}{54} \geq \frac{2}{27}$. If $(a+3, 2)$ is not in C, then $(a+2, 3)$ is in C, because $N[(a+2, 2)] \cap C \neq N[(a+1, 1)] \cap C$. Hence $(a+2, 3)$ is in B_1 and it is not generous. So its excess is at least $\frac{2}{9}$ and by (R2), it sends at least $\frac{1}{9}$ to v.
    Hence v receives at least $\frac{1}{18}$ from its left and $\frac{2}{27}$ from its right. Thus $\epsilon^*(v) \geq -\frac{11}{108} + \frac{1}{18} + \frac{2}{27} \geq 0$.
  - Assume that $t = (a+1, 2)$. By symmetry, we may assume that $(a, 3) \in U_1$. Then $(a-1, 3)$, $(a+1, 3)$ and $(a-1, 1)$ are in U and $(a+1, 1)$, $(a+2, 2)$ are in C.

Since $N[v] \cap C \neq N[w] \cap C$, necessarily $(a-2,2) \in C$. Since $N[(a-1,1)] \cap C \neq N[(a-1,3)] \cap C$, then a vertex y in $\{(a-2,1), (a-2,3)\}$ is in C. If $(a-2,2)$ is in B_0, then it sends $\frac{1}{18}$ to w by (R8); if $(a-2,2)$ is in B_1, then it sends at least $\frac{1}{18}$ to w by (R5) or (R8); if $(a-2,2)$ is in B_2, then $(a-2,2)$ sends $\frac{1}{36}$ to w by (R6). In any case, w receives at least $\frac{1}{36}$ from $(a-2,2,)$ which it redirects to v with an additional $\frac{1}{36}$ by (R10). So w sends at least $\frac{1}{18}$ to v.

Now $(a+1,1)$ has excess at least $\frac{5}{54}$ since it is adjacent to no U_1 and at least one U_3. Hence it sends at least $\frac{5}{108}$ to v by (R2).

Thus $\epsilon^*(v) \geq -\frac{11}{108} + \frac{1}{18} + \frac{5}{108} = 0$.

– Assume that v has one U_1-neighbour and two neighbours in $U_3 \cup U_4$. Then $\epsilon(v) \geq 1 - \frac{7}{18} - \frac{7}{18} - 2 \times \frac{7}{54} = -\frac{1}{27}$.
Without loss of generality, $(a-1,3)$, $(a+1,3)$ and $(a+2,2)$ are in C, and $(a-1,1)$, $(a+1,1)$ are in U. Then the vertex $(a+1,3)$ had excess at least $\frac{17}{108}$ since it is adjacent to two U_3 and no U_1. Thus by (R2) it sends at least $\frac{17}{216}$ to v. So $\epsilon^*(v) \geq -\frac{1}{27} + \frac{17}{216} > 0$.

Hence at the end, all the C-vertices have non-negative final excess and final weight at least $\frac{7}{18}$. This finishes the proof of Theorem 4.

References

1. Ben-Haim, Y., Litsyn, S.: Exact minimum density of codes identifying vertices in the square grid. SIAM J. Discrete Math. **19**, 69–82 (2005)
2. Charon, I., Honkala, I., Hudry, O., Lobstein, A.: General bounds for identifying codes in some infinite regular graphs. Electron. J. Comb. **8**(1), Research Paper 39, 21 pp. (2001)
3. Cohen, G., Gravier, S., Honkala, I., Lobstein, A., Mollard, M., Payan, C., Zémor, G.: Improved identifying codes for the grid, Comment to [4]
4. Cohen, G., Honkala, I., Lobstein, A., Zémor, G.: New bounds for codes identifying vertices in graphs. Electron. J. Comb. **6**(1), R19 (1999)
5. Cukierman, A., Yu, G.: New bounds on the minimum density of an identifying code for the infinite hexagonal grid discrete. Appl. Math. **161**, 2910–2924 (2013)
6. Daniel, M., Gravier, S., Moncel, J.: Identifying codes in some subgraphs of the square lattice. Theoret. Comput. Sci. **319**, 411–421 (2004)
7. Honkala, I.: A family of optimal identifying codes in $\mathbb{Z}^2$. J. Comb. Theor. Ser. A **113**(8), 1760–1763 (2006)
8. Junnila, V.: New lower bound for 2-identifying code in the square grid. Discrete Appl. Math. **161**(13–14), 2042–2051 (2013)
9. Junnila, V., Laihonen, T.: Optimal lower bound for 2-identifying codes in the hexagonal grid. Electron. J. Comb. **19**(2), Paper 38 (2012)
10. Karpovsky, M., Chakrabarty, K., Levitin, L.B.: On a new class of codes for identifying vertices in graphs. IEEE Trans. Inf. Theory **44**, 599–611 (1998)
11. Ray, S., Starobinski, D., Trachtenberg, A., Ungrangsi, R.: Robust location detection with sensor networks. IEEE J. Sel. Areas Commun. **22**(6), 1016–1025 (2004)
12. Slater, P.J.: Fault-tolerant locating-dominating sets. Discrete Math. **249**(1–3), 179–189 (2002)

Separating Codes and Traffic Monitoring

Thomas Bellitto[(✉)]

LaBRI, University of Bordeaux, Bordeaux, France
thomas.bellitto@labri.fr

Abstract. This paper studies the problem of traffic monitoring which consists in differentiating a set of walks on a directed graphs by placing sensors on as few arcs as possible. The problem of characterising a set of individuals by testing as few attributes as possible is already well-known but traffic monitoring presents new challenges that the previous models of separation fall short at modelling such as taking into account the multiplicity and order of the arcs in a walk. We therefore introduce a new stronger model of separation based on languages that generalises the traffic monitoring problem. We study two subproblems that we think are especially relevant for practical applications and develop methods to solve them combining integer linear programming, separating codes and language theory.

1 Introduction

Characterising objects by using as few properties as possible is a very important task in diagnosis or identification problems and has been broadly studied under different names such as separating/identifying codes or test cover. The notion of separating code finds many applications in a wide range of domains, each time we have to deliver a diagnosis with limited or expensive access to information. Notable examples include visualisation and pattern detection [3,19], routing [14] or fault detection [16] in telecommunication networks, as well as many areas of bio-informatics, such as analysis of molecular structures [8] or medical diagnosis method, where test covers are the core of diagnostic tables (see [24]) and are therefore determinant for blood sampling or bacterial identification (see [23] for a survey on the different methods). Separating codes have also been studied under the name of *sieves* in the context of logic characterisation of graphs: the size of a minimal separating code determines the complexity of the first-order logic formula required to describe a graph [12].

The problem this paper mainly addresses is called traffic monitoring. Assume that traffic is going through a network (e.g. cars in a town, packets in a telecommunication network, skiers in a ski resort...) and that we are given the possibility to install sensors on the arcs of the graph. Thus, each time an object walks in the graph and goes through an equipped arc, it activates a sensor and we know how many times and in which order each sensor was activated. We are given the set of possible walks the object can take. Our goal is to find where to place the sensors so that being able to determine exactly which route the object took from the

© Springer International Publishing Switzerland 2016
R. Dondi et al. (Eds.): AAIM 2016, LNCS 9778, pp. 89–100, 2016.
DOI: 10.1007/978-3-319-41168-2_8

information given by the sensors. This problem has been proven NP-complete in [15], even in the case of acyclic graphs. Aside from the complexity aspect, few results have been obtained on the problem. We have to take into account information such as the multiplicity and the order of the signals sent by the sensors, which place this problem beyond the expressive power of the existing models for separating codes and their resolution methods. For the special case of monitoring skiers, Meurdesoif et al. developed a solution for acyclic graphs [17]. Their algorithm is based on double-paths detection and their approach is very different than ours. In this paper, we adopt a new, more flexible approach based on separating codes that allows us to handle more general problems.

The next section provides all the necessary notions. It gives more formal definition of the separating code and traffic monitoring problems, presents how to solve the standard problem of separating codes by reducing it to integer linear programming and outlines the limitations of this model that makes it unable to handle traffic monitoring. We present in Sect. 3 a new model of separation based on language theory that overcomes these limitations and even generalises the traffic monitoring problem. The next two sections focus on particular cases of traffic monitoring. Section 4 studies the case where the set of walks to separate is finite and Sect. 5 studies the case where we want to separate every walks starting from a set of given vertices to a set of potential destinations. Such sets of walks can be infinite and would yield therefore infinitely many constraints. We study the underlying language and exhibit some properties that enable us to reformulate the problem as a standard integer linear problem (with therefore finitely many constraints). However, to fit within the page limit, some proofs are omitted in this version of the paper. Even though this paper only studies the two aforementioned problems, the model and results it presents can also be generalised to more complex sets of walks that are more relevant for some practical applications.

2 Preliminaries

2.1 Separating Codes

The separating code problem is, given a set of individuals $\mathcal{I}$ and a set of attributes $\mathcal{A}$ to find the smallest subset of attributes $\mathcal{C} \subset \mathcal{A}$ such that each individual is characterised by the attributes of $\mathcal{C}$ it possesses. Here, attributes have no value and are only properties that each individual possesses or not. This problem has been particularly studied in the case of graphs where a separating code is a subset $\mathcal{C}$ of vertices such that each vertex of the graph is characterised by the intersection of its closed neighbourhood with the code. In many practical applications, some attributes are more expensive to test than others. Let a cost be associated to each attribute. The weighted separation problem is to exhibit a separating code of minimal cost.

Given a code, we shall call *signature* of an individual i the set of attributes of the code it possesses. Thus, in the particular case of separating code on a graph, the signature of a vertex v is the intersection of its closed neighbourhood with the code. Hence, separating codes are by definition the sets of attributes such that all the individuals have different signatures.

The notion of separating code on a graph is often combined with another well-known problem of graph theory: domination. A dominating set on a graph is a set of vertices such that all the vertices of the graph are in the set or have a neighbour in the set and a set that is both separating and dominating is called an identifying code. Requiring the additional constraint that each individual should have at least one attribute in a separating code is a well-studied problem, for example in fault detection and security (see [22] for example) where domination ensures that we will notice quickly if a problem occurs and separation allows us to determine what the problem is.

While separation and domination seem unrelated at first sight, one can actually easily reduce identification to separation. Indeed, let us notice that separation consists in making the signature of the individual pairwise distinct while identification requires that the signatures are both pairwise distinct and non-empty. Therefore, all we have to do is add to our set of individuals an artificial individual with no attributes and whose signature will thus necessarily be empty, thereby forcing all the other individuals to have non-empty signatures. Hence, the identification problem is resolvable if and only if no two individuals possess exactly the same attributes and each individual possesses at least one attribute.

The problem of separating code appeared first under the name of test cover (see [18] or [3] for important examples). It first appeared on graphs in [11] together with identifying codes and has been widely studied for many subclasses of graphs. One can also find variants of identification on hypergraphs which consists in characterising the vertices by the hyperedges they belong to or in bipartite graphs which consists in separating the vertices of V_1 using only vertices of V_2 where (V_1, V_2) is a bipartition of the graph (see [4,5]). Those two problems are actually equivalent to the general problem given by a set of individuals and a set of attributes. The problem has also been studied on directed irreflexive graphs ([9,20,21]), a model whose expressiveness lies between the problem on graph and the general problem with individuals and attributes.

We now give a reformulation of the problem as an integer linear program (ILP) on which the methods developed in the next sections are based.

Let $\mathcal{I}$ be a set of individuals and $\mathcal{A}$ be a set of attributes. Let $(i, i') \in \mathcal{I}^2$ be such that $i \neq i'$. The separating set of i and i', denoted by $\mathrm{Sep}(i, i')$, is the symmetric difference of their attributes $i.e.$ the set of attributes that exactly one of them possesses. Therefore, i and i' will have distinct signatures according to a set of attributes $\mathcal{C}$ if and only if $\mathcal{C}$ possesses at least one attribute of their separating set.

For each attribute $a \in \mathcal{A}$, let x_a be a binary variable that indicates whether $a \in \mathcal{C}$. We obtain the following system:

$$\begin{cases} \forall (i, i') \in \mathcal{I}^2 \text{ with } i \neq i', \sum_{a \in \mathrm{Sep}(i, i')} x_a \geq 1 \\ \\ \text{minimise} \sum_{a \in \mathcal{A}} x_a \end{cases}$$

Polyhedra associated with identifying codes problems have been studied thoroughly in [2].

This ILP has a solution if and only if all the separating sets are non-empty *i.e.* if and only if no two individuals possess exactly the same attributes.

Notice that we have an ILP formulation of the weighted separating code problem by taking into account the costs in the linear objective function. For the sake of simplicity, we will stay with the unweigthed separating code problem in what follows but the method presented can easily be generalised to the weighted case.

2.2   The Traffic Monitoring Problem

We model a network with a directed graph $G = (V, A)$ and have the possibility to install sensors on the arcs of the graph. We want to reconstruct the route of objects walking in the graphs.

When an object walks in the graph, it activates a sensor each times it goes through an equipped arc and by moving in the graph, the object hence activates the sensors a certain number of times in a certain order. The ordered sequence of activated sensors is what we call the signature of the walk of the object.

In this problem, we are given the set $\mathcal{R}$ of potential routes the object can take and we are looking for the smallest possible set of arcs such that the signatures of all the routes of $\mathcal{R}$ are pairwise distinct. Hence, the information given by the sensors is sufficient to know exactly which route the object picked. Note that we make no assumption on the speed of the object: the time between the activation of two sensors cannot be used to determine what the object did in the meantime. We know however in which order the sensors were activated.

As before, variants where some arcs are more expensive to monitor than others or where we also want the walks of $\mathcal{R}$ to have non-empty signatures (so that we know if an object is walking in the network) are also of great practical interest.

Here again, we have to distinguish a set of individuals (here, walks in a graph) by testing as few attributes as possible (here, the arcs the walks use). However, while this problem looks close to the ones we have mentioned so far, it presents three major difficulties that we have not encountered yet and that place it beyond the expressive power of the models described previously.

Let us illustrate those difficulties on an example:

– The set of activated sensors is not sufficient to identify a walk: on the network of the Fig. 1, the walks $(1, 2, 3, 1)$ and $(1, 2, 3, 1, 2, 3, 1)$ are different but the sets of arcs they use are the same. Since they do not use these arcs as many times, we still can distinguish them - their signatures according to the sensor set $\{a, b\}$ are respectively a and aa - but this forces us to take into account the multiplicity of the attributes of our individuals, which cannot be done with the previous model.
– The number of times each sensor is activated is not sufficient to identify a walk: on the network of the Fig. 1, the walks $(1, 2, 3, 1, 4, 5, 1)$ and $(1, 4, 5, 1, 2, 3, 1)$ do not only use the same arcs but they also use them the same amount of times.

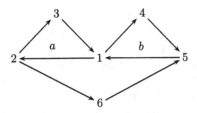

Fig. 1. An example of network $G = (V, A)$. The arcs (1,2) and (5,1) are equipped with sensors called respectively a and b.

They can still be separated: their signatures are indeed respectively ab and ba, but this requires considering the order of the attributes. This also illustrates the limit of the resolution method we showed in the previous section: indeed, we said that a code separates two individuals if and only if it contains an element that does so, but one can see here that the sensor set $\{a, b\}$ separates the walks $(1, 2, 3, 1, 4, 5, 1)$ and $(1, 4, 5, 1, 2, 3, 1)$ while neither $\{a\}$ nor $\{b\}$ can.

- The set $\mathcal{R}$ of potential walks can be infinite: if the graph contains a cycle, the number of walks in it is infinite and $\mathcal{R}$ can be any subset of it. Therefore, even checking in finite time whether a given set of sensors separates $\mathcal{R}$ is non trivial since it requires to ensure that all the walks of $\mathcal{R}$ have different signatures. A wrong intuition is that the problem can be reduced to separation on elementary paths since non-elementary walks are concatenation of elementary paths (which would be helpful since there is only a finite number of them) but this does not work. For example, assume that the set $\mathcal{R}$ we want to separate is the set of cycles starting from and leading to the vertex 1. Our set of sensors $\{a, b\}$ is not suitable since the cycles $(1, 2, 3, 1, 4, 5, 1)$ and $(1, 2, 6, 5, 1)$ both have the signature ab although all the elementary cycles $((1, 2, 3, 1), (1, 4, 5, 1)$ and $(1, 2, 6, 5, 1))$ have different signatures (respectively, a, b and ab).

3   A New Model of Separation: Separation on a Language

3.1   The Problem

We recall that an alphabet is a non-empty finite set whose elements are called letters, a word is a finite sequence of letters and a language is a set of words on a given alphabet. The reader may refer to [10] for basic notions of language and automata theory.

Let A be an alphabet and let $\mathcal{C}$ be a subalphabet of A. We call projection on $\mathcal{C}$ and we note $p_\mathcal{C}$ the function that associates to a word $u \in A^*$ its longer subword using only letters of $\mathcal{C}$. For example, $p_{\{a,b\}}(abacacb) = abaab$.

Given a language L on an alphabet A, the problem of separation on a language consists in finding the smallest subalphabet $\mathcal{C} \subset A$ such that all the words of L have different projections on $\mathcal{C}$.

For example, let $L = \{aabcc, acabc, baacb, cbaac\}$. One can immediately notice that $aabcc$ and $acabc$ use the same letters the same amount of times, so one cannot separate them with an alphabet of one letter. Plus, the projection of both those words on the subalphabet $\{a, b\}$ are aab so this subalphabet does not separate them neither. The projection of $acabc$ and $cbaac$ on $\{b, c\}$ are also the same (cbc). However, the projections of the four words of L on the subalphabet $\{a, c\}$ are respectively $aacc, acac, aac$ and $caac$ and are pairwise distinct. Hence, $\{a, c\}$ is a solution of the problem and is even the only optimal solution.

The problem of separation on a language is a generalisation of the problem of separating code. Indeed, let $\mathcal{I}$ be a set of individuals, let $\mathcal{A}$ be a set of attributes and $\leqslant$ be a total order on $\mathcal{A}$. Let us associate to each individual i a word on $\mathcal{A}$ composed of all the attributes i possesses, exactly once, in increasing order according to $\leqslant$. The subalphabet of $\mathcal{A}$ that separates the language of the words associated to the individuals of $\mathcal{I}$ are exactly the separating codes. Therefore, separation on a language is NP-complete in both the size of the language and the alphabet it is defined on.

3.2   Relation with Traffic Monitoring

Let us consider the set A of arcs of a graph as an alphabet. Since a walk on a graph is a sequence of arcs, it is a word and the set $\mathcal{R}$ of possible routes is a language on A. Given a set of sensors, the signature of a route is its projection on the subalphabet composed of the arcs that have a sensor. Hence, traffic monitoring is a particular case of separating code on a language (however, the problems are not equivalent since not all languages can be written as a set of routes on the alphabet of the arcs of a graph).

This reduction allows us to use tools arising from separating codes and language theory to address the traffic monitoring problem. However, the problem remains NP-complete in the size of the language that we want to separate. Moreover, we saw that on some instances, this language can be infinite. Hence, we do not intend to solve the problem in the general case but rather to address some classes of sets of routes of practical interest.

4   Separation of a Finite Language

The easiest place to start is the case where the language we want to separate is finite. This happens in particular when we want to solve traffic monitoring on an acyclic graph or when we can bound the length of the walks on the network. This problem already covers a wide range of applications.

Notice that if a set separates two words w and w', so do its supersets. Hence, we can look for the minimal sets of attributes that separate two words w and w': a set of attributes separates w and w' if and only if it contains one of those minimal sets. We thus define the separating set $\text{Sep}(w, w')$ of two words w and w' as follows: given two words w and w' on an alphabet A, a subalphabet $\mathcal{C}$ of A belongs to the separating set $\text{Sep}(w, w')$ of w and w' if and only if $\mathcal{C}$ separates w and w' and none of its strict subsets does. The separating set of two words is therefore a set of sets of letters. It follows from the definition that a subalphabet

$C \subset A$ separates two words w and w' if and only if there exists $C' \subset C$ such that $C' \in \mathrm{Sep}(w, w')$.

We now exhibit some properties of the structure of separating sets that are useful to compute it efficiently (Theorem 2).

Lemma 1. *A word $u \in A^*$ is characterised by its projections on the subalphabets of A of cardinality 2.*

Proof. The letter a is the first letter of a word u if and only if it is the first letter of all projections of u on the subalphabets of cardinality 2 containing a. One can then remove the a at the first position of the projection of u on the subalphabets containing a and find the second letter of u. This can be iterated until u is entirely determined.   □

For example, let $A = \{a, b, c\}$ and $u \in A^*$ such that $p_{\{a,b\}}(u) = abba$, $p_{\{a,c\}}(u) = aca$, $p_{\{b,c\}}(u) = bbc$. Since a is the first letter of the projection of u on $\{a, b\}$ and $\{a, c\}$, we know that a is the first letter of u. Thus, there exists v such that $u = av$. From the projection of u, we deduce that $p_{\{a,b\}}(v) = bba$, $p_{\{a,c\}}(v) = ca$, $p_{\{b,c\}}(v) = bbc$ and the first letter of v is therefore b. Thus, there exists w such that $u = abw$ and by iterating the process, we end up finding that $u = abbca$.

Theorem 2. *The separating set of two words contains only sets of cardinality at most 2.*

Proof. Let u and v be two words on an alphabet A and let C be a subalphabet of A of cardinality at least 3 such that no strict subalphabet of C separates u and v. Hence, for every subalphabet C' of C of cardinality 2, $p_{C'}(u) = p_{C'}(v)$. Since C' is a subalphabet of C, we know that $p_{C'}(u) = p_{C'}(p_C(u))$ and $p_{C'}(v) = p_{C'}(p_C(v))$. Hence, $p_C(u)$ and $p_C(v)$ are two words on C whose projections on every subalphabet C' of C are identical and therefore, by Lemma 1, $p_C(u) = p_C(v)$ which means that C does not separate u and v.   □

Let u and v be two words of A^*. We can build their separating set as follows:

- If a letter a does not appear as many times in u and in v, then a alone suffices to separate them. We thus add $\{a\}$ to $\mathrm{Sep}(u, v)$ and while the pairs containing a all separate u and v, they do not belong to the separating set.
- If a letter a appears neither in u nor in v, containing it would be of no help to separate u and v. Thus, the separating set contains no pair containing a. What is left to investigate are pairs $\{a, b\}$ composed of two letters appearing the same number of times in u and v. Let k be the number of occurrences of a in u and v ($k \neq 0$). Thus, there exists $u_0, \cdots, u_k, v_0, \cdots, v_k$ such that $u = u_0 a u_1 a \cdots a u_k$ and $v = v_0 a v_1 a \cdots a v_k$. The pair $\{a, b\}$ belongs to the separating set if and only if there exists $i \in [\![0, k]\!]$ such that u_i and v_i do not contain as many occurrences of the letter b. These decompositions of u and v can then be reused when investigating other pairs containing a.

We can take advantage of Theorem 2 to separate a finite language as follows: let $L \subset A^*$ be the language we want to separate. For each letter $a \in A$, let $x_{\{a\}}$ be a binary variable that indicates whether $a \in \mathcal{C}$ where $\mathcal{C}$ is the optimal separating subalphabet we want to build. For each pair of letters $\{a, b\}$, let $x_{\{a,b\}}$ be a binary variable that indicates whether both a and b belong to $\mathcal{C}$ (thus, $x_{a,b} = \min(x_a, x_b)$). We obtain the following ILP:

$$\begin{cases} \forall a \neq b \in A, 2x_{\{a,b\}} \leqslant x_{\{a\}} + x_{\{b\}} & \text{(upper bound min inequalities)} \\ \forall a \neq b \in A, x_{\{a,b\}} + 1 \geqslant x_{\{a\}} + x_{\{b\}} & \text{(lower bound min inequalities)} \\ \forall v \neq v' \in L, \displaystyle\sum_{E \in \text{Sep}(v,v')} x_E \geqslant 1 & (E \text{ can denote a singleton or a pair)} \\ \text{minimise } \displaystyle\sum_{a \in A} x_{\{a\}} \end{cases}$$

The solution is thus $\mathcal{C} = \{a \in A \mid x_{\{a\}} = 1\}$ and we do not need to look at the values of the pair-variables to read it. Notice that the pair-variables are auxiliary variables whose values are determined by singletons variables, hence the number of degrees of freedom of the problem is linear in the cardinality of A.

5   Separation of Walks from a Given Set of Starting Points to a Given Set of Destinations

In the problem that we call total separation, we are given a set of potential starting points V_I and a set of potential destinations V_F. The set $\mathcal{R}$ of routes we want to separate is the set of all the routes leading from a vertex of V_I to a vertex of V_F. If the graph contains a cycle, there will thus be infinitely many such routes.

Of course, since the number of routes to separate can be infinite, it is not feasible to compute the separating set of each pair of routes. However, notice that the separating set of two routes is included in the powerset of the set A of arcs of the graph and can therefore only take a finite number of values. Thus, while the linear program presented in the previous sections would have infinitely many constraints, only a finite number of them would be distinct. If we can determine which values of $\text{Sep}(v, v')$ are actually reached, we would therefore be able to describe the same polytope with only a finite number of constraints.

5.1   Study of the Reachable Languages

We call a language L on an alphabet A *reachable* if and only if there exists a directed graph $G = (V, A)$ (A is both the alphabet of L and the set of arcs of the graph), a set of vertices $V_I \subset V$ and a set of vertices $V_F \subset V$ such that L is the set of walks on G leading from a vertex of V_I to a vertex of V_F. We denote by $\text{Reach}(A)$ the set of reachable languages on an alphabet A.

Lemma 3. *If $L \subset A^*$ is reachable, then:*

$$\forall u, u', v, v' \in A^*, \forall a \in A, \begin{cases} uav \in L \\ u'av' \in L \end{cases} \Rightarrow \begin{cases} uav' \in L \\ u'av \in L \end{cases}$$

Proof. Omitted in this version of the paper. □

Let $\mathrm{Rat}(A)$ denote the set of rational languages on an alphabet A. We have:

Proposition 4. $\mathrm{Reach}(A) \subsetneq \mathrm{Rat}(A)$

Proof. – Given an instance of total separation, one can easily construct an automata recognising the language of all possible routes between V_I and V_F on a directed graph $G = (V, A)$: A is its alphabet, V is its states, $T = \{(u, (u, v), v) \mid (u, v) \in A\}$ is its transitions, V_I is its initial states and V_F is its final states. By the theorem of Kleene [13], this proves that reachable languages are rational.
– Note that the previous construction provides very specific automata: each letter of the alphabet labels at most one transition. There are rational languages that cannot be recognised by such automata. For example, let L be reachable and such that $ababa \in L$. Then, $(ab)a(ba) \in L$ and $\varepsilon a(baba) \in L$. Hence, according to Lemma 3, $(ab)a(baba)$ and $a(ba)$ both belong to L too. Hence, the language $\{ababa\}$ although rational, is not reachable. □

5.2 Restrictions of a Rational Language

We define a restriction of a rational language L and we denote by $\overline{L}$ the language build from a regular expression of L as follows:

- $\overline{\varnothing} = \varnothing$.
- $\forall a \in A^*, \overline{\{a\}} = \{a\}$.
- for all rational languages L_1 and L_2, $\overline{L_1 + L_2} = \overline{L_1} + \overline{L_2}$.
- for all rational languages L_1 and L_2, $\overline{L_1 L_2} = \overline{L_1}\, \overline{L_2}$.
- for all rational language L, $\overline{L^*} = \varepsilon + \overline{L} + \overline{L}^2$.

Notice that the restriction of a language L is not unique: indeed, two regular expressions can denote the same language but their associated restrictions can differ. For example, $L^{**} = L^*$ but unless $L = \varnothing$, $\overline{L^{**}} = \sum_{i=0}^{4} \overline{L}^i \neq \sum_{i=0}^{2} \overline{L}^i = \overline{L^*}$.

Proposition 5. *Every restriction $\overline{L}$ of a rational language L is finite.*

Proof. The proof by induction is immediate. Indeed, restricted languages are empty, singletons or built from other restricted languages using only finite union or concatenation, which are operations that preserve the finiteness of the language. □

5.3   Reduction Theorem and Resolution

The reason why we introduced the notion of restriction of a language is the following theorem:

Theorem 6 (reduction theorem). *For all reachable languages L on an alphabet A, for all restrictions $\overline{L}$ of L, if $A' \subset A$ separates $\overline{L}$, then it separates L.*

Proof Omitted in this version of the paper.   □

Note that the converse is obviously true since $\overline{L} \subset L$. Thus, the languages that separate $\overline{L}$ are exactly those who separate L.

Hence, given a graph, a set V_I of potential starting points and a set V_F of potential destinations, we proceed as follows to solve the traffic monitoring problem:

- we know that the language of possible routes leading from a vertex of V_I to a vertex of V_F is rational and therefore admits a regular expression. The graph directly provides us an automata which recognises it and we can use Arden's lemma [1] to find an expression of the associated reachable language that we want to separate.
- we use the regular expression of the language to determine a restriction. We know by Proposition 5 that this language is finite.
- we know by the reduction theorem that the solutions on the restricted language are exactly the solutions on the initial language. All there is left to do is use the method described in the previous section to solve the problem on the restricted language that is finite.

Finally, let us note that while the reduction theorem holds for reachable languages, it does not hold for rational languages in general. Indeed, let $L = (ab)^* + ababa$ be a rational (but not reachable, we proved in the proof of Proposition 4 that a language that contains $ababa$ and that observes Lemma 3 has to contain aba and $abababa$ too which is not the case here). A restriction of L is $\overline{L} = \varepsilon + ab + abab + ababa$. One can see that the alphabet $A' = \{a\}$ separates $\overline{L}$ while $ababa$ and $ababab$, that both belong to L, have the same image under $p_{\{a\}}$.

6   Conclusion

We studied the problem of traffic monitoring from the point of view of separating codes. To overcome the limitations of this approach (see Subsect. 2.2), we introduced a new model of separation based on language and addressed the traffic monitoring with tools stemming from language theory. The problem of separation on a language being NP-complete in the size of the language, we restricted this study to two sub-problems relevant in practice, namely finite and total separations. The strength and flexibility of our model enabled us to address the case of cyclic graphs and infinite sets of roads to separate and can also be used to set more sophisticated constraints on the set of walks to separate such as avoiding certain sequences of edges (which is especially relevant to model road networks). The expressiveness of our new model of separation on a language and the limitations it overcomes also offers hope that it could be of help in a much wider range of applications than traffic monitoring alone.

Of course, this study also opens the door to many possibilities of improvement, the two most important areas to explore being in our opinion a deeper study of the linear program our reduction leads to and the study of divide-and-conquer algorithms that would divide the initial instance in smaller subgraphs and build a solution for the initial graph from the solutions on the subgraphs.

Acknowledgements. I would like to thank my thesis advisor Arnaud Pêcher for his invaluable guidance for the writing of this paper and throughout the research that led to it.

References

1. Arden, D.N.: Delayed-logic and finite-state machines. In: Proceedings of the 2nd Annual Symposium on Switching Circuit Theory and Logical Design (SWCT 1961), FOCS 1961, pp. 133–151. IEEE Computer Society, Washington, DC (1961)
2. Argiroffo, G.R., Bianchi, S.M., Wagler, A.K.: Polyhedra associated with identifying codes. Electron. Not. Discrete Math. **44**, 175–180 (2013)
3. De Bontridder, K.M.J., Halldórsson, B.V., Halldórsson, M.M., Hurkens, C.A.J., Lenstra, J.K., Ravi, R., Stougie, L.: Approximation algorithms for the test cover problem. Math. Program. **98**(1–3), 477–491 (2003)
4. Charbit, E., Charon, I., Cohen, G.D., Hudry, O., Lobstein, A.: Discriminating codes in bipartite graphs: bounds, extremal cardinalities, complexity. Adv. Math. Comm. **2**(4), 403–420 (2008)
5. Charon, I., Cohen, G.D., Hudry, O., Lobstein, A.: Discriminating codes in (bipartite) planar graphs. Eur. J. Comb. **29**(5), 1353–1364 (2008)
6. Charon, I., Hudry, O., Lobstein, A.: Minimizing the size of an identifying or locating-dominating code in a graph is np-hard. Theoret. Comput. Sci. **290**(3), 2109–2120 (2003)
7. Garey, M.R., Johnson, D.S.: Computers and Intractability; A Guide to the Theory of NP-Completeness. W. H. Freeman & Co., New York (1990)
8. Haynes, T.W., Knisley, D.J., Seier, E., Zou, Y.: A quantitative analysis of secondary RNA structure using domination based parameters on trees. BMC Bioinform. **7**, 108 (2006)
9. Honkala, I.S., Laihonen, T., Ranto, S.M.: On strongly identifying codes. Discrete Math. **254**(1–3), 191–205 (2002)
10. Hopcroft, J.E., Motwani, R., Ullman, J.D.: Introduction to Automata Theory, Languages, and Computation - International Edition, 2nd edn. Addison-Wesley (2003)
11. Karpovsky, M.G., Chakrabarty, K., Levitin, L.B.: On a new class of codes for identifying vertices in graphs. IEEE Trans. Inf. Theory **44**(2), 599–611 (1998)
12. Kim, J.H., Pikhurko, O., Spencer, J.H., Verbitsky, O.: How complex are random graphs in first order logic? Random Struct. Algorithms **26**(1–2), 119–145 (2005)

13. Kleene, S.C.: Representation of events in nerve nets and finite automata. In: Automata Studies (1956)
14. Laifenfeld, M., Trachtenberg, A., Cohen, R., Starobinski, D.: Joint monitoring and routing in wireless sensor networks using robust identifying codes. MONET **14**(4), 415–432 (2009)
15. Maheshwari, S.: Traversal marker placement problem are np-complete. In Research report no CU-CS-092-76, Dept. of Computer Science, University of Colorado at Boulder (1976)
16. Metze, G., Schertz, D.R., To, K., Whitney, G., Kime, C.R., Russell, J.D.: Comments on "derivation of minimal complete sets of test-input sequences using boolean differences. IEEE Trans. Comput. **24**(1), 108 (1975)
17. Meurdesoif, P., Pesneau, P., Vanderbeck, F.: Meter installation for monitoring network traffic. In: International Network Optimization Conference (INOC), Spa, Belgium, 2007
18. Moret, B., Shapiro, H.: On minimizing a set of tests. SIAM J. Sci. Stat. Comput. **6**(4), 983–1003 (1985)
19. Narendra, P.M., Fukunaga, K.: A branch and bound algorithm for feature subset selection. IEEE Trans. Comput. **26**(9), 917–922 (1977)
20. Seo, S.J., Slater, P.J.: Open neighborhood locating-dominating in trees. Discrete Appl. Math. **159**(6), 484–489 (2011)
21. Seo, S.J., Slater, P.J.: Open neighborhood locating-domination for infinite cylinders. In: ACM Southeast Regional Conference, pp. 334–335 (2011)
22. Ungrangsi, R., Trachtenberg, A., Starobinski, D.: An implementation of indoor location detection systems based on identifying codes. In: Aagesen, F.A., Anutariya, C., Wuwongse, V. (eds.) INTELLCOMM 2004. LNCS, vol. 3283, pp. 175–189. Springer, Heidelberg (2004)
23. Holmes, B., Willcox, W.R., Lapage, S.P.: A review of numerical methods in bacterial identification. Antonie van Leeuwenhoek **46**(3), 233–299 (1980)
24. Willcox, W.R., Lapage, S.P.: Automatic construction of diagnostic tables. Comput. J. **15**(3), 263–267 (1972)

Know When to Persist: Deriving Value from a Stream Buffer

(Extended Abstract)

Konstantinos Georgiou[1]($\boxtimes$), George Karakostas[2], Evangelos Kranakis[3], and Danny Krizanc[4]

[1] Department of Mathematics, Ryerson University, Toronto, ON, Canada
konstantinos@ryerson.ca
[2] Department of Computing and Software, McMaster University, Hamilton, ON, Canada
karakos@mcmaster.ca
[3] School of Computer Science, Carleton University, Ottawa, ON, Canada
kranakis@scs.carleton.ca
[4] Department of Mathematics and Computer Science, Wesleyan University, Middletown, CT, USA
dkrizanc@wesleyan.edu

Abstract. We consider PERSISTENCE, a new online problem concerning optimizing weighted observations in a stream of data when the observer has limited buffer capacity. A stream of weighted items arrive one at a time at the entrance of a buffer with two holding locations. A processor (or observer) can process (observe) an item at the buffer location it chooses, deriving this way the weight of the observed item as profit. The main constraint is that the processor can only move *synchronously* with the item stream. PERSISTENCE is the online problem of scheduling the processor movements through the buffer so that its total derived value is maximized under this constraint. We study the performance of the straight-forward heuristic *Threshold*, i.e., forcing the processor to "follow" an item through the whole buffer only if its value is above a threshold. We analyze both the optimal offline and Threshold algorithms in the cases where the input stream is either a random permutation, or its items are iid valued. We show that in both cases the competitive ratio achieved by the Threshold algorithm is at least 2/3 when the only statistical knowledge of the items is the median of all possible values. We generalize our results by showing that Threshold, equipped with some minimal statistical advice about the input, achieves competitive ratios in the whole spectrum between 2/3 and 1, following the variation of a newly defined density-like measure of the input. This result is a significant improvement over the case of arbitrary input streams, where we show that no online algorithm can achieve a competitive ratio better than 1/2.

K. Georgiou, G. Karakostas, and E. Kranakis—Research supported in part by NSERC Discovery grant.

© Springer International Publishing Switzerland 2016
R. Dondi et al. (Eds.): AAIM 2016, LNCS 9778, pp. 101–112, 2016.
DOI: 10.1007/978-3-319-41168-2_9

1   Introduction

Suppose that the Automated Quality Control (AQC) of an assembly line has the ability to check all new parts as they enter the assembly line. Every such check increases our quality confidence by a certain percentage, which depends on the nature of the part itself. Now, suppose that the AQC is given the option of a second look at the same part in the next time slot, with a similar increase in our quality confidence. The downsize of this option, is that when the AQC returns to the beginning of the assembly line, it will have completely missed the part immediately following the one that was double-checked. We are looking for an algorithm to decide whether to take the option or not with every new item. Obviously, a good strategy would strive to look twice at "low-quality" items, since that would imply the greatest increases to our confidence, while "missing" only pristine-looking ones.

This problem falls within the data stream setting: a sequence of input data is arriving at a very high rate, but the processing unit has limited memory to store and process the input. Data stream algorithms have been explored extensively in the computer science literature. Typical algorithms in this area work with only a few passes (often just one) over the data input and use memory space less than linear in the input size. Applications can be found in processing cell phone calls or Internet router data, executing Web searches, etc. (cf. [17,18]).

In this work we study a new online problem in data stream processing with limited buffer capacity. An online stream of items (the parts in our AQC example) arrives (one item at a time) at a buffer with two locations L_0, L_1 (assembly points 1 and 2 respectively in the example above), staying at each location for one unit of time, in this order. A processor/observer (the AQC) can move between the two locations *synchronously*, i.e., its movements happen at the same time as the items move. This means that if the processor is processing (observing) the i-th item in time t at L_0, moving to L_1 will result in processing again the i-th item at L_1 in time $t + 1$. On the contrary, if the processor is processing the i-th item in time t at L_1, moving to L_0 will result in processing the $i + 2$-th item at L_0 in time $t + 1$; the $i + 1$-th item has already moved to L_1 and will leave the buffer without the processor ever encountering it! (just like the AQC totally missed a part). We emphasize that we restrict the processor to not even know what item it missed (i.e., cannot "see" into a location other than its current one). Processing the i-th item (either in L_0 or L_1) produces an added value or payoff. The processor has very limited (constant in our results) memory capacity, and cannot keep more than a few variables or pieces of data. The problem we address is whether such a primitive processor can have a strategy to persist and observe (if possible) mostly "good values", especially when compared to an optimal algorithm that is aware of the input stream. We call this online problem PERSISTENCE, which to the best of our knowledge is also new.

Related Work: There is extensive literature on data stream algorithms. Here the emphasis is on input data arriving at a very high rate and limited memory to store and process the input (thus stressing a tradeoff between communication and computing infrastructure). A general introduction to data stream algorithms and models can be found in [17,18]. Lower bound models for space complexity are elaborated in [3]. In the section on new directions for streaming models, [18] discusses several alternatives for data streams for permutation streaming of non-repeating items [1], windowed streaming whereby the most recent past is more important than the distant past [14], as well as reset model, distributed continuous computation, and synchronized streaming. Applications of data stream algorithms are explored extensively in the computer science literature and can be found in sampling (finding quantiles [13], frequent items [16], inverse distribution [7], and range-sums of items [2]).

Related to our study is the well-known *secretary problem* which appeared in the late 1950s and early 1960s (see [9] for a historical overview of its origins and [10] which discusses several extensions). It is concerned with the optimal strategy or stopping rule so as to maximize the probability of selecting the best job applicant assuming that the selection decision can be deferred to the end. Typically we are concerned with maximizing the probability of selecting the best job applicant; this can be solved by a maximum selection algorithm which tracks the running maximum. The problem has fostered the curiosity of numerous researchers and studied extensively in probability theory and decision theory. Several variants have appeared in the scientific literature, including on rank-based selection and cardinal payoffs [6], the infinite secretary problem in [12], secretary problem with uncertain employment in [19], the submodular secretary problem in [5], just to mention a few. The "secretary problem" paradigm has important applications in computer science of which it is worth mentioning the recent work of [4] which studies the relation of matroids, secretary problems, and online mechanisms, as well as [15] which is investigating applications of a multiple-choice secretary algorithm to online auctions. Obviously the secretary problem differs from PERSISTENCE in terms of the objective function: in our case the payoff is the sum of processing payoffs, as opposed to the maximum for the secretary problem. The two problems also differ in the synchronicity and location of arrivals, i.e., what can be accessed and how it is accessed. Nevertheless, the two problems share the inherent difficulty of having to make decisions *on the spot* while missing parts of the input altogether.

High Level Summary of Our Results and Outline of the Paper: Our primary focus is the study of the PERSISTENCE problem, which we formally define in Sect. 2.1. Our goal is to compare the performance of any primitive (online) algorithm, which is not aware of the input stream, against the optimal offline algorithm. In Sect. 2.2 we present all such possible primitive algorithms that we call *Threshold*. Subsequently, we analyze the performance of any Threshold online algorithm for deterministic input streams. Our findings indicate that simplistic primitive algorithms are actually optimal (among all online solutions), and are off no more than 1/2 the performance of an optimal (offline) algorithm that is aware of the entire input. Similar to the setting of the secretary and other online decision problems, this motivates the study of PERSISTENCE problems when the input is random, which is also our main focus.

Our main contributions are discussed in detail in Sect. 2.3. At a high level, we show that when the online observer (processor) knows the median of the possible random values that can appear in the input stream, then it is possible to perform observations in such a way that the total payoff is asymptotically at least 2/3 of the optimal offline solution (Theorem 1). Moreover, we prove that when the random input streams come from certain natural families of inputs in which the mass of possible values is concentrated in relatively few heavy items, the asymptotic performance of very primitive algorithms is nearly optimal. In fact, we parameterize the performance of online algorithms for such inputs using a proper density measure, and we show how the relative asymptotic performance changes from almost optimal (competitive ratio almost 1) to competitive ratio 2/3 (Theorem 2).

The results discussed above are just the byproduct of our main technical contributions that pertain to an analytic exposition of the performance of optimal offline and any online algorithm for random inputs, parameterized by a proper statistical density-like measure on inputs. The two random models that we study are input streams that are either random permutations (Sect. 3) or input streams whose elements assume independent and identically distributed values (Sect. 4). In each case we provide closed formulas for the performance of the optimal offline algorithm and any online algorithm (Sects. 3.1 and 4.1 respectively), which we think is interesting in its own right. Then we use the closed formulas to derive the promised asymptotic competitive analysis in Sects. 3.2 and 4.2 respectively.

We emphasize that the analysis of a size-2 buffer we provide is technically involved, and we cannot see how it could be extended to larger buffers without considerable extra effort. But even for this restricted case, the problem is interesting. Indeed, given our model of algorithms allowed (streaming algorithms with a constant-size memory that can keep only a few variable values, i.e., memoryless), the fact that the simple threshold algorithm achieves non-trivial improvements is already a rather surprising result.

Due to space limitations we omit some of the proofs. The interested reader may consult the full version of our paper [11] for the missing details.

2   Preliminaries

2.1   Model and Problem Definition

Assume that n incoming data values $v_1, v_2, \ldots, v_{n-1}, v_n$ arrive sequentially and synchronously from the left one at a time at a processing unit consisting of two registers L_0 and L_1 which are capable of storing these values instantaneously. The values pass first through location (register) L_0 and then through location L_1, before exiting. A processing unit can process (i.e., derive some payoff from or contribute some additive value to) an item either in L_0 or L_1. The value v_i derived by processing item i comes from a set of possible values $a_0 < a_1 < \cdots < a_{k-1}$, and is independent of the location that processing happened. The main constraint is that all processing is *synchronous*, i.e., at every time unit exactly one new item enters L_0 and the processor (observer) is allowed to either do some processing (observe) at the location it's already in, or perform a single move (and then do processing in) to the other location. The other important constraint is the fact that the processor has only a constant-size memory (i.e., it has space to hold at most $O(1)$ variables) as well as it is only aware of the value of the register of its location. In particular, when processor is located at one register, it is *oblivious* to the value of the other register. More formally, our model is the following:

1. At time step $t = 1$, the processor (observer) occupies position L_0, which holds value v_1.
2. At time step $t \geq 2$, the following take place in that order:
   - The processor may change the location it is about to process (observe); at the same time, locations L_0, L_1 get (new) values v_t, v_{t-1} respectively.
   - Processing is done at the location of the processor; the added value achieved at t is the value of the item in that location.

In the *online model* the observer is *not* aware of the sequence $v_1, v_2, \ldots, v_{n-1}, v_n$, rather she may only know some statistical information that requires constant memory. The limited memory implies a limited ability of keeping statistics or historical data, and, therefore, there is not much leeway for sophisticated processing policies. The (possible) movement of the observer can be determined exclusively by the current value she is observing and in particular not by the value of the location that the observer is not occupying. As a result, the only power an online algorithm has is to choose to observe a value twice in two consecutive time steps, if she thinks that this value provides high enough reward. In contrast, and in the *offline model*, the observer is aware of the entire sequence $v_1, v_2, \ldots, v_{n-1}, v_n$ in advance, and may choose to move between registers with no restrictions so as to maximize her total payoff.

Our main goal is to design PERSISTENCE strategies for the observer that maximize the total added value (or, equivalently, the *average* or *relative* added value or payoff). Our focus is to understand how the lack of information affects the performance of an oblivious online algorithm, compared to the optimal offline algorithm. The standard performance measurement that we use is the so-called *competitive ratio*, defined as the (worst case) ratio between the (expected - when the input stream is random) payoffs of an online and the optimal offline algorithm. It is immediate that for any input stream (even random), the competitive ratio of a fixed online algorithm is $ALG/OPT < 1$, where ALG, OPT are the (expected) payoff of the online and the optimal offline algorithm, respectively.

2.2   On PERSISTENCE Strategies

Given an input stream $v_1, v_2, \ldots, v_{n-1}, v_n$, the optimal solution for the offline model is straightforward; If the processor (observer) is in L_0, processing (observing) an item i with value v_i, then it moves to L_1 only if the item that follows i

has a value smaller than v_i; If the processor is in L_1, processing an item i with value v_i, then it moves to L_0 only if the item $i+2$ that will enter L_0 in the next round has a value v_{i+2} greater than the value v_{i+1} of the item $i+1$ currently in L_0. As a result, an offline and optimal observer may choose to always occupy the location (and subsequently obtain its value as a reward) that holds the maximum value that currently appears in the two locations L_0, L_1. Since at any step, an algorithm cannot have payoff more than the maximum value of the two registers, we conclude that

Observation 1. *For input stream* $v_1, v_2, \ldots, v_{n-1}, v_n$, *and at each time step* $t = 2, \ldots, n$, *the optimal solution of an offline algorithm incurs payoff equal to* $\max\{v_t, v_{t-1}\}$.

We will invoke Observation 1 later, when we will derive closed formulas for the performance of the optimal offline algorithm when the values of the input stream come from certain distributions.

Now we turn our attention to PERSISTENCE strategies in the online model. Recall that any online algorithm is oblivious, non-adaptive and with limited memory. In particular, when at register L_0, an observer has the option to process the same value for one more time in the next step, or stay put at the register and watch in the next step the (currently unknown) value which will enter L_0. If the observer is at register L_1, then the possible payoff at the next step is unknown independently of the move of the observer. Hence it is natural to move the observer back to L_0, giving her the option (in the future) to observe favorable values more than once. This primitive idea gives rise to the following *threshold algorithms*, which are determined by a choice of threshold that dictates when a register value will be observed twice in case the observer is at register L_0.

Threshold Algorithm(T)

Input: a sequence of n items with values $v_1, v_2, \ldots, v_{n-1}, v_n$

1 When the processor has finished processing an item of value τ_0 at L_0 then

   1a. **if** $\tau_0 \geq T$ **then** move to L_1

   1b. **if** $\tau_0 < T$ **then** stay at L_0

2 When the processor has finished processing an item at L_1 **then** move to L_0

Our main contribution in subsequent sections is the (competitive) analysis of Threshold algorithms for various choices of thresholds. In what follows we call the simplistic algorithm that doesn't move the processor from L_0 (or, equivalently, has a threshold greater than a_{k-1}) *Naive*.

2.3 A Summary of Our Results

In its most general version, the input stream to PERSISTENCE is chosen by an unrestricted adversary. It is not difficult to see that the threshold algorithm cannot achieve a competitive ratio better than $1/2$. This shows that in order for the threshold algorithm to perform better, we need to restrict the input instances by making assumptions about the input stream. There are two assumptions that are common in online problems such as the secretary problem [10], or resource allocation problems [8]: One is the *IID* assumption, i.e., the value of each new item is drawn independently and uniformly from the set $\{a_1, a_2, \ldots, a_{k-1}\}$. Another is the *random order* assumption, i.e., the input is a (uniformly) random permutation of n items, each with its own distinct value. In what follows, we study the threshold algorithm under these assumptions.

More formally, we study the following two random models of input streams $v_1, v_2, \ldots, v_{n-1}, v_n$: *Random Permutations:* Input sequence stream is a random permutation of values $a_0 \leq \ldots, \leq a_{n-1}$. *Independent and Identically Distributed Values:* Each v_i assumes the value a_j with probability p_j independently at random, where $j = 0, \leq k - 1$ (note that we allow that $n = \omega(k)$).

For both input families we assume that an online algorithm is oblivious, non-adaptive and with minimal memory, still we assume it has access in advance to some limited statistical information in order to determine a proper threshold. Our main technical contribution pertains to a detailed analysis of the performance of both the optimal offline and any Threshold online algorithm for any such random input. As a result we demonstrate that if the online algorithm knows the median of the set from which the input stream elements assume values, then the competitive ratio improves significantly.

Theorem 1. *For any random permutation or uniform iid input stream, the online Threshold algorithm that uses as threshold the median of the values $\{a_i\}_i$ has (asymptotic) competitive ratio 2/3.*

We emphasize that Theorem 1 is the byproduct of analytic and closed formulas that we derive for optimal offline and Threshold online algorithms, when the input stream is a random permutation (Sect. 3) or iid (Sect. 4), and not necessarily uniform. Next we ask whether it is possible for certain families of random inputs to achieve a competitive ratio better than 2/3, and given that the online algorithm has access to some statistical information. Again, we answer this in the positive by studying generic families of instances parametrized by the relative weight of their largest values.

Definition 1 (c-dense input streams). *Consider a random input stream (in either the random permutation or the iid model) whose values are chosen from $\mathcal{A} = \{a_1, \ldots, a_t\}$, with $a_i \leq a_{i+1}$. The input stream is called c-dense if the total weight of the largest $\lfloor ct \rfloor$ many values of $\mathcal{A}$, relative to the total weight of $\mathcal{A}$, is equal to $1 - c$, i.e. when $1 - c = \frac{\sum_{i=t-\lfloor ct \rfloor +1}^{t} a_i}{\sum_{i=1}^{t} a_i}$.*

Although c cannot be greater than $1/2$ by definition, when c is 0 or 1, then the left hand-side of the previous expression is 1 and 0 respectively, while the right hand-side is 0 and 1 respectively. At the same time, the two sides have different monotonicity as c increases, and as such the notion of c-dense input streams is well defined. Our main contribution pertaining to the families of random inputs which are asymptotically c-dense, for some $c \in (0, 1/2]$, is the following.

Theorem 2. *For any random permutation or uniform iid c-dense input stream, the online Threshold algorithm that uses as threshold the $\lfloor c \cdot n \rfloor$ largest value of a_i's has (asymptotic) competitive ratio $\frac{1}{2} \frac{2-c}{(1-c)(1+c)^2}$.*

Clearly, when c tends to 0, the performance of our Threshold algorithms is nearly optimal for c-dense input streams. Most notably, the worst configuration for such an input is when $c = 1/2$, inducing a competitive ratio equal to $2/3$ (and as already predicted by Theorem 1). The proof of Theorem 2 for random permutations and uniform iid inputs can be found in Sects. 3 and 4 respectively. Any omitted proofs can be found in the full version of the paper [11].

3   Random Permutation Input Streams

In this section we study the special case of inputs that are a random permutation of n items with distinct values $a_0 < a_1 < \cdots < a_{n-1}$ (with $n \geq 2$). First we find closed formulas for the performance of the optimal offline algorithm and any Threshold online algorithm for the PERSISTENCE problem, and then we conclude with the competitive analysis.

3.1   Offline and Online Algorithms for Random Permutations

Using Observation 1 we can show that

Theorem 3. *The relative expected payoff (asymptotic payoff per time step) of the optimal offline algorithm when the input is a random permutation is:* $\frac{1}{\binom{n}{2}} \sum_{i=1}^{n-1} i \cdot a_i$.

The main technical contribution of this section is the performance analysis of any Threshold online algorithm.

Theorem 4. *Let $k = k(n)$ be such that $\lim_{n \to \infty} \frac{k}{n} = c \in \Theta(1)$. Let also A^-, A^+ denote the summation of the smallest $n - k$ and largest k values respectively. Then the relative expected payoff of the Threshold algorithm (payoff per time step) when the threshold is $T := a_{n-k}$ is:* $\frac{A^-}{1+c} + \frac{2A^+}{1+c}$.

The remaining of the section is devoted in proving Theorem 4. We will need the following random variables: A_i denotes the profit of our algorithm from value a_i, or in other words, the contribution of a_i to the performance of the algorithm. Clearly, if $a_i \geq T$ then $A_i \in \{0, 2a_i\}$, and if $a_i < T$ then $A_i \in \{0, a_i\}$. $V_t \in \{a_i\}_{i=0,\ldots,n-1}$ is the value that appears in position t of the (random) permutation, where position 1 is the value that will be read first ($t = 1, \ldots, n$). Finally, O_i is the indicator random variable that equals 1 iff value a_i is observed.

Since all values a_i will appear in every permutation, we have that the expected payoff equals $\mathbb{E}\left[\sum_{i=0}^{n-1} A_i\right] = \sum_{i=0}^{n-1} \mathbb{E}[A_i]$. The contribution of each a_i clearly depends on whether the value is observed.

We can now compute the expected value of O_i given that a_i has a certain position in the permutation.

Lemma 1. $\mathbb{E}\left[O_i | V_t = a_i\right] = \begin{cases} 1 - \frac{k}{n-1} f_{n-1,k}^{t-1} \,, & \text{if } a_i < T \\ f_{n,k}^t & , \text{if } a_i \geq T \end{cases}$

where $f_{n,k}^t = \frac{1}{\binom{n-1}{k-1}} \sum_{s=0}^{\min\{t,k\}-1} (-1)^s \binom{n-1-s}{k-1-s}$.

It is clear from the previous lemma that the formulas of the payoff of online Threshold algorithms involve numerous binomial expressions, which can be simplified with proper manipulations. Given this quite technical work, we are ready to prove Theorem 4.

Proof (of Theorem 4). Let a denote some threshold value, such that $n - k$ many $a_i's$ are less than a. For every $i = 0, \ldots, n - 1$ we have

$$\mathbb{E}\left[A_i\right] = \sum_{t=1}^{n} \mathbb{P}\left[V_t = a_i\right] \mathbb{E}\left[A_i | V_t = a_i\right] = \frac{1}{n} \sum_{t=1}^{n} \mathbb{E}\left[A_i | V_t = a_i\right]. \tag{1}$$

Using the random variables O_i that indicate whether a_i is observed, we have

$$\mathbb{E}\left[A_i | V_t = a_i\right] = \begin{cases} a_i \, \mathbb{E}\left[O_i | V_t = a_i\right] & , \text{if } a_i < a \\ 2a_i \, \mathbb{E}\left[O_i | V_t = a_i\right] & , \text{if } a_i \geq a \end{cases} \quad \text{whose values are given by}$$

Lemma 1. Since the Expected payoff is $\sum_{i=0}^{n-1} \mathbb{E}\left[A_i\right]$, and by (1), we have that

$$\text{Expected Payoff} = \sum_{i=0}^{n-k-1} \frac{1}{n} \sum_{t=1}^{n} \mathbb{E}\left[A_i | V_t = a_i\right] + \sum_{i=n-k}^{n-1} \frac{1}{n} \sum_{t=1}^{n} \mathbb{E}\left[A_i | V_t = a_i\right]$$

$$= \left(1 - \sum_{s=0}^{k-1} (-1)^s \frac{\binom{n-1-s}{k-1-s}}{\binom{n}{k}}\right) A^- + 2 \left(\sum_{s=0}^{k-1} (-1)^s \frac{\binom{n-s}{k-1-s}}{\binom{n}{k-1}}\right) A^+.$$
$$\tag{2}$$

The last equality follows from binomial manipulations regarding summations of $f_{n,k}^t$. The relative performance is obtained by dividing by n. Given that $\frac{k}{n} \to c$, some technical calculations can show that the limit is exactly as claimed. ∎

3.2 Competitive Analysis for Random Permutations

We can now prove Theorems 1 and 2 pertaining to random permutations. Suppose that the Threshold algorithm chooses threshold value T equal to the $\bar{k}$ largest element of the value a_i. Denote by A the sum of all values a_i, and by $L_{\bar{k}}$ the sum of the $\bar{k}$ largest values of them. Abbreviate also $\bar{k}/n$ by c.

Theorem 4 applies with $T := a_{n-\bar{k}}$, to give (asymptotically) that $ALG = \frac{1}{1+c} \frac{1}{n} A + \frac{c}{1+c} \frac{1}{\bar{k}} L_{\bar{k}}$. At the same time, Theorem 3 implies that for the optimal offline algorithm we have that OPT is equal to

$$\frac{1}{\binom{n}{2}} \sum_{i=0}^{n-1} i \cdot a_i \leq \frac{1}{\binom{n}{2}} \left((n-k) \sum_{i=0}^{n-\bar{k}-1} a_i + n \sum_{i=n-\bar{k}}^{n-1} a_i\right) = \frac{2}{n} ((1-c)A + cL_{\bar{k}}).$$

Proof (of Theorem 1 for Random Permutations). When the Threshold value is the median, we have that $c = 1/2$. Using the bounds derived for ALG, OPT, it is straightforward that the two quantities are within $2/3$ of each other. ∎

Proof (of Theorem 2 for Random Permutations). When the input is a c-dense stream, the Threshold algorithm can choose $\bar{k}$ satisfying $1 - \bar{k}/n = 1 - c = L_{\bar{k}}/A$. But then, using the bounds for ALG, OPT, the competitive ratio becomes

$$\frac{ALG}{OPT} \geq \frac{1}{2} \cdot \frac{1}{1+c} \cdot \frac{1+\frac{L_{\bar{k}}}{A}}{(1-c)+c\frac{L_{\bar{k}}}{A}} = \frac{1}{2} \cdot \frac{2-c}{(1-c)(1+c)^2}.$$
∎

4   Random iid-Valued Input Streams

In this section we study the special case of inputs streams whose elements are iid valued. As per the description of the model in Sect. 2.3, we assume that the value v_i of the i-th input item of the stream is an independent random variable assuming a value $a_0 < a_1 < \cdots < a_{k-1}$ (with $k \geq 2$) with probability $p_0, p_1, \ldots, p_{k-1}$ respectively (i.e., $\Pr[v_i = a_j] = p_j$).

4.1   Performance of Offline and Online Algorithms for iid-Valued Streams

Using Observation 1, we can compute the asymptotic payoff of the optimal offline algorithm.

Theorem 5. *The relative expected payoff (asymptotic payoff per time step) of the optimal offline algorithm when the input is a random i.i.d. sequence is* $\sum_{i=0}^{k-1} p_i a_i + \sum_{i=0}^{k-1} \sum_{j=i+1}^{k-1} p_i p_j (a_j - a_i).$

The remaining of this section is devoted in determining the asymptotics of any Threshold algorithm.

Theorem 6. *The relative expected payoff of the Threshold algorithm (asymptotic payoff per time step) that uses threshold $T = a_r$ and when the input is a random i.i.d. is* $\frac{\sum_{i=0}^{r-1} p_i a_i + 2\sum_{i=r}^{k-1} p_i a_i}{\sum_{i=0}^{r-1} p_i + 2\sum_{i=r}^{k-1} p_i}.$

Proof. We introduce abbreviations $Avg := \sum_{i=0}^{k-1} p_i a_i$ and $P := \sum_{i=r}^{k-1} p_j$. Let also Y_i be the random variable such that $Y_i = b$ indicates that, at time i, the observer is at L_b, $b \in \{0, 1\}$. Let also $q_i := \mathbb{P}[Y_i = 0]$. By definition, $q_0 = 1$. Next we observe that $1 - q_{i+1} = \mathbb{P}[Y_{i+1} = 1] = \mathbb{P}[Y_i = 0 \,\&\, X_i \geq T] = \mathbb{P}[X_i \geq T \mid Y_i = 0] \; \mathbb{P}[Y_i = 0] = Pq_i$. A technical argument implies that $q_i = \frac{1-(-1)^i P^i}{1+P}$. Next we observe that $\mathbb{E}[X_i \mid Y_i = 0] = Avg$. If we set $Avg^+ := \sum_{s=r}^{k-1} a_s p_s$ we see that $\mathbb{E}[X_i \mid Y_i = 1] = \mathbb{E}[X_i \mid Y_{i-1} = 0 \,\&\, X_{i-1} \geq T] = \frac{Avg^+}{P}$. We now compute

$$\mathbb{E}\left[\sum_{i=1}^{n} X_i\right] = \sum_{i=1}^{n} \mathbb{E}[X_i] = \sum_{i=1}^{n} \left(\mathbb{P}[Y_i = 0]\mathbb{E}[X_i \mid Y_i = 0] + \mathbb{P}[Y_i = 1]\mathbb{E}[X_i \mid Y_i = 1]\right)$$

$$= \left(\frac{n}{1+P} + \frac{P + (-P)^{n+1}}{(1+P)^2}\right) \cdot Avg + \left(\frac{n}{1+P} + \frac{(-P)^n - 1}{(1+P)^2}\right) \cdot Avg^+.$$

Dividing the last quantity by n, and taking the limit $n \to \infty$ gives the promised formula. ∎

4.2   Competitive Analysis for Uniform iid-Valued Input Streams

Note that the formulas derived in Sect. 4.1 hold for all iid-valued input streams. In this section we provide competitive analysis for input streams that are uniformly valued, i.e. when $p_i = \frac{1}{k}$, for all $i = 0, \ldots, k-1$. That would be Theorems 1 and 2 pertaining to uniform iid-valued random input streams.

As before, denote by A the sum of all values a_i, and by $L_{\bar{r}}$ the sum of the $\bar{r}$ largest values of them. Abbreviate also $\bar{r}/n$ by c. Suppose also that the Threshold algorithm uses as threshold the $\bar{r}$-th largest value of the a_i's. We use Theorem 5 to find an upper bound for the performance of the offline algorithm:
$OPT = \frac{1}{k}A + \frac{1}{k^2}\sum_{i=0}^{k-1}\sum_{j=i+1}^{k-1}(a_j - a_i) = \frac{1}{k}A + \frac{1}{k^2}\sum_{i=0}^{k-1}(2i - k + 1)a_i \leq$
$2\left(\frac{k-\bar{r}+1/2}{k^2}A + \frac{\bar{r}}{k^2}L_{\bar{r}}\right)$. Next, using Theorem 6 (which is written for threshold
value $a_r = a_{k-1-\bar{r}}$) we obtain that for the Threshold algorithm $ALG = \frac{1}{k} \cdot \frac{A+L_{\bar{r}}}{1+\bar{r}/k}$.

Proof (of Theorem 1 for Uniform iid-Valued Streams). When the Threshold value is the median, we have that $\bar{r}/k = 1/2$. Using bounds the bounds established for ALG, OPT, it is straightfoward then to see that the two quantities are indeed within 2/3 of each other as promised. ∎

Proof (of Theorem 2 for Uniform iid-Valued Streams Random Permutations). When the input is a c-dense stream, the Threshold algorithm can choose $\bar{r}$ satisfying $1 - \bar{r}/n = 1 - c = L_{\bar{r}}/A$. Using the bounds derived previously, the
competitive ratio becomes $\frac{ALG}{OPT} \geq \frac{1}{2} \cdot \frac{1}{1+c} \cdot \frac{1+\frac{L_{\bar{r}}}{A}}{(1-c+o(c))+c\frac{L_{\bar{r}}}{A}} \rightarrow \frac{1}{2} \cdot \frac{2-c}{(1-c)(1+c)^2}$. ∎

5   Open Problems

As described in the introduction, our model can be extended in many different ways. An obvious extension is to have a bigger buffer, i.e., $k > 2$ locations $L_0, L_1, \ldots, L_{k-1}$. In this case, there are different possibilities of moving the processor within the buffer: a *single jump* model would require the processor to always jump to L_0, while a *local jump* model would allow the processor to move close to its current location. Another obvious extension would be to consider general payoffs, i.e., allowing an item to have different values in different buffer locations. Also, we leave open the potential increase in the power of the processor if it is allowed to know the item it's going to miss in L_0 (if it moves to L_0 from L_1 in the next time slot).

The threshold algorithm is probably the simplest algorithm one can use to tackle PERSISTENCE. The obvious question is whether there are better algorithms for the non-oblivious setting. Are there *upper bounds* that can be shown? The thresholds we calculated above do not apply in the oblivious setting, since we do not know the payoffs ahead of time. In that setting, it is natural to consider adaptive algorithms, probably using a prefix of the input in order to 'learn' something about it before employing a threshold-like or some other strategy.

References

1. Ajtai, M., Jayram, T.S., Kumar, R., Sivakumar, D.: Approximate counting of inversions in a data stream. In: Reif, J.H. (ed.) Proceedings of 34th Annual ACM Symposium on Theory of Computing, Montréal, Québec, Canada, 19–21 May 2002, pp. 370–379. ACM (2002)
2. Alon, N., Duffield, N., Lund, C., Thorup, M.: Estimating arbitrary subset sums with few probes. In: ACM (ed.) Proceedings of the Twenty-Fourth ACM SIGMOD-SIGACT-SIGART Symposium on Principles of Database Systems: PODS 2005, Baltimore, Maryland, 13–15 June 2005, pp. 317–325. ACM Press (2005)
3. Alon, N., Matias, Y., Szegedy, M.: The space complexity of approximating the frequency moments. J. Comput. Syst. Sci. **58**(1), 137–147 (1999)
4. Babaioff, M., Immorlica, N., Kleinberg, R.: Matroids, secretary problems, and online mechanisms. In: Bansal, N., Pruhs, K., Stein, C. (eds.) SODA, pp. 434–443. SIAM (2007)
5. Bateni, M., Hajiaghayi, M., Zadimoghaddam, M.: Submodular secretary problem and extensions. ACM Trans. Algorithms **9**(4), Art. 32, 23 (2013)
6. Bearden, J.N.: A new secretary problem with rank-based selection and cardinal payoffs. J. Math. Psychol. **50**(1), 58–59 (2006)
7. Cormode, G., Muthukrishnan, S., Rozenbaum, I.: Summarizing and mining inverse distributions on data streams via dynamic inverse sampling. In: Böhm, K., Jensen, C.S., Haas, L.M., Kersten, M.L., Larson, P.-Å., Ooi, B.C. (eds.) VLDB, pp. 25–36. ACM (2005)
8. Devanur, N.R., Jain, K., Sivan, B., Wilkens, C.A.: Near optimal online algorithms and fast approximation algorithms for resource allocation problems. In: Shoham, Y., Chen, Y., Roughgarden, T. (eds.) EC, pp. 29–38. ACM (2011)
9. Ferguson, T.S.: Who solved the secretary problem? Stat. Sci. **4**, 282–289 (1989)
10. Freeman, P.R.: The secretary problem and its extensions: a review. Int. Stat. Rev./Revue Internationale de Statistique **51**, 189–206 (1983)
11. Georgiou, K., Karakostas, G., Kranakis, E., Krizanc, D.: Know when to persist: deriving value from a stream buffer. CoRR, abs/1604.03009 (2016)
12. Gianini, J., Samuels, S.M.: The infinite secretary problem. Ann. Probab. **4**, 418–432 (1976)
13. Greenwald, M., Khanna, S.: Space-efficient online computation of quantile summaries. In: Mehrotra, S., Sellis, T.K. (eds.) SIGMOD Conference, pp. 58–66. ACM (2001)
14. Hoffman, M., Muthukrishnan, S., Raman, R.: Location streams: Models and algorithms. Technical report, DIMACS TR (2004)
15. Kleinberg, R.D.: A multiple-choice secretary algorithm with applications to online auctions. In: SODA, pp. 630–631. SIAM (2005)
16. Manku, G.S., Motwani, R.: Approximate frequency counts over data streams. In: Bernstein, P.A., et al. (eds.) VLDB 2002: Proceedings of the Twenty-Eighth International Conference on Very Large Data Bases, Hong Kong SAR, China, 20–23 August 2002, pp. 346–357. Morgan Kaufmann Publishers (2002)
17. Muthukrishnan, S.: Data Streams: Algorithms and Applications, vol. 1 (2005)
18. Muthukrishnan, S.: Data stream algorithms (notes from a series of lectures). In: The 2009 Barbados Workshop on Computational Complexity, 1–8 March 2009
19. Smith, M.: A secretary problem with uncertain employment. J. Appl. Probab. **12**, 620–624 (1975)

Algorithmic Aspects of Upper Domination: A Parameterised Perspective

Cristina Bazgan[1], Ljiljana Brankovic[2,3], Katrin Casel[3(✉)], Henning Fernau[3], Klaus Jansen[4], Kim-Manuel Klein[4], Michael Lampis[1], Mathieu Liedloff[5], Jérôme Monnot[1], and Vangelis Th. Paschos[1]

[1] PSL, University of Paris-Dauphine, CNRS, LAMSADE UMR 7243, 75775 Paris Cedex 16, France
{bazgan,michail.lampis,jerome.monnot,paschos}@lamsade.dauphine.fr
[2] School of Electrical Engineering and Computer Science, The University of Newcastle, Callaghan, NSW 2308, Australia
ljiljana.brankovic@newcastle.edu.au
[3] Fachbereich 4, Informatikwissenschaften, Universität Trier, 54286 Trier, Germany
{casel,fernau}@uni-trier.de
[4] Institut für Informatik, Universität Kiel, 24098 Kiel, Germany
{kj,kmk}@informatik.uni-kiel.de
[5] University Orléans, INSA Centre Val de Loire, LIFO EA 4022, 45067 Orléans, France
mathieu.liedloff@univ-orleans.fr

Abstract. This paper studies UPPER DOMINATION, *i.e.*, the problem of computing the maximum cardinality of a minimal dominating set in a graph, with a focus on parameterised complexity. Our main results include W[1]-hardness for UPPER DOMINATION, contrasting FPT membership for the parameterised dual CO-UPPER DOMINATION. The study of structural properties also yields some insight into UPPER TOTAL DOMINATION. We further consider graphs of bounded degree and derive upper and lower bounds for kernelisation.

1 Introduction

Domination, independence and irredundance are basic concepts in graph theory and most of the overall six respective minimisation and maximisation problems, which are related via the so-called domination chain (see [16]), are very well-studied. Especially for parameterised complexity, MINIMUM DOMINATION and MAXIMUM INDEPENDENT SET and their respective parameterised duals are sort of fundamental. With the exception of UPPER DOMINATION, all problems of the domination chain are known to be complete for either W[1] or W[2] while their corresponding parameterised dual is in FPT. This paper therefore studies the so far neglected parameter $\Gamma(G)$, which denotes the maximum cardinality of a minimal dominating set in G. More precisely, we discuss the following problems:

C. Bazgan—Institut Universitaire de France.

© Springer International Publishing Switzerland 2016
R. Dondi et al. (Eds.): AAIM 2016, LNCS 9778, pp. 113–124, 2016.
DOI: 10.1007/978-3-319-41168-2_10

Upper Domination	Co-Upper Domination
Input: A graph $G = (V, E)$, a non-negative integer k.	**Input:** A graph $G = (V, E)$, a non-negative integer ℓ.
Question: Is $\Gamma(G) \geq k$?	**Question:** Is $\Gamma(G) \geq |V| - \ell$?

Notice that Co-Upper Domination could be also addressed as Minimum Maximal Nonblocker or as Minimum Maximal Star Forest; see [1] for further discussion. From the perspective of classical complexity theory, both problems are trivially equivalent and were shown to be NP-complete quite some time ago [7]. Aside from this, very little is known, especially with respect to parameterised complexity. From this perspective, k and ℓ turn out to be the natural parameters, which turn them into dual problems in the parameterised complexity sense of this word. As we will only consider this natural parameterisation, we refrain from explicitly mentioning the parameter throughout this paper. Slightly abusing notation, we will therefore use the names Upper Domination and Co-Upper Domination to also refer to the parameterised problems.

In Sect. 2, we link minimal dominating sets to a decomposition of the vertex set that turns out to be a crucial tool for deriving our combinatorial and computational results. Section 3 then discusses properties of upper dominating sets from a parameterised point of view and reveals W[1]-hardness for Upper and Upper Total Domination. Conversely, Co-Upper Domination is shown to be in FPT, which we prove by providing both a kernelisation and a branching algorithm. In Sect. 4, we consider graphs of bounded degree and derive kernelisations for Upper and Co-Upper Domination for this restricted graph class. This section also includes an exact $O^*(1.3481^n)$-algorithm for Upper Total Domination restricted to subcubic graphs which builds on the decomposition derived in Sect. 2. We further discuss general questions of exact algorithms for Upper Domination, as well as some related questions for total domination variants (see [17]) in Sect. 5. For reasons of space, proofs and other details are omitted in this extended abstract.

Basic Notions. Throughout this paper, we only deal with undirected simple graphs $G = (V, E)$. The number of vertices $|V|$ is also known as the order of G. $N(v)$ denotes the open neighbourhood of v in a graph G, and $N[v]$ is the closed neighbourhood of v in G, i.e., $N[v] = N(v) \cup \{v\}$. These notions can be easily extended to vertex sets X, e.g., $N(X) = \bigcup_{x \in X} N(x)$. The cardinality of $N(v)$ is also known as the degree of v, denoted as $deg(v)$. The maximum degree in a graph is written as Δ. A graph of maximum degree three is called subcubic.

Given a graph $G = (V, E)$, a subset S of V is a *dominating set* if every vertex $v \in V \setminus S$ has at least one neighbour in S, i.e., if $N[S] = V$. A dominating set is called *minimal* if no proper subset of it is a dominating set. Likewise, a vertex set I is *independent* if $N(I) \cap I = \emptyset$. An independent set is maximal if no proper superset is independent. In the following we use classical notations: $\alpha(G)$ denotes the cardinality of a maximum independent set in $G = (V, E)$ and $\tau(G) := |V| - \alpha(G)$ is the cardinality of a minimum vertex cover.

For any subset $S \subseteq V$ and $v \in S$ we define the private neighbourhood of v with respect to S as $pn(v, S) := N[v] \setminus N[S \setminus \{v\}]$. Any $w \in pn(v, S)$ is called a *private neighbour of v with respect to S*. If the set S is clear from the context, we will omit the "with respect to S" part. A dominating set $S \subseteq V$ is minimal if and only if $|pn(v, S)| > 0$ for every $v \in S$. Observe that v can be a private neighbour of itself.

Parameterised Complexity. We mainly refer to the textbooks [8,10] in the area. Important notions that we will make use of include the parameterised complexity classes FPT, W[1] and W[2], parameterised reductions and kernelisation. In this area, it has also become customary not only to suppress constants (as in the O notation), but also even polynomial-factors, leading to the so-called O^*-notation.

2   Graph Decompositions for Minimal Dominating Sets

The following exposition is crucial for the development of the algorithms we derive in this paper and also for the general investigation of properties of minimal dominating sets. Any minimal dominating set D for a graph $G = (V, E)$ can be associated with a partition of the set of V into four sets F, I, P, O given by: $I := \{v \in D : v \in pn(v, D)\}$, $F := D \setminus I$, $P \in \{B \subseteq N(F) \setminus D : |pn(v, D) \cap B| = 1$ for all $v \in F\}$ and $O := V \setminus (D \cup P)$. This representation is not necessarily unique since there might be different choices for the sets P and O, but for every partition of this kind, the following properties hold:

1. Every vertex $v \in F$ has at least one neighbour in F, called a **friend**.
2. The set I is an independent set in G.
3. The subgraph induced by the vertices $F \cup P$ has an edge cut set separating F and P that is, at the same time, a perfect matching; hence, P can serve as the set of **private** neighbours for F.
4. The neighbourhood of a vertex in I is always a subset of O, which are otherwise the **outsiders** (Fig. 1).

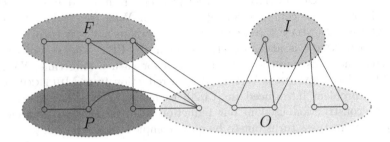

Fig. 1. Illustration of the FIPO structure imposed by minimal dominating sets

This partition is also related to a different characterisation of $\Gamma(G)$ in terms of so-called upper perfect neighbourhoods [16]. Observe two important special cases of the partition (F, I, P, O): If $F = \emptyset$, then I is an independent dominating set. If $I = \emptyset$, then F is a minimal total dominating set, *i.e.*, a set $S \subseteq V$ such that $V = N(S)$ and $N(S') \neq V$ for all $S' \subset S$. Both notions have been thoroughly studied in the literature. Observe that finding a maximum cardinality minimal dominating set for which $I = \emptyset$ holds in an (F, I, P, O) partitioning (called (F, P, O)-DOMINATION set in the following) is not equivalent to the problem UPPER TOTAL DOMINATION, which asks for a maximum cardinality minimal total dominating set. The following example illustrates the differences between optimal solutions (illustrated by the black vertices) for MINIMUM, (F, P, O)-, UPPER and UPPER TOTAL DOMINATION:

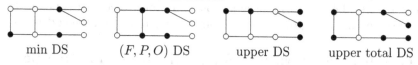

$$\text{min DS} \qquad (F, P, O) \text{ DS} \qquad \text{upper DS} \qquad \text{upper total DS}$$

From the domination chain we know $\alpha(G) \leq \Gamma(G)$ for all graphs G, which is simply due to the fact that any maximal independent set is also a minimal dominating set. Considering the partition (F, I, P, O) for a minimal dominating set S for a graph G of order $n > 0$, we immediately know that $|I| \leq \alpha(G)$. Further, we know $|F| = |P|$ and hence $|F| = 1/2(n - |I| - |O|) \leq 1/2(n - \alpha(G))$. With $|S| = |F| + |I|$, we see that $|S| \leq 1/2(n + \alpha(G))$ and since this inequality holds for all minimal dominating sets S, we can conclude:

$$\alpha(G) \;\leq\; \Gamma(G) \;\leq\; \frac{n}{2} + \frac{\alpha(G)}{2} \tag{1}$$

3   Fixed Parameter Tractability

In this section we will investigate the fixed parameter tractability of UPPER DOMINATION, its dual and related problems. The problems MINIMUM DOMINATION, MINIMUM INDEPENDENT DOMINATION and MAXIMUM INDEPENDENT SET were among the first problems conjectured not to be in FPT [9]. In fact, aside from UPPER DOMINATION, all other problems from the domination chain are now known to be complete for either W[1] or W[2] (see [2,11] for UPPER and LOWER IRREDUNDANCE respectively). It is perhaps not very surprising that UPPER DOMINATION is also unlikely to belong to FPT, and it looks rather unexpected that this question has been open for such a long time. We show that UPPER DOMINATION is W[1]-hard by a reduction from MULTICOLOURED CLIQUE, a problem introduced in [13,20] to facilitate W[1]-hardness proofs. While the construction used in our reduction itself is not very complicated, proving its correctness turns out to be quite complex and technical.

Theorem 1. UPPER DOMINATION *is W[1]-hard.*

Proof. (Sketch) Let $G = (V, E)$ be a graph with k different colour-classes given by $V = V_1 \cup V_2 \cup \cdots \cup V_k$. MULTICOLOURED CLIQUE asks if there exists a clique $C \subseteq V$ in G such that $|V_i \cap C| = 1$ for all $i = 1, \ldots, k$. For this problem, one can assume that each set V_i is an independent set in G, since edges between vertices of the same colour-class have no impact on the existence of a solution. MULTICOLOURED CLIQUE is known to be W[1]-complete, parameterised by k. We construct a graph $G' = (V', E')$ by: $V' := V \cup \{v_e : e \in E\}$ and

$$E' := \bigcup_{i=1}^{k} V_i \times V_i \ \cup \ \bigcup_{i=1}^{k}\bigcup_{j=1}^{k} \{(v_e, v_{e'}) : e, e' \in (V_i \times V_j) \cap E\}$$

$$\cup \bigcup_{i=1}^{k}\bigcup_{j=1}^{k} \{(v_{(u,w)}, x) : (u,w) \in (V_i \times V_j) \cap E, x \in ((V_i \cup V_j) \setminus \{u,w\})\}.$$

It can be shown that there exists a minimal dominating set S of cardinality $k + \frac{1}{2}(k^2 - k)$ for G' if and only if $|S \cap V_i| = 1$ for all $i = 1, \ldots, k$ and $|S \cap V_{i,j}| = 1$ for all $i \neq j$, where $V_{i,j} := \{v_e : e \in E \cap (V_i \times V_j)\}$. With this property, it is easy to see that S is minimal if and only if $S \cap V$ is a clique in the original graph; observe that if S contains two vertices v_i and v_j from V_i and V_j, respectively, which are not adjacent in G, then these already dominate all vertices of $V_{i,j}$ in G'. Overall, it can be shown that G' has an upper dominating set of cardinality $k + \frac{1}{2}(k^2 - k)$ if and only if G is a "yes"-instance for MULTICOLOURED CLIQUE, which proves W[1]-hardness for UPPER DOMINATION, parameterised by $\Gamma(G')$. $\qquad\square$

We want to point out that the above reduction also works for the restriction of UPPER DOMINATION to solutions for which I is empty:

Corollary 1. (F, P, O)-DOMINATION, *that is the restriction of* UPPER DOMI- NATION *to solutions S such that $V = N(S)$, is W[1]-hard.*

This result means that if we consider somehow splitting the problem UPPER DOMINATION into the subproblems of computing the independent vertices I and (F, P, O)-DOMINATION, we end up with two W[1]-hard problems. Considering UPPER TOTAL DOMINATION, the construction in the proof of Theorem 1 is not very helpful, since unfortunately any set S with $|S \cap V_i| = 1$ for all $i = 1, \ldots, k$ and $|S \cap V_{i,j}| = 1$ for all $i \neq j$, regardless of the structure of the original graph G, is a minimal total dominating set for G'. We can however use a much simpler construction to show W[1]-hardness for UPPER TOTAL DOMINATION, a result which cannot be inferred from the known NP-hardness of the problem, see [12].

Theorem 2. UPPER TOTAL DOMINATION *is W[1]-hard.*

Proof. (Sketch) We reduce from MULTICOLOURED INDEPENDENT SET. Let $G = (V, E)$ be a graph with k different colour-classes given by $V = V_1 \cup V_2 \cup \cdots \cup V_k$. We construct a graph $G' = (V', E')$ as follows: Starting from G, we add k vertices $C = \{c_1, \ldots, c_k\}$ and turn each vertex set $V_j \cup \{c_j\}$ into a clique. We claim that G admits a multicoloured independent set (of size k) if and only if G' has a minimal total dominating set with $2k$ vertices. $\qquad\square$

We do not know if UPPER DOMINATION belongs to W[1], but we can at least place it in W[2], the next level of the W-hierarchy. We obtain this result by describing a suitable multi-tape Turing machine that solves this problem, see [4].

Proposition 1. UPPER DOMINATION *belongs to W[2].*

Proof. Recall how MINIMUM DOMINATION can be seen to belong to W[2] by providing an appropriate multi-tape Turing machine [4]. First, the k vertices that should belong to the dominating set are guessed, and then this guess is verified in k further (deterministic) steps using n further tapes in parallel, where n is the order of the input graph. We only need to make sure that the guessed set of vertices is minimal. To this end, we copy the guessed vertices k times, leaving one out each time, and we also guess one vertex for each of the $k-1$-element sets that is not dominated by this set. Such a guess can be tested in the same way as sketched before using parallel access to the $n+1$ tapes. The whole computation takes $O(k^2)$ parallel steps of the Turing machine, which shows the claim.   □

Let us notice that very similar proofs also show membership in W[2] and hardness for W[1] for the question whether, given some hypergraph G and parameter k, there exists a minimal hitting set of G with at least k vertices. This also means that UPPER TOTAL DOMINATION belongs to W[2].

In the context of parameterised complexity, we would like to point out another difference between UPPER DOMINATION and MINIMUM DOMINATION. Despite its W[2]-hardness, there is at least a reduction-rule for MINIMUM DOMINATION, which deals with vertices of degree one, as they can be assumed not to be contained in a minimum dominating set. One might suspect that any upper dominating set would conversely always choose to contain degree-one vertices.

As the example on the right illustrates, there can not be such a rule for UPPER DOMINATION, since the degree-one vertex v is never part of a maximum solution; in fact, the black vertices form the unique optimal solution for this graph.

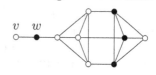

Another interesting question is to consider the dual parameter ℓ, that is to decide the existence of an upper dominating set of size at least $n - \ell$. This is in fact the natural parameterisation for CO-UPPER DOMINATION.

Theorem 3. CO-UPPER DOMINATION *is in FPT. More precisely, it admits a kernel of at most $\ell^2 + \ell$ vertices and at most ℓ^2 edges.*

Proof. Let $G = (V, E)$ be an input graph of order n. Consider a vertex $v \in V$ with $deg(v) > \ell$ and any minimal dominating set D with partition (F, I, P, O):

- If $v \in I$, all neighbours of v have to be in O which means $|O| \geq |N(v)| > \ell$.
- If $v \in F$, exactly one neighbour p of v is in P and $N[v] \setminus \{p\} \subseteq F \cup O$, which gives $|O| + |P| = |O| + |F| \geq |N[v] \setminus \{p\}| > \ell$.
- If $v \in P$, exactly one neighbour p of v is in F and $N[v] \setminus \{p\} \subseteq P \cup O$, so $|O| + |P| > \ell$.

We always have either $v \in O$ or $|O| + |P| > \ell$, which means a "no"-instance for
CO-UPPER DOMINATION. Consider the graph G' built from G by deleting the
vertex v and all its edges. For any minimal dominating set D for G with partition
(F, I, P, O) such that $v \in O$, D is also minimal for G', since $pn(w, D) \supseteq \{w\}$
for all $w \in I$ and $|pn(u, D) \cap P| = 1$ for all $u \in F$. Also, any set $D' \subset V \setminus \{v\}$
which does not dominate v has a cardinality of at most $|V \setminus N[v]| < n - \ell$, so if
G' has a dominating set D' of cardinality at least $n - \ell$, $N(v) \cap D' \neq \emptyset$; hence,
D' is also dominating for G. These observations allow us to successively reduce
(G, ℓ) to (G', ℓ') with $\ell' = \ell - 1$, as long as there are vertices v with $deg(v) > \ell$.
Any isolated vertex in the resulting graph G' originally only has neighbours in O
which means it belongs to I in any dominating set D with partition (F, I, P, O)
and can hence be deleted from G' without affecting the existence of an upper
dominating set with $|P| + |O| \leq \ell'$.

Let (G', ℓ') be the instance obtained after the reduction above with $G' = (V', E')$ and let $n' = |V'|$. If there is an upper dominating set D for G' with
$|D| \geq n' - \ell'$, any associated partition (F, I, P, O) for D satisfies $|P| + |O| \leq \ell'$.
Since G' does not contain isolated vertices, every vertex in I has at least one
neighbour in O. Also, any vertex in V', and hence especially any vertex in O,
has degree at most ℓ', which means that $|I| \leq |N(O)| \leq \ell'|O|$. Overall:

$$|V'| \leq |I| + |F| + |P| + |O| \leq (\ell' + 1)|O| + 2|P| \leq \max_{j=0}^{\ell'}\{j(\ell' + 1), 2(\ell' - j)\},$$

and hence $|V'| \leq \ell'(\ell' + 1)$, or (G', ℓ') and consequently (G, ℓ) is a "no"-instance.
Concerning the number of edges, we can derive a similar estimate. There are at
most ℓ edges incident with each vertex in O. In addition, there is exactly one
edge incident with each vertex in P that has not yet been accounted for, and, in
addition, there could be $\ell - 1$ edges incident to each vertex in F that have not
yet been counted. This shows the claim.   □

We just derived a kernel result for CO-UPPER DOMINATION, in fact a kernel
of quadratic size in terms of the number of vertices and edges. This poses the
natural question if we can do better also with respect to the question whether
the brute-force search we could perform on the quadratic kernel is the best we
can do to solve CO-UPPER DOMINATION in FPT time.

Proposition 2. CO-UPPER DOMINATION *can be solved in time* $O^*(4.3077^{\ell})$.

Proof. (Sketch) This result can be shown by designing a branching algorithm
that takes a graph $G = (V, E)$ and a parameter ℓ as input. Due to space restriction, we only describe here the rough ideas without any proof. As in Sect. 2, to
each graph $G = (V, E)$ and (partial) dominating set, we associate a partition
(F, I, P, O). We consider $\kappa = \ell - (\frac{|F|}{2} + \frac{|P|}{2} + |O|)$ as a measure of the partition and
for the running time of the algorithm. Note that $\kappa \leq \ell$. At each branching step,
our algorithm picks some vertices from R (the set of yet undecided remaining
vertices). They are either added to the current dominating set $D := F \cup I$ or
to $\overline{D} := P \cup O$. Each time a vertex is added to P (resp. to O) the value of κ

decreases by $\frac{1}{2}$ (resp. by 1). Also, whenever a vertex x is added to F, the value of κ decreases by $\frac{1}{2}$.

Let us describe the two halting rules. First, whenever κ reaches zero, we are facing a "no"-instance. Then, if the set R of undecided vertices is empty, we check whether the current domination set D is minimal and of size at least $n-\ell$, and if so, the instance is a "yes"-instance. Then, we have a simple reduction rule: whenever the neighbourhood of an undecided vertex $v \in R$ is included in $\overline{D}$, we can safely add v to I. Finally, vertices are placed to F, I or $\overline{D}$ according to three branching rules. The first one considers undecided vertices with a neighbour already in F (in such a case, v cannot belong to I). The second one considers undecided vertices with only one undecided neighbour (in such a case, several cases may be discarded as, e.g., they cannot be both in I or both in $\overline{D}$). The third branching rule considers all the possibilities for an undecided vertex and due to the previous branching rules, it can be assumed that each undecided vertex has at least two undecided neighbours (which is nice since such vertices have to belong to $\overline{D}$ whenever an undecided neighbour is added to I).   □

Of course, the question remains to what extent the previously presented parameterised algorithm can be improved on. In this context, we briefly discuss the issue of (parameterised) approximation for this parameter.

Theorem 4. Co-Upper Domination *is 4-approximable in polynomial time 3-approximable with a running time in* $O^*(1.0883^{\tau(G)})$ *and 2-approximable in time* $O^*(1.2738^{\tau(G)})$ *or* $O^*(1.2132^n)$.

Proof. First of all, observe by subtracting n from Eq. (1) that $\tau(G)$ relates to the co-upper domination number in the following way:

$$\frac{\tau(G)}{2} + 1 \leq n - \Gamma(G) \leq \tau(G) \tag{2}$$

Using any 2-approximation algorithm one can compute a vertex cover V' for G, and define $S' = V \setminus V'$. Let S be a maximal independent set containing S'. $V \setminus S$ is a vertex cover of size $|V \setminus S| \leq |V'| \leq 2\tau(G) \leq 4(n - \Gamma(G))$. Moreover, S is maximal independent and hence minimal dominating set which makes $V \setminus S$ a feasible solution for Co-Upper Domination with $|V \setminus S| \leq 4(n - \Gamma(G))$. The claimed running time for the factor-2 approximation stems from the best parameterised and exact algorithms for Minimum Vertex Cover by [6] and [19], the factor-3 approximation from the parameterised approximation in [3].   □

4   Graphs of Bounded Degree

In contrast to the case of general graphs, Upper Domination turns out to be easy (in the sense of parameterised complexity) for graphs of bounded degree.

Proposition 3. *Fix* $\Delta > 2$. Upper Domination *has a problem kernel with at most* Δk *many vertices.*

Proof. First, we can assume that the input graph G is connected, as otherwise we can apply the following argument separately on each connected component. Assume G is a cycle or a clique. Then, the problem UPPER DOMINATION can be optimally solved in polynomial time, *i.e.*, we can produce a kernel as small as we want. Otherwise, Brooks' Theorem yields a polynomial-time algorithm that produces a proper colouring of G with (at most) Δ many colours. Extend the biggest colour class to a maximal independent set I of G. As I is maximal, it is also a minimal dominating set. So, there is a minimal dominating set I of size at least n/Δ, where n is the order of G. So, $\Gamma(G) \geq n/\Delta$. If $k < n/\Delta$, we can therefore immediately answer YES. In the other case, $n \leq \Delta k$ as claimed. $\square$

With some more combinatorial effort, we obtain:

Proposition 4. *Fix* $\Delta > 2$. CO-UPPER DOMINATION *has a problem kernel with at most* $(\Delta + 0.5)\ell$ *many vertices.*

Proof. Consider any graph $G = (V, E)$. For any partition (F, I, P, O) corresponding to an upper dominating set $D = I \cup F$ for G, isolated vertices in G always belong to I and can hence be deleted in any instance of CO-UPPER DOMINATION without changing ℓ. For any graph G without isolated vertices, the set $P \cup O$ is a dominating set for G, since $\emptyset \neq N(v) \subset O$ for all $v \in I$ and $N(v) \cap P \neq \emptyset$ for all $v \in F$. Maximum degree Δ hence immediately implies $n = |N[P \cup O]| \leq (\Delta + 1)\ell$.

Since any connected component can be solved separately, we can assume that G is connected. For any $v \in P$, the structure of the partition (F, I, P, O) yields $|N[v] \cap D| = 1$, so either $|N[v]| = 1 < \Delta$ or there is at least one $w \in P \cup O$ such that $N[v] \cap N[w] \neq \emptyset$. For any $v \in O$, if $N[v] \cap F \neq \emptyset$, the F-vertex in this intersection has a neighbour $w \in P$, which means $N[w] \cap N[v] \neq \emptyset$. If $N(v) \subseteq I$ and $N[v] \neq V$, at least one of the I-vertices in $N(v)$ has to have another neighbour to connect to the rest of the graph. Since $N(I) \subseteq O$, this also implies the existence of a vertex $w \in O$, $w \neq v$ with $N[w] \cap N[v] \neq \emptyset$. Finally, if $N[v] \not\subseteq I \cup F$, there is obviously a $w \in P \cup O$, $w \neq v$ with $N[w] \cap N[v] \neq \emptyset$.

Assume that there is an upper dominating set with partition (F, I, P, O) such that $|P \cup O| = l \leq \ell$ and let $v_1, \ldots, v_l$ be the $l > 1$ vertices in $P \cup O$. By the above argued domination-property of $P \cup O$, we have:

$$n = |\bigcup_{i=1}^{l} N[v_i]| = \tfrac{1}{2} \sum_{i=1}^{l} |N[v_i] \setminus \bigcup_{j=1}^{i-1} N[v_j]| + \tfrac{1}{2} \sum_{i=1}^{l} |N[v_i] \setminus \bigcup_{j=i+1}^{l} N[v_j]|$$

Further, by the above argument about neighbourhoods of vertices in $P \cup O$, maximum degree Δ yields for every $i \in \{1, \ldots, l\}$ either $|N[v_i] \setminus \bigcup_{j=1}^{i-1} N[v_j]| \leq \Delta$ or $|N[v_i] \setminus \bigcup_{j=i+1}^{l} N[v_j]| \leq \Delta$ which gives:

$$n = \tfrac{1}{2} \sum_{i=1}^{l} |N[v_i] \setminus \bigcup_{j=1}^{i-1} N[v_j]| + |N[v_i] \setminus \bigcup_{j=i+1}^{l} N[v_j]| \leq \tfrac{1}{2} l(2\Delta + 1) \leq (\Delta + 0.5)\ell.$$

Any graph with more than $(\Delta + 0.5)\ell$ vertices is consequently a "no"-instance which yields the stated kernelisation, as the excluded case $|P \cup O| = 1$ (or in other words $N[v] = V$ for some $v \in O$) can be solved trivially. $\square$

This implies that we have a $3k$-size vertex kernel for UPPER DOMINATION, restricted to subcubic graphs, and a 3.5ℓ-size vertex kernel for CO-UPPER DOMINATION, again restricted to subcubic graphs. With [5, Theorem 3.1], we can conclude the following consequence:

Corollary 2. *Unless P equals NP, for any $\varepsilon > 0$, UPPER DOMINATION, restricted to subcubic graphs, does not admit a kernel with less than $(1.4 - \varepsilon)k$ vertices; neither does CO-UPPER DOMINATION, restricted to subcubic graphs, admit a kernel with less than $(1.5 - \varepsilon)\ell$ vertices.*

Exact Algorithms

Let us recall one important result on the pathwidth of subcubic graphs from [15].

Theorem 5. *Let $\epsilon > 0$ be given. For any subcubic graph G of order $n > n_\epsilon$, a path decomposition proving $pw(G) \le n/6 + \epsilon$ is computable in polynomial time.*

This result immediately gives an $O^*(1.2010^n)$-algorithm for solving MINIMUM DOMINATION on subcubic graphs. We will take a similar route to prove moderately exponential-time algorithms for UPPER DOMINATION.

Proposition 5. UPPER DOMINATION *on graphs of pathwidth p can be solved in time $O^*(7^p)$, given a corresponding path decomposition.*

We are considering all partitions of each bag of the path decomposition into 6 sets: F, F^*, I, P, O, O^*, where

- F is the set of vertices that belong to the upper dominating set and have already been matched to a private neighbour;
- F^* is the set of vertices that belong to the upper dominating set and still need to be matched to a private neighbour;
- I is the set of vertices that belong to the upper dominating set and is independent in the graph induced by the upper dominating set;
- P is the set of private neighbours that are already matched to vertices in the upper dominating set;
- O is the set of vertices that are not belonging neither to the upper dominating set nor to the set of private neighbours but are already dominated;
- O^* is the set of vertices not belonging to the upper dominating set that have not been dominated yet.

The upper bound on the running time can be improved for graphs of a certain maximum degree to $O^*(6^p)$ so that we can conclude:

Corollary 3. UPPER DOMINATION *on subcubic graphs of order n can be solved in time $O^*(1.3481^n)$, using the same amount of space.*

We like to point out that the idea from the pathwidth algorithm above can be adapted to work for treewidth.

Proposition 6. UPPER DOMINATION *on graphs of treewidth p can be solved in time $O^*(11^p)$, given a corresponding nice tree decomposition.*

5   Discussions and Open Problems

The motivation to study UPPER DOMINATION (at least for some of the authors) was based on the following observation based on enumeration; see [14].

Proposition 7. UPPER DOMINATION *can be solved in time* $O^*(1.7159^n)$ *on general graphs of order* n.

It is of course a bit nagging that there seems to be no better algorithm (analysis) than this enumeration algorithm for UPPER DOMINATION. Recall that the minimisation counterpart can be solved in better than $O^*(1.5^n)$ time [18,21]. As this appears to be quite a tough problem, it makes a lot of sense to study it on restricted graph classes. This is what we did above for subcubic graphs, see Corollary 3. We summarise some open problems.

- Is UPPER DOMINATION in W[1]? Or, hard for W[2]?
- Can we improve on the 4-approximation of CO-UPPER DOMINATION?
- Can we find smaller kernels for UPPER or CO-UPPER DOMINATION on degree-bounded graphs?
- Can we find exact (*e.g.,* branching) algorithms that beat the enumeration or pathwidth-based ones for UPPER DOMINATION, at least on cubic graphs?

Also for UPPER TOTAL DOMINATION, the best exact algorithm seems to be based on enumeration. The $O^*(1.7159^n)$ bound from [14] is achieved by a branching algorithm that enumerates all minimal set covers of an instance $(\mathcal{U}, \mathcal{S})$, where $\mathcal{S}$ is a collection of subsets over a universe $\mathcal{U}$ and then uses a simple reduction from a dominating set instance to a set cover instance. It is implicit from the analysis (see Sect. 4 of [14]) that a SET COVER instance has at most $1.156154^{|\mathcal{U}|+2.720886|\mathcal{S}|}$ minimal set covers which can be enumerated in time $O^*(1.156154^{|\mathcal{U}|+2.720886|\mathcal{S}|})$. As an easy consequence, minimal total dominating sets of a graph $G = (V, E)$ can be enumerated in time $O^*(1.7159^n)$, by picking as the universe $\mathcal{U} = V$ and $\mathcal{S} = \{N(v) : v \in V\}$. This allows to conclude that UPPER TOTAL DOMINATION can be solved in the same time. Similarities to UPPER DOMINATION continue to some extent; however, the general picture is not very clear and still needs some research.

Acknowledgements. We thank our colleagues Serge Gaspers, David Manlove and Daniel Meister for some discussions on (total) upper domination. Part of this research was supported by Deutsche Forschungsgemeinschaft, grant FE 560/6-1.

References

1. Abu-Khzam, F.N., Bazgan, C., Chopin, M., Fernau, H.: Data reductions and combinatorial bounds for improved approximation algorithms. J. Comput. Syst. Sci. **82**(3), 503–520 (2016)
2. Binkele-Raible, D., Brankovic, L., Cygan, M., Fernau, H., Kneis, J., Kratsch, D., Langer, A., Liedloff, M., Pilipczuk, M., Rossmanith, P., Wojtaszczyk, J.O.: Breaking the 2^n-barrier for irredundance: two lines of attack. J. Discrete Algorithms **9**, 214–230 (2011)

3. Brankovic, L., Fernau, H.: A novel parameterised approximation algorithm for minimum vertex cover. Theor. Comput. Sci. **511**, 85–108 (2013)
4. Cesati, M.: The turing way to parameterized complexity. J. Comput. Syst. Sci. **67**, 654–685 (2003)
5. Chen, J., Fernau, H., Kanj, I.A., Xia, G.: Parametric duality and kernelization: lower bounds and upper bounds on kernel size. SIAM J. Comput. **37**, 1077–1108 (2007)
6. Chen, J., Kanj, I.A., Xia, G.: Improved upper bounds for vertex cover. Theor. Comput. Sci. **411**(40–42), 3736–3756 (2010)
7. Cheston, G.A., Fricke, G., Hedetniemi, S.T., Jacobs, D.P.: On the computational complexity of upper fractional domination. Discrete Appl. Math. **27**(3), 195–207 (1990)
8. Cygan, M., Fomin, F., Kowalik, Ł., Lokshtanov, D., Marx, D., Pilipczuk, M., Pilipczuk, M., Saurabh, S.: Parameterized Algorithms. Springer, Switzerland (2015)
9. Downey, R.G., Fellows, M.R.: Fixed parameter tractability and completeness. Congressus Numerantium **87**, 161–187 (1992)
10. Downey, R.G., Fellows, M.R.: Fundamentals of Parameterized Complexity. Texts in Computer Science. Springer, London (2013)
11. Downey, R.G., Fellows, M.R., Raman, V.: The complexity of irredundant set parameterized by size. Discrete Appl. Math. **100**, 155–167 (2000)
12. Fang, Q.: On the computational complexity of upper total domination. Discrete Appl. Math. **136**(1), 13–22 (2004)
13. Fellows, M.R., Hermelin, D., Rosamond, F., Vialette, S.: On the parameterized complexity of multiple-interval graph problems. Theor. Comput. Sci. **410**(1), 53–61 (2009)
14. Fomin, F.V., Grandoni, F., Pyatkin, A.V., Stepanov, A.A.: Combinatorial bounds via measure and conquer: bounding minimal dominating sets and applications. ACM Trans. Algorithms **5**(1), 1–17 (2008)
15. Fomin, F.V., Høie, K.: Pathwidth of cubic graphs and exact algorithms. Inf. Process. Lett. **97**, 191–196 (2006)
16. Haynes, T.W., Hedetniemi, S.T., Slater, P.J.: Fundamentals of Domination in Graphs. Monographs and Textbooks in Pure and Applied Mathematics, vol. 208. Marcel Dekker, New York (1998)
17. Hennings, M., Yeo, A.: Total Domination in Graphs. Springer, New York (2013)
18. Iwata, Y.: A faster algorithm for dominating set analyzed by the potential method. In: Marx, D., Rossmanith, P. (eds.) IPEC 2011. LNCS, vol. 7112, pp. 41–54. Springer, Heidelberg (2012)
19. Kneis, J., Langer, A., Rossmanith, P.: A fine-grained analysis of a simple independent set algorithm. In: Kannan, R., Narayan Kumar, K. (eds.) IARCS Annual Conference on Foundations of Software Technology and Theoretical Computer Science. LIPIcs, FSTTCS 2009, vol. 4, pp. 287–298. Schloss Dagstuhl – Leibniz-Zentrum für Informatik (2009)
20. Pietrzak, K.: On the parameterized complexity of the fixed alphabet shortest common supersequence and longest common subsequence problems. J. Comput. Syst. Sci. **67**(4), 757–771 (2003)
21. van Rooij, J.M.M., Bodlaender, H.L.: Exact algorithms for dominating set. Discrete Appl. Math. **159**(17), 2147–2164 (2011)

On Network Formation Games with Heterogeneous Players and Basic Network Creation Games

Christos Kaklamanis, Panagiotis Kanellopoulos[✉], and Sophia Tsokana

Department of Computer Engineering and Informatics, Computer Technology
Institute and Press "Diophantus", University of Patras, 26504 Rio, Greece
{kakl,kanellop,tsokana}@ceid.upatras.gr

Abstract. We consider two variants of the network formation game that aims to capture the impact of selfish behavior, on behalf of the network administrators, on the overall network structure and performance. In particular, we study basic network creation games, where each player aims to minimize her distance to the remaining players, and we present an improved lower bound on the graph diameter of equilibria of this game. We also consider network formation games with a large number of heterogeneous players and monetary transfers, and prove tight bounds on the price of anarchy under realistic assumptions about the cost function. Finally, we argue about the setting where these heterogeneous players must be connected with additional node-disjoint paths to mitigate the impact of failures.

1 Introduction

The advent and widespread adoption of the Internet has given rise to questions like how such networks are formed and sustained. Since such a network mainly relies on the voluntary participation of its members, the need to understand the motives behind network administrators' actions arises quite naturally. Being part of a large-scale network offers great benefits to the participating entities, who, however, would prefer to enjoy these benefits with the minimum possible cost. The cost that each entity suffers may be due to several reasons, for instance, being part of such a network requires investing in infrastructure (e.g., in the form of links), while network administrators desire to be connected via short paths to the remaining (sub)network.

Such tradeoffs, between being well-connected on the one hand and paying as less as possible on the other, have been studied under the lens of algorithmic game theory with an emphasis on understanding the network administrator objectives and analyzing their impact on the global network performance. We view each administrator as a player that only cares about maximizing her utility, which, in turn, is a function of connectivity, equipment cost, and distance to other players. In other words, players make decisions according to their best interest, observe the decisions of other players, and respond, if necessary, to

© Springer International Publishing Switzerland 2016
R. Dondi et al. (Eds.): AAIM 2016, LNCS 9778, pp. 125–136, 2016.
DOI: 10.1007/978-3-319-41168-2_11

such decisions. This process eventually ends when all players are satisfied by the resulting network, i.e., no player has an incentive to unilaterally change her decision, given that all other players do not change theirs. The resulting stable networks are also called equilibria. A natural question, then, is to study how such stable networks look like, and how close are they to an optimal network that maximizes a global utility function. The notion of the *price of anarchy* [9,17] is used to quantify this efficiency loss due to the selfish behavior on behalf of the network administrators, by comparing the optimal network to the worst equilibrium.

Related work: Fabrikant et al. [7] initiated the study of network formation games where the objective of each player is to be connected to the network with the minimum possible cost. Each edge has a (uniform) creation cost and the total player cost depends on the number of links that a player has to pay for, to ensure connectivity and proximity to other networks of her choice, and the sum of distances to the other players. Apart from defining the model, Fabrikant et al. also present upper and lower bounds on the price of anarchy. A series of papers improve these bounds for different ranges of the edge cost function; to the best of our knowledge, the most recent results are presented in [13].

Aiming for a simpler model that would be independent of the edge cost and would allow a player to compute her best response in polynomial time, Alon et al. [2] introduced the class of basic network creation games, where no additional edges can be bought and no existing edges can be deleted. However, a player is allowed to exchange a single edge for another; this is called an *edge swap*. A player is satisfied with a graph if there is no profitable edge swap available, i.e., any such swap does not decrease the sum of distances from this player (this is the SUM version of the game) or the maximum distance from this player (this is the MAX version). A natural question, then, is how large can such graphs be, when all players are satisfied. It was shown in [2] that, in the SUM version, every swap equilibrium graph has diameter at most $2^{O(\sqrt{\log n})}$ while there exists such a swap equilibrium graph with diameter 3; the proof for the lower bound appears in [3]. Following [2], Nikoletseas et al. [15] (see also [16]) consider necessary conditions under which the diameter can be polylogarithmic for the SUM version. Furthermore, they defined a local cost network creation game where the utility of a player depends only the degree of her neighbors. In a similar attempt to stress the impact of locality on the game, Bilò et al. [5] consider network creation games but impose restrictions on the players' knowledge about the graph. In particular, each player can observe the behavior of players that are at distance at most k from the player. Mihàlak and Schlegel [14] consider asymmetric swap equilibria where, in contrast to [2,3,15], each edge is owned by a player and each player can only swap an edge she owns, and provide upper and lower bounds on the diameter of equilibria graphs. Ehsani et al. [6] study a network formation game where each player has a fixed budget to establish edges to other players and prove bounds on the price of anarchy for both versions.

Another direction towards generalizing the model in [7] was presented in [11] where players were classified in core and peripheral, in an attempt to model in a more realistic manner the actual status of Autonomous Systems on the Internet. In particular, there is a smaller group of Autonomous Systems that are well-interconnected and essentially form the Internet core, and a larger group that is connected to the Internet core via tree-like structures. Apart from presenting upper bounds on the price of anarchy, [11] also considers the setting where a player may pay a neighboring node in order to reach a joint decision. Similarly, Álvarez et al. [1] consider a setting where players sign contracts in order to exchange traffic and, among other results, provide a characterization of topologies that are pairwise stable for a given traffic matrix.

Finally, Arcaute et al. [4] as well as Lenzner [10] focus on game dynamics of network formation and creation games.

Our contribution: We first present an improved lower bound of 4 on the diameter of swap equilibria for the SUM version of basic network creation games. This is the first improvement over a lower bound of 3 due to Alon et al. [3]. We then consider network formation games with a large number of heterogeneous players and, in particular, a setting where a player may compensate another player if the total cost for both players is reduced. We present an asymptotic upper bound of 1 on the price of anarchy of such games under realistic assumptions on the cost function. Furthermore, we also show that, in general, the price of anarchy is at most 5/4 and this is tight, as there exists a graph with price of anarchy arbitrarily close to 5/4. Our work can be cast in the framework of studying the performance deterioration in games played by a large number of players, e.g., see the very recent work in [8].

Roadmap: The remainder of the paper is structured as follows. We begin, in Sect. 2, by formally introducing the problems and by presenting the necessary definitions. Then, in Sect. 3, we present the improved lower bound of 4 on the diameter of sum equilibria of basic network creation games. We continue in Sect. 4 by considering large network formation games with heterogeneous players and monetary transfers and present improved bounds on the price of anarchy. Finally, we conclude with open problems in Sect. 5.

2   Preliminaries

We provide the necessary definitions used throughout the paper for basic network creation games and network formation games with heterogeneous players. In both games, the players are the nodes of a graph and each player cares only about minimizing her cost. Given a graph $G = (V, E)$, we denote by $d(u, v)$ the distance between nodes u and v in G and by $\deg(u)$ the degree of node u. In case there is no path connecting u and v in G, then we set $d(u, v) = +\infty$ and, hence, we can restrict our attention to graphs with a single connected component.

A *basic network creation game* $\Gamma(n)$ is played by n players. Given a graph $G = (V, E)$ with $|V| = n$, the cost of player u is defined as $\text{cost}_u(G) = \sum_{v \neq u} d(u, v)$. Players may not add new edges or delete existing ones, and their strategy space is confined in swapping a single adjacent edge. We say that G is a *sum equilibrium graph* if for every edge $(u, v) \in E$ and any player $w \in V$ such that $(u, w) \notin E$, creating the graph G' by swapping the edge (u, v) with (u, w) does not decrease the total sum of distances from v to all other players, i.e., $\text{cost}_u(G) \leq \text{cost}_u(G')$. In such games, we are interested in upper-bounding the *diameter* of a sum equilibrium graph, i.e., the maximal distance between any pair of players.

A *network formation game with heterogeneous players* $\Gamma(\alpha, \beta)$ consists of two *types* of players, namely there is a set T_A of central players, where $|T_A| = \alpha$ and a set T_B of minor players, where $|T_B| = \beta$; we also use the terms *type A players* and *type B players*. For any type A player, each adjacent edge incurs cost c_A, while, for any type B player, each adjacent edge incurs cost c_B, with $c_A \leq c_B$. Given a graph $G = (V, E)$, the cost of a type A player u is defined as $\text{cost}_u^A(G) = \deg(u)c_A + A\sum_{v \in T_A} d(u, v) + \sum_{v \in T_B} d(u, v)$, where $A > 1$ denotes the importance of proximity to type A players compared to proximity to type B players. Similarly, the cost of type B player u is defined as $\text{cost}_u^B(G) = \deg(u)c_B + A\sum_{v \in T_A} d(u, v) + \sum_{v \in T_B} d(u, v)$. The *social cost* of a graph G is the sum of the players' costs, i.e., $\text{cost}(G) = \sum_{u \in T_A} \text{cost}_u^A(G) + \sum_{u \in T_B} \text{cost}_u^B(G)$. We sometimes use $\text{cost}_u(G)$ to denote player u's cost, e.g., when her type is unknown or not relevant. As argued in [11], when $c_A \leq A$ any pair of type A nodes is connected, and, since this corresponds to the behavior observed on the Internet, in the following we impose this restriction. Similarly, we consider only the case where $c_A + c_B \leq 2A$. We find it convenient to use $c = \frac{c_A + c_B}{2}$ and remark that if $c \leq \frac{A+1}{2}$, then the optimal graph contains a clique of $|T_A|$ players (these are the type A players) and any type B player is directly connected to all type A players. Otherwise, when $c > \frac{A+1}{2}$, the optimal graph contains the clique of the type A players and all type B players are connected to a particular type A player. The optimal cost in the last case is $\text{cost}(G_{opt}) = |T_A|(|T_A - 1)(c_A + A) + 2|T_B|(|T_B| - 1) + |T_B|(2|T_A| - 1)(A + 1) + 2c|T_B|$.

For the case of network formation games with heterogeneous players, we consider the setting with *monetary transfers*, i.e., players are allowed to reimburse another player if this is to their best joint interest. We remark that these transfers are limited so that any player can compensate only a neighboring player and only for either creating a new edge between them or for deleting the existing edge that connects them. Consider a graph G' obtained from G either by adding a new edge or removing an existing edge and let (u, v) be the relevant edge. We say that G is a *pairwise-stable equilibrium* (or pairwise-stable graph) if $\text{cost}_u(G) + \text{cost}_v(G) \leq \text{cost}_u(G') + \text{cost}_v(G')$, for any graph G' differing by a single edge (u, v) from G. Monetary transfers essentially allow player u with $\text{cost}_u(G) < \text{cost}_u(G')$ to compensate player v with $\text{cost}_v(G) > \text{cost}_v(G')$ as long as the total cost of the two players in G is at most the corresponding total cost in G'.

The *price of anarchy* denotes the worst-case performance deterioration over all possible pairwise-stable equilibria. Given a game $\Gamma(\alpha, \beta)$, and its corresponding set of equilibrium graphs G_{eq}, the price of anarchy for the game $\Gamma(\alpha, \beta)$ is formally defined as $\text{PoA}(\Gamma(\alpha, \beta)) = \max_{G \in G_{eq}} \frac{\text{cost}(G)}{\text{cost}(G_{opt})}$, where G_{opt} is the graph of minimum cost. Similarly, given a class Γ of games, the price of anarchy is defined as $\text{PoA}(\Gamma) = \max_{\Gamma(\alpha, \beta) \in \Gamma} \text{PoA}(\Gamma(\alpha, \beta))$.

3   A New Lower Bound for Sum Equilibria of Basic Network Creation Games

In this section we present a swap equilibrium graph with a diameter of 4 for the SUM version of basic network creation games. This improves upon a previous lower bound of 3 stated in [2] and proven in [3].

Theorem 1. *There is a diameter-4 sum equilibrium graph.*

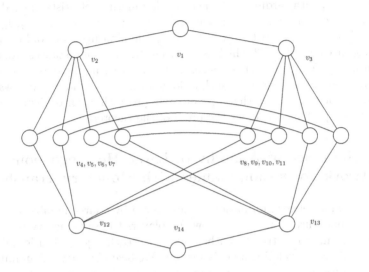

Fig. 1. A diameter-4 sum equilibrium graph.

Proof. The graph G is presented in Fig. 1. Note that, due to symmetry, it suffices to prove that nodes v_1, v_2 and v_4 have no incentive to swap an adjacent edge. We begin by proving that node v_1 (respectively, v_{14}) has no incentive to swap an adjacent edge. Without loss of generality, due to symmetry, we assume that v_1 swaps edge (v_1, v_2). Given graph G with edge (v_1, v_2) removed, we can observe, by considering all possible edge additions that can connect v_1 to another node, that the best possible edge addition, i.e., the one that minimizes the sum of distances from v_1, is to reinclude edge (v_1, v_2).

We now prove that node v_2 (respectively, v_3, v_{12}, and v_{13}) has no incentive to swap an adjacent edge. Since in a swap equilibrium the degree cannot change, v_2 must have 5 nodes at distance 1. If there is one (or more) node at distance 3, the remaining nodes must have distance 2, and the cost cannot be better than the current cost. Note that, in G, there is a single node at distance 3 from v_2, i.e., node v_{14}, and, so, any edge swap should reduce this distance in order to reduce the sum of distances from v_2. That is, v_2 can decrease the sum of distances only if it can be connected to one of v_{12}, v_{13}, v_{14} without this edge swap inducing another node at distance 3. Swapping edge (v_2, v_1) will increase the distance from v_3 to 3, while swapping (v_2, v_4) will increase the distance either from v_{11} or from v_{12} to at least 3; similar reasoning applies also to (v_2, v_5), (v_2, v_6) and (v_2, v_7). Hence, there is no edge swap that will connect v_2 to one of v_{12}, v_{13}, or v_{14} without increasing the distance to another node to at least 3.

Finally, we now prove that node v_4 (respectively, v_5, v_6, v_7, v_8, v_9, v_{10}, and v_{11}) has no incentive to swap an adjacent edge. Consider the graph G_{-4} obtained by removing v_4 and its adjacent edges. Note that v_4 has degree 3 in G and that the optimal solution for v_4 would be to connect its three edges to a dominating set of graph G_{-4} with cardinality 3, if such a dominating set exists. Indeed, there exist two such dominating sets that are symmetric; these are $\{v_1, v_{12}, v_{13}\}$ and $\{v_2, v_3, v_{14}\}$. Connecting v_4 to any of these two dominating sets would lead to a sum of distances equal to 23 which is less than the sum of distances from v_4 in G, which is 24. Since, however, v_4 is connected in G with only one node of each dominating set, connecting v_4 to such a dominating set would require swapping two edges. Hence, it follows that v_4 cannot improve the sum of distances by a single edge swap. This concludes the proof of the theorem.   □

4   On the Price of Anarchy of Large Heterogeneous Network Formation Games with Monetary Transfers

We now consider network formation games played by a large number of heterogeneous players. Furthermore, we allow the players to use monetary reimbursements as a means to motivate another player to reach a joint decision, such as adding a new edge or deleting an existing one. We begin by showing an improved upper bound on the price of anarchy of such games and we remark that for the most natural choices of parameter values, this upper bound in fact tends to 1. In case we allow large parameter values, then this bound becomes 5/4. In the last case, we also present a matching lower bound that demonstrates that our analysis is tight. Recall that we assume that $c_A \leq c_B$ and $c_A + c_B \leq 2A$. Furthermore, since we are motivated by networks such as the Internet, we assume that $|T_B| \gg |T_A| \gg 1$.

4.1   The Upper Bound

We prove an upper bound of 5/4 on the price of anarchy of large network formation games with heterogeneous players and monetary transfers, whenever $c = \frac{c_A + c_B}{2} \leq A$, thus improving upon the 3/2 upper bound of Meirom et al. [11]. Furthermore, we show that for a natural and wide range of values for parameters c_A, c_B, and A, the price of anarchy tends to 1. We begin this section by presenting our arguments for the latter claim, i.e., that the price of anarchy of large network formation games with heterogeneous players tends to 1 for certain parameter values. We slightly abuse notation and use T_A (T_B, respectively) to denote both the set of type A nodes (type B nodes, respectively) and its cardinality $|T_A|$ ($|T_B|$, respectively).

First, it is shown in [11] that whenever $c_B < 1$ the only pairwise-stable graph is the clique, which is also the graph with lowest cost, so, in this case, the price of anarchy is 1. Also, when $c_B \geq 1$ and $c < \frac{A+1}{2}$, the lowest cost graph contains all edges except those among type B nodes as it is beneficial to connect all type A players in a clique and to connect any type B player to any type A player. In this case, where $c_A \leq A$, $c_B \geq 1$ and $c < \frac{A+1}{2}$, it holds that any pairwise-stable graph contains at least the same set of edges as the optimal graph, and additional edges connecting two type B nodes only in case $c_B = 1$. These additional edges, however, do not increase the cost, as the edge cost is compensated by the reduced distance cost. Hence, in this case, also, the price of anarchy is 1. Therefore, in the following we assume that $\frac{A+1}{2} \leq c \leq A$. In this case, the lowest cost graph consists of the clique of the type A nodes while all type B nodes are connected to a unique type A node. Recall that the cost in this case is

$$\text{cost}(G_{opt}) = T_A(T_A - 1)(c_A + A) + T_B(c_A + c_B) + (A+1)(2T_A - 1)T_B$$
$$+ 2T_B(T_B - 1). \tag{1}$$

We now argue about structural properties of pairwise-stable graphs and how these can be exploited. In particular, Meirom et al. [11] argue that, when $T_B \gg T_A \gg 1$, then the diameter is at most 3 with any type B node having at least one direct link to a type A node, while the maximal distance of a type B node to a type A node is 2.

Theorem 2 [11]. *The maximal distance of a type B node from a type A node in a pairwise-stable graph is* $\max\{\lfloor \sqrt{A^2|T_A|^2 + 4cA|T_A|} - A|T_A| \rfloor, 2\}$.

Corollary 1. *Whenever $T_A \gg 1$ and $c = o(A|T_A|)$, the maximal distance of a type B node from a type A node in a pairwise-stable graph is 2.*

The next theorem considers the case where parameters c_A, c_B and A are independent of T_A and T_B and states that, in this case, the price of anarchy tends to 1 as the number of players increases. The main idea in the proof is that, in any pairwise-stable graph, all type B players are directly connected to a specific node, hence any pairwise-stable graph resembles the structure of the graph of lowest cost.

Theorem 3. *The price of anarchy of large network formation games with heterogeneous players and monetary transfers tends to 1 whenever $T_B \gg T_A \gg 1$ and $c \leq A$ with $A \cdot T_A \in o(T_B)$.*

Proof. Consider a pairwise-stable graph G and a type B node j. Let ξ_j be the number of type B nodes at distance 3 from j. Also, let $\xi^* = \max_j \xi_j$ and let $j^* \in \arg\max_j \xi_j$. Note that these ξ^* nodes do not share a type A neighbor with j^*, as in that case their distance from j^* would be 2. Among the type A neighbors of these ξ^* nodes, there exists at least one node i^* with $\xi^*/(T_A - 1)$ neighbors of type B. Since j^* and i^* are not connected by an edge in G, we obtain that $c_A + c_B \geq A + 1 + \frac{\xi^*}{T_A - 1}$, since establishing the edge (i^*, j^*) increases the joint cost of i^* and j^* by $c_A + c_B$, decreases the distance cost of i^* due to j^* by A and decreases the distance cost of j^* due to i^* and its adjacent type B players by at least $1 + \frac{\xi^*}{T_A - 1}$.

Since $A + 1 + \frac{\xi^*}{T_A - 1} \le 2c \le 2A$, we obtain that $A \ge 1 + \frac{\xi^*}{T_A - 1}$. Since $A \cdot T_A \in o(T_B)$, then $\xi^* \in o(T_B)$ and, subsequently, the total contribution of the type B nodes that are at distance 3 from T_B nodes to the distance cost is $o(T_B^2)$. Therefore, the leading term in the cost of any pairwise-stable graph is the cost due to the distance between pairs of type B nodes which is 2 for almost all such pairs. Thus, the leading term tends to $2T_B^2$ and the theorem follows since this matches the leading term in the cost of the optimal graph. □

So far, we have established conditions that lead to a price of anarchy that is asymptotically 1. The following theorem considers the case where parameters c_A, c_B, and A can obtain rather large values.

Theorem 4. *The price of anarchy of network formation games with a large number of heterogeneous players and monetary transfers is at most $5/4$ whenever $c \le A$.*

4.2   A Lower Bound of 5/4 for Large Values of c and A

We complement the discussion in the previous section by presenting a lower bound on the price of anarchy for the case where c_A, c_B and A can take large values. Our construction exhibits a simple structure and consists of a graph with diameter 3.

Theorem 5. *There exists a network formation game with a large number of heterogeneous players and monetary transfers with price of anarchy $5/4 - \epsilon$ with $\epsilon > 0$.*

Proof. Consider a graph G with $|T_A| = \kappa$ players of type A and $|T_B| = \kappa\lambda$ players of type B. Let $c = c_A = c_B$ be the edge cost. Players of type A form the clique $K_{|T_A|} = K_\kappa$, while each such player is connected to λ players of type B.

Recall that, since $c \le A$, any pair of type A nodes always prefer to be connected via an edge, and, therefore, type A nodes form a clique. We begin by proving that for appropriate values of c and A, graph G is a pairwise-stable graph, that is, no pair of players has an incentive to jointly deviate from their current strategy by either adding or deleting an edge between them. Note that any edge deletion results either in a disconnected graph (when we delete an edge connecting a node of type A to a node of type B) or, when the edge connects two type A nodes, does not reduce the costs of the type A nodes involved, since $c \le A$. Therefore, it suffices to consider only the case of edge additions. There are three such cases for an edge:

Adding an edge from $i \in T_A$ to $j \in T_B$: In this case, the degree of both nodes i and j increases by 1. Edge (i, j) reduces the distance cost of node i to node j from 2 to 1 and the distance cost of j to i from $2A$ to A. Also, the total distance of j to nodes of type B that are neighbors with node i decreases by λ, while any other distance remains the same. The cost difference is, thus, $2c - A - \lambda - 1$ and, hence, it must hold that $c \geq \frac{A+\lambda+1}{2}$ in order for G to be pairwise-stable.

Adding an edge from $j_1 \in T_B$ to $j_2 \in T_B$ where j_1 and j_2 have a common type A neighbor: In this case, the cost difference is $2c - 2$, since the new edge incurs cost c to each involved player and reduces their distance from 2 to 1. Therefore, in order for G to be pairwise-stable it must hold that $c \geq 1$.

Adding an edge from $j_1 \in T_B$ to $j_2 \in T_B$ where j_1 and j_2 have no common type A neighbor: The cost of both nodes j_1 and j_2 increases by c, while their distance is reduced by 2; all other distances are unaffected. Hence, the cost difference is $2c - 4$, thus it should hold that $c \geq 2$ in order for G to be pairwise-stable.

We conclude that, when $\max\{2, \frac{A+\lambda+1}{2}\} \leq c \leq A$, graph G is pairwise-stable.

We now proceed to prove the lower bound on the price of anarchy. We begin by computing the total cost in G. The cost of a type A node u is $\text{cost}_u^A(G) = (\kappa + \lambda - 1)c + (\kappa - 1)A + 2\kappa\lambda - \lambda$, since each such node has degree $(\kappa + \lambda - 1)$ and is at distance 1 from the remaining $\kappa - 1$ type A nodes, at distance 1 from its λ neighbors of type B and at distance 2 from the remaining $(\kappa - 1)\lambda$ type B nodes. Similarly, the cost of a type B node v is $\text{cost}_v^B(G) = c + (2\kappa - 1)A + 3\kappa\lambda - \lambda - 2$, since such a node has degree 1 and is at distance 1 from a single type A node, at distance 2 from the remaining $\kappa - 1$ type A nodes, at distance 2 from the $\lambda - 1$ type B nodes that share the same type A neighbor, and at distance 3 from the remaining $(\kappa - 1)\lambda$ nodes of type B. Since in total there are κ nodes of type A and $\kappa\lambda$ nodes of type B, we obtain that the total cost is:

$$\text{cost}(G) = \kappa \sum_{u \in T_A} \text{cost}_u^A(G) + \kappa\lambda \sum_{v \in T_B} \text{cost}_v^B(G)$$
$$= \left(\kappa^2 - \kappa + 2\kappa\lambda\right)c + \left(\kappa^2 - \kappa + 2\kappa^2\lambda - \kappa\lambda\right)A + \kappa\lambda(3\kappa\lambda + 2\kappa - \lambda - 3).$$

Consider now the graph G' where all $\kappa\lambda$ players of type B are connected to the same node of type A. The cost in this case is:

$$\text{cost}(G') = \left(\kappa^2 - \kappa + 2\kappa\lambda\right)c + \left(\kappa^2 - \kappa + 2\kappa^2\lambda - \kappa\lambda\right)A + 2\kappa^2\lambda^2 + 2\kappa^2\lambda - 3\kappa\lambda.$$

Trivially, for the optimal cost it holds that $\text{cost}(G_{opt}) \leq \text{cost}(G')$. We set $A = c = \lambda + 1$; note that these values satisfy both $c \leq A$ and $c \geq \frac{A+\lambda+1}{2}$, and recall that $|T_A| = \kappa$, $|T_B| = \kappa\lambda$. We can now obtain a lower bound on the price of anarchy, i.e.,

$$\text{PoA} \geq \frac{5\kappa^2\lambda^2 + 6\kappa^2\lambda + 2\kappa^2 - 4\kappa\lambda - 2\kappa}{4\kappa^2\lambda^2 + 6\kappa^2\lambda + \kappa\lambda^2 + 2\kappa^2 - 4\kappa\lambda - 2\kappa} \geq \frac{5}{4} - \epsilon,$$

where ϵ is an arbitrarily small positive number. This concludes the proof of the theorem. $\quad\square$

4.3   2-Reliable Network Formation Games

We consider the setting introduced by Meirom et al. in [12] where we require that any pair of nodes is connected by at least two node-disjoint paths. This model captures the case of failures where a backup path is required and uses two additional parameters $\delta \in [0,1]$ and $\tau \in \{0,1\}$ that will be detailed in the following. The cost of a type A node player is

$$\text{cost}_u^A(G) = \deg(u)c_A + \frac{A}{1+\delta} \sum_{v \in T_A} (d(u,v) + \delta d'(u,v))$$

$$+ \frac{\tau}{1+\delta} \sum_{v \in T_B} (d(u,v) + \delta d'(u,v)) + (1-\tau) \sum_{v \in T_B} d(u,v),$$

while the cost of a type B node player is

$$\text{cost}_u^B(G) = \deg(u)c_B + \frac{A}{1+\delta} \sum_{v \in T_A} (d(u,v) + \delta d'(u,v))$$

$$+ \frac{\tau}{1+\delta} \sum_{v \in T_B} (d(u,v) + \delta d'(u,v)) + (1-\tau) \sum_{v \in T_B} d(u,v).$$

For any pair u, v of nodes, $d(u,v)$ and $d'(u,v)$ are the lengths of a pair of node-disjoint paths that minimizes the cost function. Parameter δ relates to the frequency with which the second path will be used. For example, in a setting where failures are frequent, then δ would be close to 1, while if failures are seldom, then δ would be closer to 0. Parameter τ defines whether the requirement for backup paths is only relevant for connectivity to all nodes (and then $\tau = 1$) or only to type A nodes (and then $\tau = 0$).

Meirom et al. [12] argue that in order for the type A nodes to establish a clique, which is consistent to the topology observed in practice, it should hold that $c_A < \frac{A}{1+\delta}$. Furthermore, they show that when $c_A + c_B < \frac{A+1}{1+\delta}$ and $\tau = 1$, then the optimal graph contains all edges except those where both endpoints are type B nodes. We now present our first result for this setting which improves upon an upper bound of 3/2 due to [12] and remark that it holds for any value of T_A and T_B.

Theorem 6. *Whenever $c_A + c_B < \frac{A+1}{1+\delta}$, the price of anarchy of 2-reliable network formation games with heterogeneous players and monetary transfers is 1.*

Proof. Since $c_A < \frac{A}{1+\delta}$, there exists an edge connecting any pair of type A nodes. Furthermore, since $c_A + c_B < \frac{A+1}{1+\delta}$, there is an edge connecting any type B node j to any type A node i, since, if there exist nodes $i \in T_A$, $j \in T_B$ where the edge (i,j) does not exist, then establishing this edge is profitable for the pair i, j as it increases the edge cost by $c_A + c_B$ and decreases the distance cost by at least $\frac{A+1}{1+\delta}$ when $\tau = 1$ and by at least $\frac{A+1+\delta}{1+\delta} \geq \frac{A+1}{1+\delta}$ when $\tau = 0$.

Note that when $\tau = 1$, by the discussion above and due to [12], it holds that the optimal graph has the same structure, while when $\tau = 0$, if there are nodes $i \in T_A$ and $j \in T_B$ that are not adjacent, then since $c_A + c_B < \frac{A+1}{1+\delta} \leq \frac{A+1+\delta}{1+\delta}$, adding edge (i,j) reduces the total cost.

Hence, we conclude that any pairwise-stable graph contains all edges that are also present in the optimal graph and the theorem follows. $\qquad\square$

Whenever $c_A + c_B \geq \frac{A+1}{1+\delta}$, the optimal graph contains the clique of the type A nodes while all type B nodes have degree 2 and are all connected to the same pair of type A nodes. Then, for any pair of nodes (i, j) it holds that, when i and j are both type A nodes, then $d(i, j) = 1$ and $d'(i, j) = 2$, when i and j are both type B nodes, then $d(i, j) = d'(i, j) = 2$, when $i \in T_A$ and $j \in T_B$ and the edge (i, j) does not exist, then $d(i, j) = d'(i, j) = 2$, while, finally, when $i \in T_A$ and $j \in T_B$ and the edge (i, j) exists, then $d(i, j) = 1$ and $d'(i, j) = 2$. The optimal cost, in this case, is

$$\text{cost}(G_{opt}) = T_A(T_A - 1)\left(\frac{A(1 + 2\delta)}{1 + \delta} + c_A\right) + 2T_B(c_A + c_B)$$
$$+ 2(A + 1)T_B\left(T_A - \frac{1}{1 + \delta}\right) + 2T_B(T_B - 1). \tag{2}$$

The proof of the following lower bound on the price of anarchy is in the same spirit as Theorem 5. Again, the lower bound relies on fairly large values for c_A, c_B and A.

Theorem 7. *There exists a 2-reliable large network formation game with heterogeneous players and monetary transfers with price of anarchy $7/6 - \epsilon$.*

5   Conclusions

We have presented improved bounds on the price of anarchy of network formation games with a large number of heterogeneous players that resemble the structure of Internet-like networks and monetary transfers. For fairly natural choices of cost functions, we essentially show that any pairwise-stable graph is very similar to the graph of lowest cost. Moreover, we have shown an improved lower bound on the diameter of a sum equilibrium graph for basic network creation games.

A natural question is to establish the correct bound on the diameter of such sum equilibria. There is still a large gap between the lower bound of 4 presented in this paper and the upper bound of $2^{O(\sqrt{\log n})}$ due to [2]. Sufficient conditions such that the diameter is polylogarithmic have already been identified in [2,15]. The fact that the smallest sum equilibrium graph order is 2 when the diameter is 1, is 3 when the diameter is 2, is 8 when the diameter is 3 and is at most 14 when the diameter is 4 suggests that further progress can be also made with respect to the lower bound.

Concerning network formation games with heterogeneous players, extending the problem to include additional reliability constraints is also an interesting research direction. The setting where two node-disjoint paths are required between any pair of nodes was recently introduced in [12]. It seems natural to allow for greater reliability guarantees by requiring k node-disjoint paths between pairs of nodes, and especially so when type A nodes are involved.

References

1. Àlvarez, C., Serna, M., Fernàndez, A.: Network formation for asymmetric players and bilateral contracting. Theor. Comput. Syst. forthcoming. doi:10.1007/s00224-015-9640-6
2. Alon, N., Demaine, E., Hajiaghayi, M., Leighton, T.: Basic network creation games. SIAM J. Discrete Math. **27**(2), 656–668 (2013)
3. Alon, N., Demaine, E., Hajiaghayi, M., Kanellopoulos, P., Leighton, T.: Correction: basic network creation games. SIAM J. Discrete Math. **28**(3), 1638–1640 (2014)
4. Arcaute, E., Dyagilev, K., Johari, R., Mannor, S.: Dynamics in tree formation games. Games Econ. Behav. **79**, 1–29 (2013)
5. Bilò, D., Gualà, L., Leucci, S., Proietti, G.: Locality-based network creation games. In: Proceedings of the 26th ACM Symposium on Parallelism in Algorithms and Architectures (SPAA), pp. 277–286 (2014)
6. Ehsani, S., Fadaee, S., Fazli, M., Mehrabian, A., Sadeghabad, S., Safari, M., Saghafian, M.: On a bounded budget network creation game. ACM Trans. Algorithms **11**(4), 34 (2015)
7. Fabrikant, A., Luthra, A., Maneva, E., Papadimitriou, C.H., Shenker, S.: On a network creation game. In: Proceedings of the 22nd Annual Symposium on Principles of Distributed Computed (PODC), pp. 347–351 (2003)
8. Feldman, M., Immorlica, N., Lucier, B., Roughgarden, T., Syrgkanis, V.: The price of anarchy in large games. In: Proceedings of the 48th ACM Symposium on Theory of Computing (STOC) (2016)
9. Koutsoupias, E., Papadimitriou, C.: Worst-case equilibria. In: Meinel, C., Tison, S. (eds.) STACS 1999. LNCS, vol. 1563, pp. 404–413. Springer, Heidelberg (1999)
10. Lenzner, P.: On dynamics in basic network creation games. In: Persiano, G. (ed.) SAGT 2011. LNCS, vol. 6982, pp. 254–265. Springer, Heidelberg (2011)
11. Meirom, E., Mannor, S., Orda, A.: Network formation games with heterogeneous players and the Internet structure. In: Proceedings of the 15th ACM Conference on Economics and Computation (EC), pp. 735–752 (2014)
12. Meirom, E., Mannor, S., Orda, A.: Formation games of reliable networks. In: Proceedings of the 2015 IEEE Conference on Computer Communications (INFOCOM), pp. 1760–1768 (2015)
13. Mamageisvhili, A., Mihalák, M., Müller, D.: Tree Nash equilibria in the network creation game. Internet Math. **11**(4–5), 472–486 (2015)
14. Mihalák, M., Schlegel, J.C.: Asymmetric swap-equilibrium: a unifying equilibrium concept for network creation games. In: Rovan, B., Sassone, V., Widmayer, P. (eds.) MFCS 2012. LNCS, vol. 7464, pp. 693–704. Springer, Heidelberg (2012)
15. Nikoletseas, S., Panagopoulou, P., Raptopoulos, C., Spirakis, P.G.: On the structure of equilibria in basic network formation. Theor. Comput. Sci. **590**, 96–105 (2015)
16. Panagopoulou, P.N.: Efficient equilibrium concepts in non-cooperative network formation. In: Zaroliagis, C., et al. (eds.) Spirakis Festschrift. LNCS, vol. 9295, pp. 384–395. Springer, Heidelberg (2015). doi:10.1007/978-3-319-24024-4_22
17. Papadimitriou, C.H.: Algorithms, games and the Internet. In: Proceedings of the 33rd Annual ACM Symposium on Theory of Computing (STOC), pp. 749–753 (2001)

Parameterized Complexity of Team Formation in Social Networks

Robert Bredereck[1]([✉]), Jiehua Chen[1], Falk Hüffner[1], and Stefan Kratsch[2]

[1] Institut für Softwaretechnik und Theoretische Informatik,
Technische Universität Berlin, Berlin, Germany
{robert.bredereck,jiehua.chen,falk.hueffner}@tu-berlin.de
[2] Institut für Informatik I, Universität Bonn, Bonn, Germany
kratsch@cs.uni-bonn.de

Abstract. Given a task that requires some skills and a social network of individuals with different skills, the TEAM FORMATION problem asks to find a team of individuals that together can perform the task, while minimizing communication costs. Since the problem is NP-hard, we identify the source of intractability by analyzing its parameterized complexity with respect to parameters such as the total number of skills k, the team size l, the communication cost budget b, and the maximum vertex degree Δ. We show that the computational complexity strongly depends on the communication cost measure: when using the weight of a minimum spanning tree of the subgraph formed by the selected team, we obtain fixed-parameter tractability for example with respect to the parameter k. In contrast, when using the diameter as measure, the problem is intractable with respect to any single parameter; however, combining Δ with either b or l yields fixed-parameter tractability.

1 Introduction

Assembling teams based on required skills is a classic management task. Recently, it has been suggested to take into account not only the covering of the required skills, but also the expected communication costs (see Lappas et al. [11] for a survey). This cost can be estimated based on a given edge-weighted social network, where a low weight value on an edge between two individuals indicates a low communication cost. For example, edge weights can reflect distance in an organizational chart or the number of joint projects completed.

Lappas et al. [10] formalized the setting as the optimization problem of minimizing the communication cost and studied two cost measures: the diameter (DIAM) and the weight of a minimum spanning tree (MST). For our complexity analysis, we formulate it as a decision problem by fixing the maximum team size.

DIAM-TEAM FORMATION
Input: An undirected graph $G = (V, E)$ with edge-weight function $w \colon E \to \mathbb{N}$, a set T of k skills, a skill function $S \colon V \to 2^T$, a team size $l \in \mathbb{N}$, and a budget $b \in \mathbb{N}$.
Question: Is there a subset $V' \subseteq V$ with $|V| \leq l$ such that $\bigcup_{v \in V'} S(v) = T$ and the w-weighted diameter of the induced subgraph $G[V']$ is at most b?

© Springer International Publishing Switzerland 2016
R. Dondi et al. (Eds.): AAIM 2016, LNCS 9778, pp. 137–149, 2016.
DOI: 10.1007/978-3-319-41168-2_12

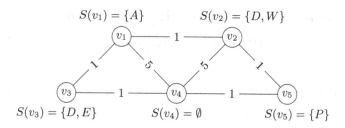

Fig. 1. A TEAM FORMATION example: a social network of five potential team members and five skills, "algorithms" (A), "data bases" (D), "software engineering" (E), "programming" (P), and "web programming" (W). When minimizing the weight of a minimum spanning tree of the subgraph induced by a team (MST), the team with members v_1, v_2, v_3, v_5 has the lowest cost, 3, (one can build a path with weight one at each edge). However, when minimizing the diameter of the subgraph induced by a team (DIAM), it is worthwhile to add the individual v_4—who has no specific skill—to reduce the diameter of 3 in $G[\{v_1, v_2, v_3, v_5\}]$ to the diameter of 2 in $G[\{v_1, v_2, v_3, v_4, v_5\}]$.

Here, the *diameter* of an edge-weighted graph G, denoted as DIAM(G), is the maximum distance between any two vertices in the input graph and the *distance* between two vertices is the minimum sum of the weights of the edges along any path between these two vertices. Our formulation of the team formation problem allows to choose individuals (vertices) that do not contribute any skills, but serve as intermediate vertices to lower overall communication costs. We further assume w.l.o.g. that no individual has a skill that is not in the request set T.

The *weight of a minimum spanning tree* of graph G, MST(G), is the smallest sum of the weights of the edges in a spanning tree of G. We define the corresponding MST-TEAM FORMATION problem by replacing "diameter" in the definition of DIAM-TEAM FORMATION with "weight of a minimum spanning tree".

Figure 1 illustrates an example for the DIAM-TEAM FORMATION and MST-TEAM FORMATION problems. Lappas et al. [10] showed that both problems are NP-complete. Experiments on DIAM-TEAM FORMATION , MST-TEAM FORMATION and similar team formation problems so far use heuristic algorithms [1,4,8, 10,12]. However, it might be that instances encountered in practice are actually easier than a one-dimensional complexity analysis suggests, and can be solved optimally. For example, it might be reasonable to assume that only a small number of skills is required. Thus, we try to identify the sources of intractability by a parameterized complexity analysis.

Optimization Variant. There are two natural ways to define approximate solutions for our team formation problem. First, one allows to find solutions with larger communication costs. This leads to the MINCOST-ζ-TEAM FORMATION problem, ζ being either DIAM or MST, which asks for a vertex subset $V' \subseteq V$ with $|V'| \leq l$ such that $\bigcup_{v \in V'} S(v) = T$ and the communication cost $\zeta(G[V'])$ is minimized. Second, one allows to find solutions with larger teams. This leads to the MinTeamSize-ζ-Team Formation problem, which asks for a minimum vertex subset $V' \subseteq V$ such that $\bigcup_{v \in V'} S(v) = T$ and $\zeta(G[V']) \leq b$.

Cost Measure "Diameter". Arkin and Hassin [2] studied MinCost-Diam-Team Formation with unlimited team size l under the name Multiple-Choice Cover. They showed that even when no skill is allowed to be covered by more than three team members the problem cannot be approximated with a constant-factor error guarantee, unless P = NP. However, when the weights satisfy the triangle inequality, a 2-approximation is possible; this bound is sharp [2].

Cost Measure "Minimum Spanning Tree". As already mentioned by Lappas et al. [10], the MinCost-Mst-Team Formation problem with an unlimited team size l is equivalent to the Group Steiner Tree problem: given an undirected edge-weighted graph $G = (V, E)$ and vertex subsets (*groups*) $g_i \subseteq V$, $1 \leq i \leq k$, find a subtree $T = (V_T, E_T)$ of G such that $V_T \cap g_i \neq \emptyset$ for all $1 \leq i \leq k$ and the cost $\sum_{e \in E_T} w(e)$ is minimized. Clearly, each group of Group Steiner Tree corresponds to a subset of vertices in Mst-Team Formation that have a particular skill. From this relation to Group Steiner Tree and an inapproximability result of Halperin and Krauthgamer [9], we obtain that it is unlikely that MinCost-Mst-Team Formation can be approximated to a factor of $O(\log^{2-\epsilon} k)$ for any $\epsilon > 0$, where k is the number of skills to be covered.

Despite the polylogarithmic inapproximability result, we can obtain fixed-parameter tractability for the parameter "number k of skills to be covered". First, we reduce the Mst-Team Formation problem with limited team size l to the Mst-Team Formation problem with an unlimited team size by adding a large weight W (for example the sum over all edge weights) to each edge weight and adding $l \cdot W$ to the budget. Then, by the relation between Mst-Team Formation and Group Steiner Tree, we can think of the resulting instance as a Group Steiner Tree instance, which can be solved by reducing it to Steiner Tree: introduce a new vertex for each group and connect it to each vertex contained in this group by an edge with very high weight. The resulting Steiner Tree instance can be solved using inclusion–exclusion in $O^*(2^k)$ time and polynomial space when the edge weights are integers [13] (the O^* notation omits factors polynomial in the input size); for arbitrary weights, it can be solved in $O^*(3^k)$ time and exponential space by dynamic programming [6].

Further Related Work. Our team formation problem can be generalized in different ways. First, we can require each skill to be covered by a given number of team members instead of once. Li and Shan [12] proposed three heuristics for this problem; Gajewar and Sarma [8] studied it with the objective of maximizing the *collaborative compatibility*, an alternative to Diam and Mst. Here, the collaborative compatibility is the sum of the weights of all edges in the subgraph induced by the team divided by the team size (the number of vertices in the subgraph). They showed that this version is also NP-hard and provided a 1/3-approximation algorithm. Second, we can additionally require the workload to be balanced within the team [1,4].

A number of experimental studies examine the validity of these models, using data for example from bibliography databases [1,8,10,12] or the GitHub programming collaboration platform [4].

DIAM-TEAM FORMATION and MST-TEAM FORMATION have also applications in keyword search in relational databases [15]: the vertices in the graph correspond to tables, edges represent foreign key relationships, and skills model keywords that match the table. A subgraph covering all keywords with small communication costs helps to create efficient SQL queries.

Parameterized Complexity. Parameterized algorithmics analyzes problem difficulty not only in terms of the input size, but also for an additional parameter, typically an integer p. Thus, formally, an instance of a parameterized problem is a tuple of the unparameterized instance I and the parameter p. A parameterized problem with parameter p is *fixed-parameter tractable* (FPT) if there is an algorithm that decides each instance (I, p) in $f(p) \cdot |I|^{O(1)}$ time, where f is a computable function depending only on p; we call this algorithm a *fixed-parameter algorithm*. In such case, we say that our problem can be solved in FPT-time for the parameter p. Clearly, if the problem is NP-hard, we must expect f to grow superpolynomially.

There are parameterized problems for which there is good evidence that no fixed-parameter algorithms exist. Analogously to the concept of NP-hardness, the concept of W[1]-*hardness* was developed. It is widely assumed that a W[1]-hard problem cannot have a fixed-parameter algorithm (hardness for the classes W[t], $t \geq 2$ has the same implication). To show that a problem is W[t]-hard, a *parameterized reduction* from a known W[t]-hard problem can be used. This is a reduction that runs in FPT-time and maps the parameter p to a new parameter p' that is upper-bounded by some function $g(p)$. We refer to recent text books [3,5,7,14] for details on parameterized complexity theory and W[t]-complete problems.

Contributions. We focus on the parameterized complexity of DIAM-TEAM FORMATION, which has to the best of our knowledge not been considered before. We consider parameters that are related to the communication cost and to the input graph: the number k of skills to be covered, the cost budget b, the maximum vertex degree Δ, and the team size l.

For the parameter l, DIAM-TEAM FORMATION is W[2]-hard even with either constant budget b or constant maximum degree Δ (Proposition 1). For the parameter k, while MST-TEAM FORMATION is fixed-parameter tractable, DIAM-TEAM FORMATION is W[1]-hard even on graphs of maximum vertex degree three and with unrestricted team size l (Theorem 1). For the combined parameter $l + k$, DIAM-TEAM FORMATION is W[1]-hard even if the cost budget b is two (Theorem 2). Concerning the parameter maximum degree Δ, we find that the problem is NP-hard even if the graph is a caterpillar with maximum degree $\Delta = 3$ (Proposition 1), where a caterpillar is a tree in which all the vertices are within distance one of a central path. Our results rule out fixed-parameter tractability for all considered single parameters and several parameter combinations.

By our parameterized hardness reductions, we can obtain that MINCOST-DIAM-TEAM FORMATION is inapproximable even when we allow for a super-polynomial running time factor in the team size l, even on complete graphs or on stars (Corollary 1). MINTEAMSIZE-DIAM-TEAM FORMATION is inapproximable even when we allow for a superpolynomial running time factor in the number k of skills, even on graphs with maximum degree $\Delta = 3$ (Corollary 2).

Geared towards robustness, we also consider the situation where the subgraph induced by the team is two-connected (that is, between each two team members, there are at least two edge-disjoint paths). We find that unless $\mathsf{FPT} = \mathsf{W}[1]$, it is unlikely that there is an algorithm that forms a team of size at most l, covering all k skills and inducing a two-connected subgraph, in $f(k + l) \cdot |I|^{O(1)}$ time, where $|I|$ denotes the size of our input instance (Theorem 3).

On the positive side, we provide some tractability results: DIAM-TEAM FORMATION can be solved in $O^*(\Delta^{\Delta^b} \cdot \text{dcheck})$ time and in $O^*(\Delta^l \cdot \text{dcheck})$ time (Theorem 4), where dcheck denotes the running time of checking whether a subgraph has diameter most b, which can for example be solved in $O(\Delta \cdot n^2 \cdot \log(n))$ time by Dijkstra's algorithm. Finally, if the input graph is a tree, then we obtain that DIAM-TEAM FORMATION is fixed-parameter tractable for parameter k (Theorem 5).

2   Hardness Results

Throughout this section, we assume each edge in the input graph to have weight one; thus, we omit the introduction of the edge weight function w. We will see that our DIAM-TEAM FORMATION problem is already hard in this setting. First, to get a feeling for the computational hardness of our TEAM FORMATION model we start with a simple observation which basically says that DIAM-TEAM FORMATION with an unbounded number k of skills is basically at least as hard as the SET COVER problem, even on simple graph classes.

SET COVER
Input: A set family $\mathcal{F} = \{F_1, \ldots, F_\alpha\}$ over a universe $U = \{u_1, \ldots, u_\beta\}$ and a non-negative integer h.
Question: Is there a *set cover* of size at most h, that is, a subfamily $\mathcal{F}' \subseteq \mathcal{F}$ with $|\mathcal{F}'| \leq h$ such that $\bigcup_{F \in \mathcal{F}'} F = U$?

Observation 1. *For edge weight one,* DIAM-TEAM FORMATION *parameterized by the team size l generalizes* SET COVER *parameterized by the set cover size h, even on simple graph classes such as (1) complete graphs, (2) stars, and (3) caterpillars with maximum vertex degree three.*

Proof. Given a SET COVER instance $(\mathcal{F}, U, h)$, we construct a DIAM-TEAM FORMATION instance $(G = (V, E), T, S, l, b)$ for each of the settings as follows.

(1) Define the skill set $T := U$, and for each set $F_i \in \mathcal{F}$, create one vertex v_i and define $S(v_i) := F_i$. Add an edge between each pair of vertices to obtain a complete graph. Finally, define the team size $l := h$ and let the cost budget b be an arbitrary integer at least one.

(2) Define the skill set $T := U$, and for each set $F_i \in \mathcal{F}$, create one vertex v_i and define $S(v_i) := F_i$. Add a special skill 0 to T, add a center vertex v_r to V, and define $S(v_r) := \{0\}$. Construct a star graph with center v_r by adding an edge between each vertex v_i and v_r, $1 \le i \le \alpha$. Finally, define the team size $l := h$, and let the cost budget b be an arbitrary integer at least two.

(3) Define the skill set $T := U \uplus \{0, 1, 2\}$, and for each set $F_i \in \mathcal{F}$, create two vertices u_i and v_i. Define $S(v_i) := F_i$ for all $1 \le i \le \alpha$, $S(u_1) := \{0\}$ and $S(u_\alpha) := \{2\}$, and $S(u_i) := \{1\}$ for all $1 < i < \alpha$. Add an edge between u_i and u_{i+1} for all $1 \le i < \alpha$ and and edge between u_i and v_i for all $1 \le i \le \alpha$. Finally, define the team size l to be at least $h + \alpha$, and the cost budget $b := \alpha + 2$.

It is easy to verify that the constructed instances are yes-instances if and only if $(\mathcal{F}, U, h)$ is a yes-instance. □

In terms of parameterized and classical complexity analysis, Observation 1 yields the following hardness results.

Proposition 1. *Even when each edge has weight one, the following holds. (1)* DIAM-TEAM FORMATION *parameterized by the team size l is* W[2]*-hard even if the budget b is one and the graph is complete. (2)* DIAM-TEAM FORMATION *parameterized by the team size l is* W[2]*-hard even if the budget b is two and the graph is a star. (3)* DIAM-TEAM FORMATION *is* NP*-hard even on caterpillar graphs with maximum degree three.*

We note that the budget in the proof of Statements (1)–(2) in Observation 1 as well as the team size in the proof of Statements (3) may have extremely large values that effectively do not upper-bound the communication costs or the team size. Since SET COVER is NP-complete and W[2]-complete when parameterized by h, in terms of minimizing the communication cost or team size, we have the following inapproximability result.

Corollary 1. *Unless all problems in* W[2] *are fixed-parameter tractable,* MIN-COST-DIAM-TEAM FORMATION *is inapproximable even in* FPT*-time for the parameter team size l, even on complete graphs or on stars. Unless* P = NP, MINTEAMSIZE-DIAM-TEAM FORMATION *is inapproximable even in polynomial time, even on caterpillar graphs with maximum degree three.*

Consequently, to identify tractable cases one should start with cases where SET COVER is tractable. A very well-motivated restriction for DIAM-TEAM FORMATION is to assume that there are not too many skills to cover, that is, the number k of skill is (part of) the parameter. Our next result, however, shows that this assumption (alone) does not make the problem fixed-parameter tractable.

Theorem 1. DIAM-TEAM FORMATION *parameterized by the number k of skills is* W[1]*-hard, even on graphs with maximum degree three and with each edge weight one, and when the team size is unrestricted.*

Proof. We give a parameterized reduction from the W[1]-complete problem MULTICOLORED CLIQUE parameterized by the clique size h to TEAM FORMATION parameterized by k on graphs of maximum degree three.

MULTICOLORED CLIQUE
Input: An undirected graph $G = (V, E)$, a non-negative integer $h \in \mathbb{N}$, and a vertex coloring $\phi \colon V \to \{1, 2, \ldots, h\}$.
Question: Does G admit a colorful h-clique, that is, a size-h vertex subset $Q \subseteq V$ such that the vertices in Q are pairwise adjacent and have pairwise distinct colors?

Let (G, ϕ, h) be a MULTICOLORED CLIQUE instance. We construct in FPT-time an equivalent DIAM-TEAM FORMATION instance $(G' = (V', E'), T, S, l, b)$. Without loss of generality, we assume that in G all edges are between vertices of different colors (according to ϕ) since they could be deleted without changing presence of a colorful h-clique. Let $n := |V|$, let $V = \{v_1, \ldots, v_n\}$ be an arbitrary ordering of the vertices, and let y be the smallest integer with $n \leq 2^y$.

Construction. We construct the graph G' by adding to V' all vertices in V (as an independent set). We connect the vertices in V by the following three steps:

1. Attach to each v_i a path of length s (to be determined later) whose other endpoint is denoted w_i; i.e., the distance between v_i and w_i shall be s. We call this path, including v_i and w_i the *path of v_i*.
2. Make each w_i the root of a newly added complete binary tree of height y, i.e., with 2^y leaves. Arbitrarily pick any n of its leaves and assign them names $x_{i,1}, \ldots, x_{i,n}$. Thus, $x_{i,j}$ will be the jth leaf in the binary tree attached by a path of length s (via w_i) to vertex v_i. We call the binary tree with root w_i and leaves $x_{i,1}, \ldots, x_{i,n}$ the *binary tree of v_i*.
3. Finally, we encode the adjacency from G as follows: If v_i and v_j are vertices that are adjacent in G, which implies by our assumption that they have different colors, i.e., $\phi(v_i) \neq \phi(v_j)$, then add an edge between $x_{i,j}$ and $x_{j,i}$. Thus, a leaf in the binary tree of v_i (namely $x_{i,j}$) is now adjacent to a leaf in the binary tree of v_j (namely $x_{j,i}$). Note that the naming convention of leaves prevents using each leaf for more than one adjacency.

This completes the construction of the graph G'. The graph can be constructed in FPT-time; indeed, it can be computed in polynomial time. Its maximum degree is three. Observe that if v_i and v_j are adjacent in G, then they have distance at most $2s + 2y + 1$ in G'. We set $s := 4n$ (with the intention of having vertices in two binary trees with adjacent leaves be at distance at most $4n$).

To complete the construction define the skill set $T := \{1, 2, \ldots, h\}$ and the skill function $S \colon V \to 2^T$ such that $S(v) := \{\phi(v)\}$ for all vertex $v \in V \subseteq V'$, i.e., for all vertices of the input graph G the skill equals the color according to ϕ, and $S(v) = \emptyset$ for all further vertices of $V' \setminus V$. The budget b (the diameter) is set to $b := 2s + 2y + 1$. Finally, the team size l is set to $|V'|$, i.e., the team size is effectively unbounded. We return instance $(G' = (V', E'), T, S, l, b)$.

The equivalence of (G, ϕ, h) and (G', T, S, l, b) is omitted due to space.   □

We know from Proposition 1 that DIAM-TEAM FORMATION is W[2]-hard for the parameter team size l and from Theorem 1 that DIAM-TEAM FORMA-TION is W[1]-hard for the parameter number k of skills (in both cases even if the maximum degree Δ is a small constant). This invokes the question whether our problem becomes tractable for the combined parameter $l + k$, that is, for cases where both the team size and the number of skills are small. In the following, we obtain W[1]-hardness for $l + k$ even for a constant cost budget b. We will see later (Theorem 4 in Sect. 3) that our problem becomes tractable when both values, the maximum vertex degree Δ and the cost budget b, are small.

Theorem 2. DIAM-TEAM FORMATION *parameterized by the combined parameter $l + k$ is* W[1]*-hard, even if the cost budget is two.*

Proof. We provide a parameterized reduction from MULTICOLORED CLIQUE parameterized by the clique size h. Let (G, ϕ, h) be an instance of MULTI-COLORED CLIQUE; w.l.o.g. there are no edges $\{u, v\}$ with $\phi(u) = \phi(v)$. For $i \in \{1, \dots, h\}$ define $V_i := \phi^{-1}(i)$, i.e., the set of all vertices of color i in G.

We create a graph G' from G as follows. First, subdivide all edges of G using new vertices. We use $V_{i,j}$ with $1 \le i < j \le k$ for the set of all newly introduced vertices that subdivide an edge between V_i and V_j. Now, we turn all subdividing vertices into a single large clique by adding edges. We define the skill set $T := \{1, 2, \dots, h\}$. We assign each vertex v in a set V_i the skill set: $S(v) := \{i\}$, and each vertex v in a set $V_{i,j}$ an empty skill set: $S(v) := \emptyset$. We set the team size l to be $h + \binom{h}{2}$ and set the budget b (the diameter) to be two. This completes the construction which can clearly be done in FPT-time; indeed it can be computed in polynomial time. The correctness proof is omitted due to space.   □

Observe that in the proofs of Theorem 1 and Theorem 2 we have made no use of the upper bound on the team size. Any team with all k skills and diameter at most b was proved to lead directly to a k-clique. Thus, the minimum team size is strongly inapproximable in the sense that even finding any feasible team respecting the cost budget is W[1]-hard with respect to k.

Corollary 2. *Unless all problems in* W[1] *are fixed-parameter tractable,* MIN-TEAMSIZE-DIAM-TEAM FORMATION *is inapproximable even in* FPT*-time for the parameter number k of skills, even either on graphs with maximum degree three or with cost budget two.*

Finally, we show that the W[1]-hardness for the combined parameter $l + k$ still holds if we require that the graph induced by the team is only two-vertex-connected (resp. two-edge-connected) instead of requiring a small diameter, that is, requiring robustness instead of low communication costs.

Theorem 3. *Finding a team of size at most l, covering all k skills, such that the subgraph induced by the team is two-connected is W[1]-hard with respect to the combined parameter parameters $l + k$.*

Proof. We provide a parameterized reduction from MULTICOLORED CLIQUE parameterized by the clique size h to our problem parameterized by $l + k$ where l denotes the team size and k the number of skills. Given a MULTICOLORED CLIQUE instance (G, ϕ, h) construct a graph G' by again subdividing all edges. Just as in the proof of Theorem 2, use vertex sets $V_1, \ldots, V_h$ and let $V_{(i,j)}$ contain the subdividing vertices of (former) edges between V_i and V_j, $i < j$. Define the skill set $T := \{1, \ldots, h\} \cup \{(i,j) \mid 1 \leq i < j \leq h\}$. Assign the skills according to membership in sets V_i and $V_{(i,j)}$, i.e., a vertex v has skill $x \in T$ if and only if it is contained in set V_x. Finally, set the team size l to $h + \binom{h}{2}$. Note that the parameter value $l + k$ is upper-bounded by $O(h^2)$, which is sufficient. This completes the construction which can clearly be done in FPT-time; indeed it can also be computed in polynomial time. Equivalence of (G, ϕ, h) and the constructed instance is omitted due to space.   □

3   Tractability Results

In contrast to our hardness results, which all hold even for unit weights, we identify tractable cases with arbitrary positive integer weights. The first case models the situation where each potential team member is connected only to few others in the social network (that is, the maximum vertex degree Δ) and either the budget (that is, the diameter b) or the desired team size l are small.

Theorem 4. DIAM-TEAM FORMATION *can be solved in $O^*(\Delta^l \cdot dcheck)$ time and in $O^*(\Delta^{\Delta^b} \cdot dcheck)$ time where Δ is the maximum vertex degree of the input graph, b is the communication cost budget (the diameter), l is the team size, and checking whether the diameter of a subgraph is at most b takes dcheck time.*

Proof. Let $(G = (V, E), w, T, S, l, b)$ be our DIAM-TEAM FORMATION instance. Without loss of generality, we assume that the team $V' \subseteq V$ which we search for induces a connected subgraph. Given that each vertex has at most Δ neighbors, we build a search tree algorithm that branches into selecting one of the Δ neighbors of a potential team member (vertex) adding it to our partial solution (team). Since the team can have at most l members, the depth of our search tree is upper-bounded by l. In each node of the search tree we need to check whether the subgraph induced by the partial solution has diameter at most b (regarding the edge weight function w); this check runs in polynomial time. We omit the details due to space.   □

Our second tractable case models situations where the team members are organized in a hierarchical tree structure.

Theorem 5. *If the input social network is a tree, then* DIAM-TEAM FORMA-
TION *can be solved in* $O(2^k \cdot n \cdot b^2 \cdot B_k)$ *time, where* k *denotes the number of
skills,* n *denotes the number of individuals in the network,* b *denotes the target
diameter, and* B_k *denotes the* k*th Bell number.*

Proof. We describe a dynamic programming algorithm to solve DIAM-TEAM
FORMATION on trees. Let $I = (G = (V, E), w, T, S, l, b)$ be an instance of
DIAM-TEAM FORMATION with the input graph G being a tree. The basic idea
is to store for each vertex $v \in V$ of the tree and each subset $T' \subseteq T$ of skills
whether T' can be covered within the subtree rooted at v. To this end, we
assume that $G = (V, E)$ is an arbitrarily rooted tree and denote the subtree
rooted at $v \in V$ by subtree(v). We denote the set of children of each vertex
$v \in V$ by children(v).

We define the dynamic programming table A as follows. For each subset $T' \subseteq
T$ of skills, each vertex $v \in V$, each cost budget b', $b' \in \{0, 1, \ldots, b\}$, and each
depth bound z, $z \in \{0, 1, \ldots, b\}$, the entry $A(T', v, b', z)$ stores the size of a
smallest team $V' \subseteq V$ which fulfills the following requirements:

(a) V' covers T'.
(b) V' consists of vertex v and vertices only from subtree(v).
(c) The subgraph induced by V' is a tree with diameter at most b' and depth at
    most z. (That is, the largest weight of a shortest path between two arbitrary
    vertices of the tree $G[V']$ is at most b' and the largest weight of a shortest
    path between v and an arbitrary vertex $v \in V'$ is at most z.)

It is easy to see that there is a yes-instance if and only if $\min_{v \in V} A(T, v, b, b) \le l$.

We fill the table entries following the tree from the leaves to the root. We
initialize the entries concerning the set $V_L \subseteq V$ of leaves of the tree as follows.

$$\forall T' \subseteq T; v \in V_L; b' \in \{0, \ldots, b\}; z \in \{0, \ldots, b\}:$$

$$A(T', v, b', z) = \begin{cases} 1 & \text{if } T' \subseteq S_v \\ \infty & \text{otherwise} \end{cases}$$

Now, we consider some non-leaf v of the tree. The key question is which subtree
rooted at some child of v contributes to the team. Clearly, in a *smallest* team,,
each of such subtrees must cover at least some skill uniquely. Observe that there
are at most B_k partitions of T' where B_k is the kth Bell number. That is, there
are at most B_k possibilities of having at most k subtrees, each of which is rooted
at some child of v and contributes to some smallest team covering T'. The idea is
to consider for each part of the partition only the "cheapest" subtree covering it
while fulfilling the diameter and depth requirements. Another crucial observation
is that the diameter of the subtree rooted at $v \in V$ is the maximum of

(1) the largest diameter of all subtree(v'), $v' \in$ children(v), and
(2) the length of a longest path in subtree(v) containing v.

To calculate the value in (2), we need to know the length z_1 of a longest path from v to a leaf of the subtree(v') rooted at a child v' of v, and the length z_2 of a longest path from v to a leaf of the subtree(v'') rooted at a child v'' of v with $v'' \neq v'$.

Using these ideas, updating the entries bottom up works as follows. To handle the diameter costs that come from two different subtrees in subtree(v), we fix a partition of T' and the part of skills to be covered by the child v' of v such that subtree(v') has the largest depth.

$$\forall T' \subseteq T; v \in V; b', z \in \{0, \ldots, b\}: A(T', v, b', z) =$$
$$\min_{\substack{1 \leq i' \leq k' \leq k \\ T' = S_v \uplus T'_1 \uplus T'_2 \uplus \cdots \uplus T'_{k'}}} \text{cheapestCover}(T'_1, \ldots, T'_{k'}, i', v, b', z),$$

where "cheapestCover()" denotes the size of a smallest team covering T' in the following way.

(i) Each disjoint subset T'_i of skills is covered by the vertices of the subtree rooted at one child of v.
(ii) The team that covers $T'_{i'}$ induces a subtree with the largest depth.
(iii) The overall team (including v and covering T') has diameter at most b', and depth at most z.

It can be computed as follows.

$$\text{cheapestCover}(T'_1, \ldots, T'_{k'}, i', v, b', z) =$$
$$1 + \min_{\substack{z_2 \leq z_1 \leq z \\ z_1 + z_2 \leq b'}} \left\{ \min_{\substack{v' \in \text{children}(v) \\ z_1 \geq w(\{v, v'\})}} A(T'_{i'}, v', b', z_1 - w(\{v, v'\})) + \right.$$
$$\left. \sum_{\substack{1 \leq i \leq k' \\ i \neq i'}} \min_{\substack{v'' \in \text{children}(v) \\ z_2 \geq w(\{v, v''\})}} A(T'_i, v'', b', z_2 - w(v, v'')) \right\}$$

For the correctness of our algorithm, if $\min_{v \in V} A(T, v, b, b) \leq l$, then there is indeed a set $V' \subseteq V$ with at most l vertices such that $\bigcup_{v \in V'} S(v) = T$ and $\text{DIAM}(G[V']) \leq b$, which can be constructed by standard backtracking of our dynamic programming algorithm. However, it is not obvious that our algorithm considers all possible solutions, since it assumes that each part $T'_i, 1 \leq i \leq k'$ of the partition $T' = S_v \uplus T'_1 \uplus T'_2 \uplus \cdots \uplus T'_{k'}$ is covered by a distinct cheapest subtree. To see that this is no real restriction, consider some fixed partition with $T' = S_v \uplus T'_1 \uplus T'_2 \uplus \cdots \uplus T'_{i_1} \uplus T'_{i_2} \uplus \cdots \uplus T'_{k'}$. Of course, it may happen that the cheapest subtrees for T'_{i_1} and T'_{i_2} are identical. In this case, the value of "cheapestCover()" might be higher than the size of the corresponding team and one might think that a smaller team may not be identified. However, as this also means that $T'_{i''} := T'_{i_1} \uplus T'_{i_2}$ can also be covered within the same subtree, the team will be found using some partition with $T' = S_v \uplus T'_1 \uplus T'_2 \uplus \cdots \uplus T'_{i''} \uplus \cdots \uplus T'_{k'}$, that is, replacing the two parts T'_{i_1} and T'_{i_2} with $T'_{i''}$. Summarizing, for every team there is always a partition of T' such that no two parts are covered within the same cheapest subtree.

Finally, the size of the two tables is upper-bounded by some function in $O(2^k \cdot n \cdot b^2)$. The initialization phase takes $O(2^k \cdot n \cdot b^2)$ time. The update phase takes $O(2^k \cdot n \cdot b^2 \cdot B_k)$ time.   □

Finally we conjecture that the fixed-parameter tractability result from Theorem 5 can be extended to hold even for the combined parameter "number k of skill" and "treewidth t". However, showing this certainly requires extensive technical details going beyond the scope of this work.

Acknowledgement. This work was started at the research retreat of the TU Berlin Algorithms and Computational Complexity group held in April 2014.

References

1. Anagnostopoulos, A., Becchetti, L., Castillo, C., Gionis, A., Leonardi, S.: Online team formation in social networks. In: Proceedings of the 21st International Conference on World Wide Web, WWW 2012, pp. 839–848. ACM (2012)
2. Arkin, E.M., Hassin, R.: Minimum-diameter covering problems. Networks **36**(3), 147–155 (2000)
3. Cygan, M., Fomin, F.V., Kowalik, L., Lokshtanov, D., Marx, D., Pilipczuk, M., Pilipczuk, M., Saurabh, S.: Parameterized Algorithms. Springer, Switzerland (2015)
4. Datta, S., Majumder, A., Naidu, K.: Capacitated team formation problem on social networks. In: Proceedings of the 18th ACM SIGKDD International Conference on Knowledge Discovery and Data Mining, KDD 2012, pp. 1005–1013. ACM (2012)
5. Downey, R.G., Fellows, M.R.: Fundamentals of Parameterized Complexity. Springer, London (2013)
6. Dreyfus, S.E., Wagner, R.A.: The Steiner problem in graphs. Networks **1**(3), 195–207 (1972)
7. Flum, J., Grohe, M.: Parameterized Complexity Theory. Springer, Heidelberg (2006)
8. Gajewar, A., Sarma, A.D.: Multi-skill collaborative teams based on densest subgraphs. In: Proceedings of the 12th SIAM International Conference on Data Mining, SDM 2012, pp. 165–176. SIAM/Omnipress (2012)
9. Halperin, E., Krauthgamer, R.: Polylogarithmic inapproximability. In: Proceedings of the 35th Annual ACM Symposium on Theory of Computing, STOC 2003, pp. 585–594. ACM (2003)
10. Lappas, T., Liu, K., Terzi, E.: Finding a team of experts in social networks. In: Proceedings of the 15th ACM SIGKDD International Conference on Knowledge Discovery and Data Mining, KDD 2009, pp. 467–476. ACM (2009)
11. Lappas, T., Liu, K., Terzi, E.: A survey of algorithms and systems for expert location in social networks. In: Social Network Data Analytics, pp. 215–241. Springer (2011)
12. Li, C., Shan, M.: Team formation for generalized tasks in expertise social networks. In: Proceedings of the 2nd IEEE International Conference on Social Computing, SocialCom 2010, pp. 9–16. IEEE 2010

13. Nederlof, J.: Fast polynomial-space algorithms using Möbius inversion: Improving on steiner tree and related problems. In: Albers, S., Marchetti-Spaccamela, A., Matias, Y., Nikoletseas, S., Thomas, W. (eds.) ICALP 2009, Part I. LNCS, vol. 5555, pp. 713–725. Springer, Heidelberg (2009)
14. Niedermeier, R.: Invitation to Fixed-Parameter Algorithms. Oxford University Press, Oxford (2006)
15. Park, J., Lee, S.: Keyword search in relational databases. Knowl. Inf. Syst. **26**(2), 175–193 (2010)

Reconstructing Cactus Graphs
from Shortest Path Information
(Extended Abstract)

Evangelos Kranakis[1], Danny Krizanc[2(✉)], and Yun Lu[3]

[1] School of Computer Science, Carleton University, Ottawa, ON, Canada
[2] Department of Mathematics and Computer Science, Wesleyan University,
Middletown, CT, USA
dkrizanc@wesleyan.edu
[3] Department of Mathematics, Kutztown University, Kutztown, PA, USA

Abstract. Imagine a disaster in which all edges of a network go down simultaneously. Imagine further that the routing tables of the nodes were not destroyed but are still available. Can one "reconstruct" the network from the routing tables? This question was first asked by Kranakis, Krizanc and Urrutia in 1995 as part of an attempt to answer the question of how much information about the network is stored in the node routing tables. In this paper, we answer this question in the affirmative if the underlying network is a cactus graph by providing an algorithm that returns all node-labelled cacti consistent with a given set of routing tables. This is the first such result for a class of non-geodetic graphs.

Keywords: Graph reconstruction · Shortest path information · Cactus graph

1 Introduction

Imagine a disaster in which all edges of a network go down simultaneously. Imagine further that the routing tables from each of the nodes were not destroyed but are still available. Can one "reconstruct" the network from the routing tables? This question was first asked by Kranakis et al. [19] in 1995 as part of an attempt to answer the question of how much information about the network is stored in the node routing tables. To be a bit more precise, assume we know the degree of each node, i, and for each other node, j, we know one of the edges adjacent to i which is on a shortest path from i to j. (If more than one edge is the first edge on a shortest path from i to j then one is chosen arbitrarily.) Is this "shortest path information" sufficient to reconstruct the underlying graph of the network?

The problem is similar in nature to many other "reconstruction" problems studied in graph theory. An early result along these lines is by Erdős and Gallai [9] who characterize those sequences that are realized as the degree sequence of a

Research supported in part an by NSERC grant.

© Springer International Publishing Switzerland 2016
R. Dondi et al. (Eds.): AAIM 2016, LNCS 9778, pp. 150–161, 2016.
DOI: 10.1007/978-3-319-41168-2_13

graph by providing an algorithm for reconstructing such a graph if one exists. The question of characterizing degree sequences for limited classes of graphs and finding graphs realizing a given sequence has a long history including for such classes as trees (folklore), split graphs [14, 23], 4-cycle-minor free graphs [27], unicyclic graphs [3], Halin graphs [5], 2-trees [7], and k-trees [21].

Perhaps the most famous of these reconstruction problems is the conjecture due to Kelly [18] and Ulam [28]: Given a multiset of n induced subgraphs generated by deleting each vertex of an n-node graph, can one reconstruct the graph? (Harary [15] posed the same question for edge deletions.) The problem remains open in general but a series of papers have solved the problem for a number of different classes of graphs starting with trees and regular graphs [16] and continuing with (among others) outerplanar graphs [12], unit interval graphs [29] and random graphs [6].

Perhaps most closely related to our reconstruction problem is that of reconstructing a graph from the set system of the neighborhoods of the graph's nodes (referred to as a star system) introduced by Sabidussi and Sos [13]. For this problem it is known that for general graphs the problem is NP-complete [2, 20] and for some classes of graphs it remains NP-complete but for others it becomes solvable in polynomial time [10]. Other related (but distinct from our version) reconstruction problems include those of reconstructing trees from edge splits [8] (with applications in phylogentics), reconstructing graphs from distance queries [22] (with applications in network tomography), reconstructing graphs from embedding requests [26] (with applications to security) and reconstructing graphs with a betweenness oracle [1].

In [19], Kranakis et al. show that their reconstruction problem is solvable for unique shortest path graphs (also known as geodetic graphs) by giving a polynomial time algorithm to construct such a node-labelled graph from its shortest path information. In fact, they show that in the "full information" model, where all the first edges on a path from i to j are provided for each i and j, there is a unique graph satisfying the set of constraints this provides. In unique shortest path graphs there is only one such edge, so the full information model is equivalent to having the routing tables (with a single first edge) and their algorithm for the full information model answers the question for this class of graphs. They also show that if the underlying graph includes an even length cycle, it is possible that a given set of routing tables may be realized by more than one labelled graph. A simple 4-cycle is sufficient to show this. Building on this example, they point out that a given set of routing tables (referred to as a shortest path set system below), may have as many as 2^k different labelled graphs consistent with it, if the graph contains k "independent" 4-cycles (sharing no edges).

In this paper, we provide an algorithm for reconstructing all possible node labelled graphs consistent with a given shortest path set system assuming the graphs are from the class of cactus graphs. A *cactus* graph is a connected graph in which any two simple cycles have at most one node in common. This is the first such result for graphs with potentially more than one shortest path between pairs of nodes, i.e., non-geodetic graphs. Interestingly, the degree sequences of cacti have been characterized [17] and it is one of the classes for which the Ulam's reconstruction problem is known to be solvable [11]. Cacti have been studied in their own right for applications in traffic flow [30], electrical circuits [24], wireless sensor networks [4] and genomics [25].

A *shortest path set system*, $S = \{(i_1, i_2, ..., i_n), F_{i_1}, F_{i_2}, ..., F_{i_n}\}$, (or set system, for short), is a set containing an n-tuple along with n sets indexed by the elements of the n-tuple. We refer to n as the *size* of S, denoted $|S|$. Each of the sets, F_i, is made up of one or more subsets of $V(S) = \{i_1, i_2, \ldots, i_n\}$ which we refer to as the *vertices* (or *nodes*) of S. I.e., each $F_i = \{F_{i,1}, F_{i,2}, \ldots, F_{i,k}\}$ for some $k > 0$. We refer to k as the $degree(i)$ and to the $F_{i,j}$'s as the *edge sets* of i. We say a set system is *valid* if for $1 \leq i \leq n$, we have

- $F_{i,l} \cap F_{i,k} = \emptyset$ for $1 \leq l \neq k \leq degree(i)$, and
- $\cup_{1 \leq l \leq degree(i)} F_{i,l} = \{i_1, i_2, ..., i_n\} \setminus \{i\}$.

In other words, the F_i's form a partition of $V(S) \setminus \{i\}$. For a valid set system we denote by F_i^j the unique $F_{i,t}$ such that $j \in F_{i,t}$, where $j \neq i$.

We say an n-node graph G with vertex set $V(G) = \{i_1, i_2, \ldots, i_n\}$ is *consistent* with a valid set system $S = \{(i_1, i_2, ..., i_n), F_{i_1}, F_{i_2}, ..., F_{i_n}\}$ if and only, for all $i \in V(S)$, there is a one-to-one mapping, ϕ_i, of the edge sets of F_i to the edges adjacent to node i in G, with the following property: $\phi(F_i^j)$ is the first edge on a shortest path from i to j in G. In this way we can think of the set system as encoding a possible set of routing tables for the nodes of G, providing the first edge on a shortest path to each possible destination node. (Note: our graphs are unweighted, i.e., the length of a path is the number of edges in it.) Given a valid set system, we are interested in finding those node-labelled graphs that are consistent with it, if any.

As was pointed out earlier, [19] showed that even length cycles can cause problems when trying to reconstruct a graph from its shortest path set system. In particular, there is a set system that is consistent with two different node labelled 4-cycles (see Fig. 1). It turns out, if the underlying graph must be a cactus, it is only the 4-cycles that can cause such problems. Odd cycles do not introduce multiple shortest paths between nodes and we show below that even cycles of length 6 or more can be uniquely determined from their shortest path information. Building upon this, we are able to provide an algorithm that given a shortest path set system, reconstructs all of the node-labelled cacti consistent with the system or reports that none exists. The cycles in cacti are all edge-disjoint so that each restriction of a set system to 4 nodes consistent with a 4-cycle has the potential to double the number of cacti consistent with a given set system. If we let k be the number of 4-cycles in a cactus, the number of node-labelled cacti consistent with a set system derived from that cactus can be as large as 2^k depending upon how nodes are assigned to edge sets in the case where multiple shortest paths exist. We show our algorithm runs in time $O(n^6 + 2^k n^3)$ if the resulting reconstructed cacti have k 4-cycles. While this is not polynomial in general it is polynomial in the worst case size of the output.

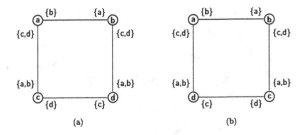

Fig. 1. Two node-labelled 4-cycles consistent with one set system.

If we limit ourselves to the class of cacti with a fixed number of 4-cycles the algorithm becomes worst-case polynomial.

In the next section, we present our algorithm. This is followed by a proof of its correctness and an analysis of its runtime. We end with some conclusions and open problems.

2 The Algorithm

In this section we give an overview of the algorithm, followed by detailed pseudocode for the main procedure. Details of the sub-procedures are provided in an appendix. The input to our algorithm is a valid shortest path set system potentially representing a cactus graph, $S = \{(i_1, i_2, ..., i_n), F_{i_1}, F_{i_2}, ..., F_{i_n}\}$ and the output is the set of all n-node-labelled cacti that are consistent with S, if any. Before we begin our algorithm description we introduce a little more notation.

It is well-known that the nodes of a cactus can be partitioned into three types. *Cycle* nodes are nodes of degree two that appear on exactly one cycle. *Non-cycle* nodes are nodes that are not included in any cycle. The remaining nodes are called *hinges*. Observe that a hinge is of degree greater than 2 and appears on at least one cycle. We refer to non-cycle nodes of degree one as *leaves* and a cycle containing exactly one hinge as a *leaf cycle* (see Fig. 2).

The following lemma is easily established from the definitions.

Lemma 1. *Let G be a cactus graph. One of the following must hold:*

1. *G consists of a single vertex and no edges.*
2. *G consists of a simple cycle of length three or more.*
3. *G has a leaf.*
4. *G has a leaf cycle.*

We denote by $S \setminus A$, where $A \subset \{i_1, i_2, ..., i_n\}$, the set system formed by removing the vertices associated with A wherever they appear in S. Note that if S is valid then $S \setminus A$ is valid as well. By S_A, where $A \subset \{i_1, i_2, ..., i_n\}$, we mean the set system S restricted to only vertices in A, i.e., $S \setminus (\{i_1, i_2, ..., i_n\} \setminus A)$. We represent a leaf of a cactus graph by the directed edge (i, j) from the parent i to the leaf j. Given a graph G and a leaf (i, j), where $i \in V(G)$ and $j \notin V(G)$, the

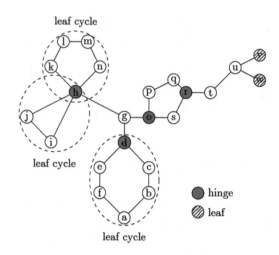

Fig. 2. Cactus graph with leaves, leaf cycles and hinges.

graph $G + (i, j)$ is G with the vertex j added to $V(G)$ and the edge $\{i, j\}$ added to $E(G)$. We represent a leaf cycle of a cactus graph by a pair (h, C) where h is the hinge of the cycle and C is the cycle through h. Given a graph, G, and a hinge (h, C), where $h \in V(G)$ and $V(G) \cap V(C) = \{h\}$, the graph $G + (h, C)$ is G with the cycle C added to h.

Based on Lemma 1 our algorithm is recursive. Their are two base cases:

1. the graph consistent with S consists of a single node, i.e., $G = (\{i\}, \{\{\}\})$ or
2. it consists of a simple cycle of length 3 or more.

Otherwise, if S represents a set system of a cactus, G, one of the following must hold:

3. G contains a leaf node or
4. G contains a leaf cycle.

Using the degree sequence implied by S, the algorithm checks for each of these cases in turn. If all of the nodes are of degree 2 then the only (connected cactus) consistent with S is a cycle of length 3 or more and the algorithm checks to see if S satisfies the necessary constraints. Note: if the set system is of size 4, then it is possible that two cycles may be consistent with S. In this case, both cycles are returned. If there are nodes of degree one then these must be leaves of the cactus. For a given leaf, (i, j), the algorithm recursively finds the set of cacti, $\mathcal{C}$, consistent with $S \setminus \{j\}$ and returns the set of $G + (i, j)$, such that $G \in \mathcal{C}$ and $G + (i, j)$ is consistent with S. If there are no leaves and there are nodes of degree greater than 2, the algorithm checks for a potential leaf cycle. If a leaf cycle (h, C) is found, the algorithm recursively finds the set of cacti, $\mathcal{C}$, consistent with $S \setminus (V(C) \setminus \{h\})$ and returns the set of $G + (h, C)$, such that $G \in \mathcal{C}$ and $G + (h, C)$ is consistent with S. Note: if the leaf cycle is of length 4 and there are two cycles consistent with it, both are checked against all the cacti returned after its removal.

The main complications in the above come from (1) checking if S is consistent with a cycle and finding that cycle and (2) finding a leaf cycle if one exists.

To check if a set system, S, (with all F_i of size 2) is consistent with being a cycle comes down to three cases:

1. $|S| = 4$,
2. $|S|$ is odd or
3. $|S| > 4$ and even.

If $|S| = 4$ then each F_i consists of 2 sets, one a singleton, the other of size 2. Let $S = \{(a,b,c,d), \{A_1, A_2\}, \{B_1, B_2\}, \{C_1, C_2\}, \{D_1, D_2\}, \}$. Observe that each singleton determines an edge that must exist if S is to be consistent with a cycle of length 4. E.g., if $A_1 = \{c\}$ then the edge $\{a, c\}$ must be part of the cycle. If $A_1 = \{c\}$ and $C_1 = \{a\}$ they both determine the same edge, $\{a, c\}$. Regarding the determined edges there are only 3 possible cases:

1. The determined edges include a triangle. In this case, S is not consistent with a 4-cycle and $\emptyset$ is returned.
2. The number of determined edges is 3 or 4 (without a triangle). In this case, there is a unique 4-cycle consistent with S and it is returned (see Fig. 3).
3. The number of determined edges is 2. In this case, there are two 4-cycles consistent with S and both are returned (see Fig. 2).

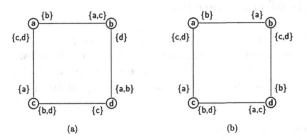

(a)   (b)

Fig. 3. Examples of set systems with a unique consistent 4-cycle.

If $|S|$ is odd, the cycle is a unique shortest path graph and using techniques from [19] we can determine if S is consistent with a cycle as well as determine the edges of the unique such cycle.

For the last case, we have $|S| = 2k$ with $k > 2$. In this case we pick an arbitrary vertex of S, say s, with corresponding edge sets $F_s = \{A, B\}$ where $|A| = k - 1$ and $|B| = k$. When S is restricted to the smaller of these along with s, it must form a path starting at s, a unique shortest path graph, and again by [19] we can determine the complete path if it exists. The second node in this path must introduce either one or two nodes of B into its edge set not containing s. If one node, x, is introduced then it must be at distance k from s and we can determine the path starting at s and containing the nodes $B \setminus \{x\}$, if it exists. If two nodes, x and y, are introduced we can determine which, if either, is the node at distance k from s by checking which is adjacent to the last node in the path formed by the vertices in A. Say it is x. We can again determine the path starting at s and containing the nodes of $B \setminus \{x\}$ using the fact they must form a unique shortest path graph. Finally we can confirm that the two paths starting at s, along with x, form a cycle and if so, return the cycle, otherwise the empty set.

Once all of the leaves are dealt with, if what remains is not a single vertex or a cycle, it must contain a leaf cycle. The first step is to find all of the potential hinges and check each in turn to see if it is a hinge of leaf cycle. A node is a potential hinge if it has degree greater than 2 and it contains two edge sets of the same size (odd cycle) or that differ by one (even cycle) all of whose vertices are of degree 2. For each potential pairing of edge sets (the same size or differing by one) we check if they, together with the potential hinge, form a cycle. If a cycle is found, the cycle along with its hinge is returned as a potential leaf cycle. If the cycle is of length 4 and consistent with two different cycles, both potential leaf cycles are returned. If no leaf cycles are found, the empty set is returned.

The detailed pseudocode of the main procedure appears as Algorithm 1. The pseudocode for the remaining subroutines may be found in the full paper. In the next section we prove the algorithm's correctness and analyze its run time. Proofs omitted due to space constraints may be found in the full paper.

3   Analysis of the Algorithm

In order to prove our main result we will need some lemmas. The following obvious fact is used implicitly in many of the arguments below: Say G is consistent with S. Let H be an induced connected subgraph of G with vertex set $V(H)$. Then H is consistent with $S_{V(H)}$. We encapsulate its contrapositive in the following lemma:

Lemma 2. *Say that $\mathcal{G} = \{G_1, G_2, \ldots, G_t\}$ is the set of all connected graphs consistent with some S_A, $A \subseteq V(S)$. Further assume G is consistent with S. Then if the subgraph induced by A in G is connected it is in $\mathcal{G}$.*

Observe that if G is consistent with S, the mappings ϕ_i are implied by the correspondence of the vertex sets of G and S as, if $\{i, j\}$ is an edge then $j \in \phi_i^{-1}(\{i, j\})$ (also, $i \in \phi_j^{-1}(\{i, j\})$) and thus we know the edge set corresponding to this adjacency. Further observe that given a graph G and a potential set system S for G, it is straightforward to verify if G is consistent with S. The following lemma is easily established:

Lemma 3. *CheckGraph(G, S), returns "true" if G is consistent with S, "false" otherwise. Furthermore it runs in $O(mn)$ time where m is the number of edges in G.*

Algorithm 1. ConstructCactus(S)

1: Compute the degree sequence implied by S
2: $Cacti = \emptyset$
3: **if** $S = \{(i), \{\{\}\}\}$ **then**
4:     **return** $\{G\}$ where $G = (\{i\}, \{\})$
5: **else if** every vertex has degree 2 **then**
6:     $C = CheckCycle(S)$
7:     **for** $c \in C$ **do**
8:         **if** $CheckGraph(c, S)$ **then**
9:             $Cacti = Cacti \cup \{c\}$
10:     **return** $Cacti$
11: **else if** there exists a vertex of degree 1 **then**
12:     $leaf = FindLeaf(S)$
13:     **if** $leaf$ **then**
14:         $G = ConstructCactus(S \setminus leaf)$
15:         **for** $g \in G$ **do**
16:             **if** $CheckGraph(g + leaf, S)$ **then**
17:                 $Cacti = Cacti \cup \{g + leaf\}$
18:     **return** $Cacti$
19: **else**
20:     $leafcycle = FindLeafCycle(S)$
21:     **if** $leafcycle$ **then**
22:         **for** $l \in leafcycle$ **do**
23:             $G = ConstructCactus(S \setminus l)$
24:             **for** $g \in G$ **do**
25:                 **if** $CheckGraph(g + l, S)$ **then**
26:                     $Cacti = Cacti \cup \{g + l\}$
27:     **return** $Cacti$

The following observation follows from the discussion in [19] and proves the correctness of a function that is used extensively in our algorithm.

Lemma 4. *If $CheckEdge(a, b, S)$, returns "false" then $\{a, b\}$ is not an edge in any G consistent with S. If it returns "true", then $\{a, b\}$ is potentially an edge in some G consistent with S but need not be.*

The remaining lemmas establish the correctness of our sub-procedures whose pseudo-code may be found in the full version of the paper.

Lemma 5. *Given a set system S, a distinguished vertex, s, of degree 1 in S, and the edge set A of s, $FindPath(s, A, S)$, returns the unique path consistent with S, starting at s and containing all of the nodes of A, if such a path exists. Furthermore, the algorithm runs in $O(n^2)$ time.*

Lemma 6. *Given a set system S with all vertices of degree 2, $CheckCycle(S)$, returns all cycles that are potentially consistent with S. If $|S| = 4$, there may be two such cycles. Otherwise, the potential cycle is unique. Furthermore, the algorithm runs in $O(n^2)$ time.*

Lemma 7. *Given a set system S with at least one vertex of degree 1, the procedure $FindLeaf(S)$, returns a potential leaf, if one exists. In any graph consistent with S, the returned vertex must be a leaf with the given parent. Furthermore, the algorithm runs in $O(n^2)$ time.*

Lemma 8. *Given a set system S and a potential hinge (a node of degree greater than 2 with at least two edge sets containing nodes all of degree 2) h, the procedure $CheckHinge(h, S)$, returns a potential leaf cycle with hinge h, if one exists. If there are two such cycles, both are returned. In any graph consistent with S, the returned cycle(s) must be a leaf cycle with the given hinge. Furthermore, the algorithm runs in $O(n^4)$ time.*

Lemma 9. *Given a set system S with no vertices of degree 1 and at least one vertex of degree 3, $FindLeafCycle(S)$, returns a potential leaf cycle, if one exists. If there are two such cycles, both are returned. In any graph consistent with S, the returned cycle(s) must be a leaf cycle with the given hinge. Furthermore, the algorithm runs in $O(n^5)$ time.*

We are now ready to prove our main result:

Theorem 1. *Given a valid set system S, $ConstructCactus(S)$, outputs all node-labelled cacti consistent with S. Furthermore, if the resulting cacti have at most k 4-cycles the algorithm runs in time $O(\max\{n^6, 2^k n^3\})$.*

PROOF. First observe that before returning any graph, G, the algorithm calls $CheckGraph(G, S)$ which by Lemma 3 correctly verifies that G is consistent with S. I.e., all graphs returned by the algorithm are guaranteed to be consistent with S. We need only show that all cacti consistent with S are returned.

The proof of this is divided into two cases: S is consistent with simple cycle or not. If S is consistent with a simple cycle then $|S| > 2$ and all of its vertices have degree 2. In this case, $ConstructCactus(S)$ will call $CheckCycle(S)$ which by Lemma 6 correctly returns all possible cycles consistent with S, either 0, 1 or 2. These are checked for consistency and those that are consistent are returned.

The remaining case is shown by induction on the size of S. The case $|S| = 1$ follows trivially. Assume the algorithm correctly outputs all cacti consistent with set systems of up to size $n - 1$ (including simple cycles by the first case). We show that it works correctly for set systems of size n, where $n > 1$.

Since S is not consistent with a cycle of length n, by Lemma 1 we know that to be consistent with any cacti, it must either have a leaf or a leaf cycle. If it has a leaf, then it must have a vertex of degree 1. In this case, the procedure $FindLeaf(S)$ is called and by Lemma 7, if there is a potential leaf in S it is returned. If there is no potential leaf but S has a vertex of degree 1 then there are no cacti consistent with S and the empty set is returned. If there is a potential leaf, $leaf$, then it is removed from S forming a set system S' of size $n - 1$. By induction, calling $ConstructGraph(S')$ will find all cacti consistent with S'. In any cactus consistent with S, $leaf$ must be a leaf of the cactus and that cactus without the leaf must be consistent with S'. Therefore, by Lemma 2, inserting the leaf into each cactus returned and verifying the result is consistent with S produces all of the cacti consistent S.

If S (of size greater than 1) has no leaves (and is not consistent with a cycle) it must contain a leaf cycle. In this case, the algorithm will call $FindLeafCycle(S)$ and by Lemma 9, if there is a restriction of S consistent with a leaf cycle, it will be found. (If it is of size 4, possibly two will be found.) Let L be the set of leaf cycles returned by $FindLeafCycle(S)$. If $|L| = 0$ then there can be no cactus consistent with S and the empty set is returned. Otherwise, the algorithm makes recursive calls on S' equal to S with the nodes of the leaf cycle (but not the hinge) removed, finding, by induction, all cacti consistent with the smaller set system. Note: if $|L| = 2$, both cycles are on the same vertex set and removing one results in the same S as removing the other. In any cacti consistent with S, the potential leaf cycle must be a leaf cycle and its removal must be consistent with S'. Therefore, by Lemma 2, inserting the leaf cycle (or both, separately, if there are two) into each cactus returned and verifying the result is consistent with S produces all cacti consistent with S.

The algorithm makes at most $n - 1$ recursive calls. The cost of such a call is dominated by either the cost of finding a leaf cycle ($O(n^5)$) or the cost of checking all of the possible graphs returned for consistency. Since each 4-cycle at most doubles the number of potential graphs this is bounded by $O(2^k n^2)$. (Note: cacti are planar and therefore have $O(n)$ edges). ∎

4   Conclusions

We have shown the first reconstruction result from shortest path information for a class of graphs that potentially includes graphs with multiple shortest paths between pairs of nodes, namely cacti. The runtime of our reconstruction algorithm is not polynomial in the size of the input but is polynomial in the potential worst case size of the output. Unfortunately, the final size of the output may be much smaller than the total number of graphs investigated by the algorithm. Ideally, the algorithm would be truly *output sensitive*, running in time polynomial in the actual number of graphs consistent with the input. On the other hand, it would be straightforward to modify our algorithm to solve the problem for the class of cacti with a bounded number of 4-cycles resulting in a fully polynomial algorithm in that case. The modified algorithm would keep track of how many 4-cycles it has found and exit as soon as this exceeded the bound. For example, such algorithm with a bound of 0 would run in $O(n^6)$ time and would reconstruct the unique cactus among the class of 4-cycle-free cacti consistent with a given set system, if one exists. With judicious use of preprocessing and clever data structures, the runtime of the algorithm could no doubt be significantly reduced but our goal here is only to establish it to be polynomial.

Using Lemma 3, one can show that the problem of deciding if there exists a graph consistent with a given set system is in NP. It would be interesting to determine if the problem is NP-complete or if there is a polynomial time algorithm to answer this question for general graphs or for other classes of graphs besides geodetic graphs or 4-cycle-free cacti.

References

1. Abrahamsen, M., Bodwin, G., Rotenberg, E., Stöckel, M.: Graph reconstruction with a betweenness oracle. In: 33rd Symposium on Theoretical Aspects of Computer Science (STACS 2016). Leibniz International Proceedings in Informatics (LIPIcs), vol. 47, pp. 5:1–5:14 (2016)
2. Aigner, M., Triesch, E.: Reconstructing a graph from its neighborhood lists. Comb. Probab. Comput. **2**, 103–113 (1993)
3. Beineke, L., Schmeichel, E.: Degrees and cycles in graph. In: 2nd International Conference on Combinatorial Mathematics, pp. 64–70 (1979)
4. Ben-Moshe, B., Dvir, A., Segal, M., Tamir, A.: Centdian computation in cactus graphs. J. Graph Algorithms Appl. **16**(2), 199–224 (2012)
5. Bıyıkoğlu, T.: Degree sequences of Halin graphs, and forcibly cograph-graphic sequences. Ars Combinatoria **75**, 205–210 (2005)
6. Bollobas, B.: Almost every graph has reconstruction number three. J. Graph Theor. **14**, 1–4 (1990)
7. Bose, P., Dujmović, V., Krizanc, D., Langerman, S., Morin, P., Wood, D., Wuhrer, S.: A characterization of the degree sequences of 2-trees. J. Graph Theor. **58**(3), 191–209 (2008)
8. Buneman, P.: The recovery of trees from measures of dissimilarity. In: Mathematics in the Archaeological and Historical Sciences, pp. 387–395. Edinburgh University Press (1971)
9. Erdos, P., Gallai, T.: Graphs with prescribed degrees of vertices. Mat. Lapok **11**, 264–274 (1960)
10. Fomin, F., Kratochivil, J., Lokshtanov, D., Mancini, F., Telle, J.A.: On the complexity of reconstructing H-free graphs from their star systems. J. Graph Theor. **68**, 113–124 (2011)
11. Geller, D., Manvel, B.: Reconstruction of cacti. Can. J. Math. **21**(6), 1354–1360 (1969)
12. Giles, W.: The reconstruction of outerplanar graphs. J. Comb. Theor. Ser. B **16**(3), 215–226 (1974)
13. Hajnal, A., Sos, V. (eds.): Combinatorics. In: Colloquia Mathematica Societatis Janos Bolya 18, vol. II. North-Holland (1978)
14. Hammer, P., Simeone, B.: The splittance of a graph. Combinatorica **1**, 275–284 (1981)
15. Harary, F.: On the reconstruction of a graph from a collection of subgraphs. In: Theory of Graphs and its Applications (Proceedings Symposium Smolenice, 1963), pp. 47–52 (1964)
16. Harary, F.: A survey of the reconstruction conjecture. In: Bari, R.A., Harary, F. (eds.) Graphs and Combinatorics. Lecture Notes in Mathematics, vol. 406, pp. 18–28. Springer, Heidelberg (1974)
17. Havel, V.: Eine Bemerkung über die Existenz der endlichen Graphen. Casopis Pest. Mat. **80**, 477–480 (1955)
18. Kelly, P.: A congruence theorem for trees. Pac. J. Math. **7**, 961–968 (1957)
19. Kranakis, E., Krizanc, D., Urrutia, J.: Implicit routing and shortest path information (Extended Abstract). In: Colloquium on Structural Information and Communication Complexity, pp. 101–112 (1995)
20. Lalonde, F.: Le problem d'toiles pour graphes est NP-complet. Discrete Math. **33**, 271–280 (1981)

21. Lotker, Z., Majumdar, D., Narayanaswamy, N.S., Weber, I.: Sequences characterizing k-trees. In: Chen, D.Z., Lee, D.T. (eds.) COCOON 2006. LNCS, vol. 4112, pp. 216–225. Springer, Heidelberg (2006)

22. Mathieu, C., Zhou, H.: Graph reconstruction via distance oracles. In: Fomin, F.V., Freivalds, R., Kwiatkowska, M., Peleg, D. (eds.) ICALP 2013, Part I. LNCS, vol. 7965, pp. 733–744. Springer, Heidelberg (2013)

23. Merris, R.: Split graphs. Eur. J. Comb. **24**, 413–430 (2003)

24. Nishi, T.: On the number of solutions of a class of nonlinear resistive circuits. In: International Symposium on Circuits and Systems, pp. 766–769. IEEE (1991)

25. Paten, B., Diekhans, M., Earl, D., John, J., Ma, J., Suh, B., Haussler, D.: Cactus graphs for genome comparisons. J. Comput. Biol. **18**(3), 469–481 (2011)

26. Pignolet, Y., Schmid, S., Tredan, G.: Adversarial VNet embeddings: a threat for ISPs? In: INFOCOM, pp. 415–419. IEEE (2013)

27. Rao, R.: Degree sequences of cacti. In: Rao, S. (ed.) Combinatorics and Graph Theory. Lecture Notes in Mathematics, vol. 885, pp. 410–416. Springer, Heidelberg (1981)

28. Ulam, S.: A Collection of Mathematical Problems. Wiley, New York (1960)

29. von Rimscha, M.: Reconstructibility and perfect graphs. Discrete Math. **47**, 283–291 (1983)

30. Zmazek, B., Žerovnik, J.: Estimating the traffic on weighted cactus networks in linear time. In: Ninth International Conference on Information Visualisation, pp. 536–541. IEEE (2005)

Near-Optimal Dominating Sets via Random Sampling

Martin Nehéz[✉]

Institute of Information Engineering, Automation and Mathematics,
Faculty of Chemical and Food Technology,
Slovak University of Technology in Bratislava,
Radlinského 9, 812 37 Bratislava, Slovak Republic
martin.nehez@stuba.sk

Abstract. A minimum dominating set (MDS) of a simple undirected graph G is a dominating set with the smallest possible cardinality among all dominating sets of G and the MDS problem represents the problem of finding the MDS in a given input graph.

Motivated by the transportation, social and biological networks from a control theory perspective, the main result of this paper is the assertion that a random sampling is usable to find a near-optimal dominating set in an arbitrary connected graph. Our result might be of significance in particular contexts where exact algorithms cannot be run, e.g. in distributed computation environments. Moreover, the analysis of the relationship between the time complexity and the approximation ratio of the corresponding sequential algorithm exposes the counterintuitive behavior.

1 Introduction

A dominating set is a structure in graph theory which have been originally examined regarding optimization problems in operations research. Namely, the optimal placement of key facilities in traffic networks is an instance of network location problems known for decades [8]. The computation of the minimum dominating set (abbreviated MDS) is a classical discrete optimization problem which has become fairly pervasive in various other areas, e.g. in sociology [10], mobile ad-hoc networks or sensor networks [1]. Recently it has found applications in the network controlling theory being exploited for viral marketing in social networks, in biology and medicine of cancer therapy [13,14,18,19].

The MDS problem is one of the first to be classified as NP-hard [4] and to approximate it to within a logarithmic factors is NP-hard as well [16]. If $P \neq NP$ then there is a logarithmic lower bound for the approximability of such a problem even for the class of power-law graphs when the scaling exponent $\gamma > 2$ [5]. The k-dominating set problem was the first one to be shown W[2]-complete in terms of designing FPT algorithms [3]. The hardness results imply that the corresponding optimal (i.e. exact) algorithms are all of the (moderately) exponential-time complexity and fast polynomial heuristics very often compute outputs which are far from optimum; see [11,14] for short surveys. Yet the fastest exact algorithm was

© Springer International Publishing Switzerland 2016
R. Dondi et al. (Eds.): AAIM 2016, LNCS 9778, pp. 162–172, 2016.
DOI: 10.1007/978-3-319-41168-2_14

invented by van Rooij and Bodlaender [17] and fast optimal algorithm for sparse graphs is based on the integer linear programming (shortly ILP) which uses the branch-and-bound method [14,18]. However, the ILP-optimization seems to be unusable for dense graphs due to high memory consumption [14]. Nevertheless, it was shown in [15] that there are randomized algorithms which are able to compute, with high probability, a near-optimal dominating sets with a constant approximation ratio in polynomial time for Erdös-Rényi random graphs $G(n, 1/2)$.

The main result of this paper is the lower bound on the probability that a random set with a given number of vertices (greater than the size of the corresponding MDS) is a dominating set. It implies that a simple Monte Carlo randomized algorithm can compute a near-optimal dominating set with high probability for an arbitrary class of undirected connected graphs (i.e. it is not restricted for e.g. random graphs only). The proof of this result essentially relies on the "replacing argument" which could be explained as the fact that some vertices of a MDS may be replaced by other suitable vertices outside the MDS. Our emphasis is also focused on the analysis of the algorithm's time complexity[1]. Namely, it is moderately exponential, i.e. $O^*(t^n)$, where $1 \leq t \leq 2$ and t depends both on the constant approximation ratio ρ and the domination number γ of an input graph. As regards the time complexity, it is comparable with the $O(1.4969^n)$ exact algorithm [17] and for some particular values of γ and ρ it is even better. Although the time complexity is not polynomial, our result has more significance for unsequential computation, e.g. in distributed environments.

2   Definitions and Preliminaries

2.1   Graph Theory

Let $G = (V, E)$ be a simple undirected graph which consists of a non-empty n-element set $V(G)$ of *vertices* (or *nodes*) and a finite set $E(G)$ of unordered pairs of distinct vertices called *edges*. For a vertex $v \in V(G)$, its degree (the number of adjacent vertices of v) is denoted by $\deg(v)$. The maximum (minimum) degree of a graph G, denoted by Δ (δ), is the maximum (minimum) degree over all its vertices. Note that the inequality $0 \leq \delta \leq \Delta \leq n - 1$ is true for any G. The distance of two vertices $u, v \in V(G)$ is denoted by $\mathrm{dist}(u, v)$. The diameter of G is the maximum distance over all pairs of nodes in G. The diameter is denoted by $\mathrm{diam}(G)$. For any $v \in V(G)$, let $N(v) = \{ w \mid \mathrm{dist}(w, v) = 1 \}$ be the *open neighborhood* of a vertex v. Note that it holds $|N(v)| = \deg(v)$ for each $v \in V(G)$. Let $N[v] = N(v) \cup \{v\}$ be the *closed neighborhood* of $v \in V(G)$. The closed neighborhood is called the *ball of radius 1* (or *1-ball*) *centered at v* as well. Its extension to sets of nodes is defined as:

$$N[S] = \bigcup_{u \in S} N[u],$$

where $S \subseteq V(G)$.

---

[1] For the time complexity, it is used a modified big-Oh notation O^* throughout this paper. For functions f and g, we write $f(n) = O^*(g(n))$ if $f(n) = O(g(n)poly(n))$, where $poly(n)$ is a polynomial.

Given a graph G, a set $V_D \subseteq V(G)$ is said to be a *dominating set* of G if every vertex $v \in V$ is either in V_D or adjacent to some vertex in V_D, i.e. it holds $N[V_D] = V(G)$. The *domination number* $\gamma(G)$ is the minimum cardinality of a dominating set of G. For subsets S and T of $V(G)$, we say that S *dominates* T if S is a dominating set of the subgraph induced by $S \cup T$.

In the *Minimum Dominating Set* problem, one is given a graph G and is asked to compute a dominating set of G of minimum cardinality. This problem is NP-hard [4], thus we cannot expect to find a polynomial time algorithm for it.

2.2   Approximation Algorithms and Their Performance

According to [6], an optimization problem Π is a set of tuples $\{(I, Sol(I), m_I, Opt(I))\}$ over all possible instances I, where:

- $Sol(I)$ is the set of feasible solutions to I,
- $m_I : Sol(I) \to \mathbb{R}$ is the measure function associated with I, and
- $Opt(I) \subseteq Sol(I)$ are the feasible solutions with optimal measure.

In case of the MDS problem, instances are all graphs $G = (V, E)$, feasible solutions are all possible subsets of vertices in a graph G, m_I is a cardinality of an element of $Sol(I)$ and finally, elements of $Opt(I)$ are all subsets of vertices of G with the minimum cardinality. Let A be an algorithm which feasibly solves the MDS problem, i.e. for a given instance $G \in$ MDS, $A(G) \in Sol(G)$. The *approximation ratio* $\rho(A)$ of A is:

$$\rho(A) = \min_{G \in MDS} \frac{|A(G)|}{|Opt(G)|},$$

where $|Opt(G)| = \gamma(G)$.

2.3   The Monte Carlo Algorithms

The common framework for the Monte Carlo algorithms is described as follows.

Input: I, a positive integer k
Output: $\mathcal{A}(I) \in Sol(I)$

1. for $i = 1$ to k do
      - randomly select one element out of all possible elements;
      - if a result is found **then** write the result to output;
2. endfor

The Monte Carlo algorithm $\mathcal{A}$.

The following property is valid for the algorithm $\mathcal{A}$.

Lemma 1. *Let $\mathcal{A}$ be a Monte Carlo randomized algorithm described above. Let π_1 $(0 < \pi_1 < 1)$ be a probability that a single iteration of $\mathcal{A}$ in step 1 finds a solution for the input I. Let $\varphi : \mathbb{N} \to \mathbb{N}$ be an arbitrary constant or slowly increasing function such that $\varphi(n) = \Omega(1)$. If the number of iterations of the algorithm $\mathcal{A}$ is set to $k = \varphi(n)/\pi_1$ then:*

$$\Pr[\, \mathcal{A} \text{ finds a solution after } k \text{ iterations} \,] \geq 1 - e^{-\varphi(n)}.$$

3   The Main Result

For any simple connected graph $G = (V, E)$, and r such that $\gamma(G) \leq r \leq n$, let $V_D \subseteq V(G)$ be an r-node set in which each its element is selected uniformly at random out of $V(G)$. Let us consider that the selection of any V_D is independent of each other. Such a way, V_D is a result of a random sampling over $V(G)$. Let us define the following property:

$$A_r \;=\; \text{``}V_D \text{ is a dominating set of } G\text{''}.$$

The fundamental part of this section contains the estimation of the probability $\Pr[A_r]$. The main idea of our analysis is based on the *replacing argument* which is explained in proof of the following lemma.

Lemma 2. *Let $G = (V, E)$ be a graph with n vertices and let $\gamma = \gamma(G)$ be the domination number of G. If*

$$\gamma \geq s \geq \frac{r - \gamma}{\Delta - 1}$$

then:

$$\Pr[A_r] \geq \frac{\binom{\gamma}{s}}{\binom{n}{r}}.$$

Proof. Let V_D be a randomly selected r-set of vertices from $V(G)$ and let D_{opt} be an optimal dominating set of G, i.e. $|D_{opt}| = \gamma$. Let us denote the difference of sets D_{opt} and $(D_{opt} \cap V_D)$ by W, i.e.,

$$W = D_{opt} \setminus (D_{opt} \cap V_D).$$

(It is not necessary to distinguish whether $D_{opt} \cap V_D = \emptyset$ or not.) In order to evaluate the probability $\Pr[A_r]$ we shall estimate the number of all positive events which is, in our setting, the number of all dominating sets with the cardinality r in G. To do so, let us propose the *replacing argument* which is based on the following idea: if some vertex $w \in W$ is not in V_D then w must be replaced in V_D by such set of vertices which dominates at least the same part of G as w. Roughly speaking, if $w \in W$ is not in V_D then it can be replaced by such suitable vertices outside W which belong to V_D, see Fig. 1. Such a way, for each $w \in W$ all vertices from the open neighborhood $N(w)$ should be occurred in V_D. Note that if $N(w) \subseteq V_D$ then $N(w)$ dominates at least the same part of G as the vertex w. In particular, all vertices from $N(w)$ dominate w, see Fig. 1. In the

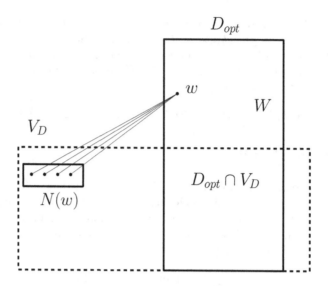

Fig. 1. The sets V_D, D_{opt} and $W = D_{opt} \setminus (D_{opt} \cap V_D)$, respectively. Their cardinalities are: $|V_D| = r = \rho\gamma$, $|D_{opt}| = \gamma$ and $|W| = s$, respectively.

worst case, for each $w \in W$ it is necessary to include all vertices from $N(w)$ into V_D. It means that for each $w \in W$ the cardinality of V_D is increased to at most $|N(w)| = \deg(w) \leq \Delta$. Since the replacing argument does not involve vertices from the intersection $D_{opt} \cap V_D$, the following inequality holds:

$$|V_D| \leq |D_{opt} \cap V_D| + \sum_{w \in W} \deg(w). \qquad (1)$$

Let $|W| = s$. By the definition of W, W and $(D_{opt} \cap V_D)$ are disjoint and thus

$$|D_{opt} \cap V_D| + |W| = |D_{opt}| = \gamma.$$

It turns out that $|D_{opt} \cap V_D| = \gamma - s$ and by substitution into (1), it holds:

$$r \leq \gamma - s + \sum_{w \in W} \deg(w). \qquad (2)$$

For all $w \in W$, we use the inequality $\deg(w) \leq \Delta$ and thus in Eq. (2) is rewritten into the following form:

$$r \leq \gamma - s + s \cdot \Delta.$$

This is used to claim the lower bound on s:

$$s \geq \frac{r - \gamma}{\Delta - 1}. \qquad (3)$$

Recall that if in Eq. (3) holds then we may use the replacing argument to determine V_D as a dominating set of G. We shall count the number of all positive events for a random selection of V_D. The vertices of V_D are unambigously determined by the choice of the set W. Clearly, if W is chosen then its complement in D_{opt}, i.e. the set $(D_{opt} \cap V_D)$, is chosen as well. The remaining vertices of V_D are determined by the replacing argument. Consequently, the number of positive events is at least the number of possibilities how to choose vertices in W out of D_{opt}. It is:

$$\binom{\gamma}{s}.$$

The number of all possibilities is

$$\binom{n}{r},$$

because it is the number of ways how to choose a set V_D with r vertices out of all vertices from $V(G)$. The resulting lower bound on $\Pr[A_r]$ is the ratio of the above binomial coefficients. □

The main result is the following assertion.

Theorem 1. *Let G be an arbitrary connected graph with n vertices. Let $\lambda = \gamma(G)/n$ and let $\rho = r/\gamma(G)$. It holds:*

$$\Pr[A_r] \geq \left[(\lambda\rho)^{\lambda\rho}(1 - \lambda\rho)^{(1-\lambda\rho)} \right]^n \rho^{1/2}(1 - \lambda\rho)^{1/2}.$$

Proof. According to [7], for each connected graph G with n vertices the upper bound on its domination number is $\gamma(G) \leq n/2$. Clearly, $0 < \lambda \leq 1/2$. By Lemma 2 and by the Stirling's approximation[2] of $n!$, it holds:

$$\Pr[A_r] \geq \frac{\gamma! \, r! \, (n-r)!}{n! \, s! \, (\gamma - s)!} \approx \sqrt{\frac{\gamma r(n-r)}{ns(\gamma - s)}} \cdot \frac{\gamma^\gamma r^r (n-r)^{n-r}}{s^s n^n (\gamma - s)^{\gamma - s}}$$

$$= \left(\frac{r(1 - r/n)}{s(1 - s/\gamma)} \right)^{1/2} \cdot \gamma^s r^r s^{-s} n^{-r} \left(1 - \frac{s}{\gamma} \right)^{s-\gamma} \left(1 - \frac{r}{n} \right)^{n-r}.$$

We obtain the following lower bound.

$$\Pr[A_r] \geq \left(\frac{r}{s} \right)^{1/2} \left(\frac{\gamma}{s} \right)^s \left(\frac{r}{n} \right)^r \left(1 - \frac{s}{\gamma} \right)^{s-\gamma-1/2} \left(1 - \frac{r}{n} \right)^{n-r+1/2}$$

$$= \left(\frac{rn}{sn} \right)^{1/2} \left(\frac{\gamma}{s} \right)^s \left(\frac{r}{n} \right)^r \left(1 - \frac{s}{\gamma} \right)^{s-\gamma-1/2} \left(1 - \frac{r}{n} \right)^{n-r+1/2} \quad (4)$$

Recall the assumptions of Lemma 2, i.e. $n \geq \gamma \geq s$ which yields:

$$\left(\frac{rn}{sn} \right)^{1/2} \left(\frac{\gamma}{s} \right)^s \geq \left(\frac{r}{n} \right)^{1/2} \left(\frac{\gamma}{s} \right)^{s+1/2} \geq \left(\frac{r}{n} \right)^{1/2} 1^{s+1/2} \geq \left(\frac{r}{n} \right)^{1/2}.$$

---

[2] We use the Stirling's formula in the form $n! \approx \sqrt{2\pi n}(n/e)^n$.

Further, it is sufficient to replace the lower bound $s \geq (r - \gamma)/(\Delta - 1)$ by the weaker one, i.e. $s > 0$. Hence,

$$\left(1 - \frac{s}{\gamma}\right)^{s-\gamma-1/2} \geq \left(1 - \frac{s}{\gamma}\right)^{-1/2} \geq 1.$$

Thus, in Eq. (4) is simplifying accordingly.

$$\Pr[A_r] \geq \left(\frac{r}{n}\right)^{r+1/2} \left(1 - \frac{r}{n}\right)^{n-r+1/2}$$

If $\lambda = \gamma/n$ and $\rho = r/\gamma$, then the equation $r = \lambda\rho n$ is substituted into the former inequality and the resulting inequality follows,

$$\Pr[A_r] \geq (\lambda\rho)^{\lambda\rho n+1/2}(1 - \lambda\rho)^{(1-\lambda\rho)n+1/2}$$
$$= \left[(\lambda\rho)^{\lambda\rho}(1 - \lambda\rho)^{(1-\lambda\rho)}\right]^n \rho^{1/2}(1 - \lambda\rho)^{1/2}.$$

The proof is complete.   □

4   The Time Complexity Analysis of the Sequential Randomized Algorithm

We propose the sequential randomized approximation algorithm *Rand_DS* which probabilistically finds a dominating set in a given connected graph. Its inputs are k and r, where k is a number of iterations and r is the number of vertices in resulting DS. Choice of r can be determined by the binary searching similarly as in [14], it is omitted in the description of the algorithm.

Input: $G = (V, E)$: a connected graph with n vertices,
                k: a positive integer,
                r: a positive integer such that $\gamma(G) \leq r \leq n/2$
Output: a dominating set V_D of a graph $G = (V, E)$

1. $i = 1$;   *dominated* := false;
2. while $i \leq k$ and not *dominated*
        – randomly select an r-node subset $V_D \subseteq V(G)$;
        – if $N[V_D] = V(G)$ then   *dominated* := true;
        – $i = i + 1$;
3. endwhile

Algorithm *Rand_DS*: Randomized finding of DS in G.
The following lemma is a consequence of Theorem 1.

Lemma 3. *Let $\rho \geq 1$ be a constant. There is a randomized approximation algorithm which for each connected graph G with n vertices computes a dominating set of G such that its approximation ratio is at most ρ. The probability of each feasible solution of the algorithm is at least $1 - n^{-\tau}$ for an arbitrary constant $\tau > 0$ and the time complexity of the algorithm is at most $O^*(t^n)$, where:*

$$t = t(\lambda, \rho) = (\lambda\rho)^{-\lambda\rho}(1 - \lambda\rho)^{\lambda\rho - 1} \qquad and \qquad \lambda = \gamma(G)/n.$$

Proof. Let G be a connected graph with n vertices and m edges. Note that for each such G, it is always possible to express λ by $\gamma(G)/n$. Recall that $0 < \lambda \leq 1/2$ (see [7]). It is easy to see that algorithm *Rand_DS* probabilistically computes a dominating set for an input graph G. By Lemma 2, the approximation ratio of *Rand_DS* is at most ρ. Let $\tau > 0$ and $\rho > 1$ be constants. In *Rand_DS*, let us set:

$$k = \tau \ln n \left[(\lambda\rho)^{\lambda\rho}(1 - \lambda\rho)^{(1-\lambda\rho)} \right]^{-n} \rho^{-1/2}(1 - \lambda\rho)^{-1/2}.$$

Since *Rand_DS* is the algorithm of the same type as the Monte Carlo one $\mathcal{A}$, it is possible to apply Lemma 1. By Lemma 1 and Theorem 1,

$$\Pr[\ Rand\_DS \text{ finds a feasible solution after } k \text{ iterations}\ ] \geq 1 - n^{-\tau},$$

since $\varphi(n) = \tau \ln n$. The time complexity of a single iteration of *Rand_DS* is at most $O(m \log n)$ and the overall time complexity of *Rand_DS* is $k \cdot O(m \log n)$. Note that $\rho^{-1/2}(1 - \lambda\rho)^{-1/2} = O(1)$ and the time complexity of *Rand_DS* is

$$O^* \left(\left[(\lambda\rho)^{-\lambda\rho}(1 - \lambda\rho)^{(\lambda\rho - 1)} \right]^n \right).$$

$\square$

The plot of the function $t(\lambda, \rho)$ is shown in Fig. 2. It follows that the time complexity of *Rand_DS* is moderately exponential, comparable with the $O(1.4969^n)$ exact algorithm [17].

Dependence of $t(\lambda, \rho)$ on its variable ρ is examined in more details due to the fact that it represents a trade-off relationship between the time complexity and the approximation ratio of an algorithm in question. In our setting, it leads to quite surprising conclusion. The result is based on the following lemma. Meaning of λ and ρ is the same as in Lemma 3, i.e. λ is a relative size of the MDS and ρ is an approximation ratio of the algorithm *Rand_DS*.

Lemma 4. *Let ρ_{MAX} be a constant such that $1 < \rho_{MAX} \ll n$. The maximum time complexity of the algorithm Rand_DS is $O^*(2^n)$ and it is achieved iff $(\lambda, \rho) \in \hat{L}$ such that:*

$$\hat{L} = \left\{ (\lambda, \rho) \mid \lambda\rho = \frac{1}{2};\ 0 < \lambda \leq 0.5;\ 1 \leq \rho \leq \rho_{MAX} \right\}.$$

Proof. Sketch. We shall find a maximum of the function $t(\lambda, \rho)$ for all plausible pairs (λ, ρ). To do so, the standard method based on partial derivatives is used. $\square$

According to Fig. 3, the time complexity of *Rand_DS* is an increasing function with respect to ρ if $\rho < 1/(2\lambda)$, further, it attains the maximum for $\rho = 1/(2\lambda)$ and then, it is decreasing for $\rho > 1/(2\lambda)$. The increasing behavior of the time complexity with respect to increasing approximation ratio is counterintuitive,

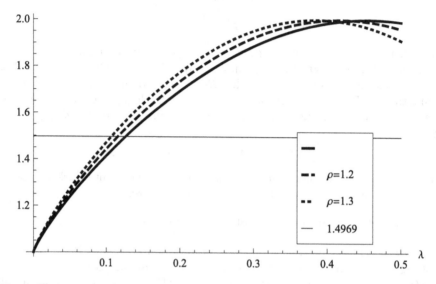

Fig. 2. Function $t(\lambda, \rho)$ for $\lambda \in [0, 0.5]$ and for three choices of ρ. The horizontal line represents the base of the exponential-time complexity for the fastest exact algorithm known which is 1.4969 [17].

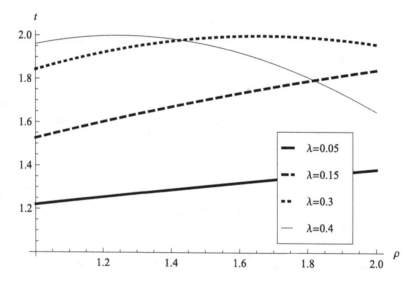

Fig. 3. The dependence of the *Rand_DS* time complexity, i.e. the function $t(\lambda, \rho)$, on the approximation ratio $\rho \in [1, 2]$ for 4 various choices of λ.

however, the same property was also achieved for the near-optimal DS algorithm presented in [15], see Theorem 4, p. 9. Our algorithm and the algorithm in [15] are ones from only few instances with the similar time complexity functions.

5   Conclusions

We have estimated a probability that a randomly selected subset of vertices represents a dominating set in an arbitrary connected graph. The corresponding sequential Monte Carlo randomized algorithm can compute a dominating set with high probability, with a constant approximation ratio and moderately exponential time complexity. The algorithm can be easily extend for disconnected graphs by applying it for each connected component separately.

Regarding the essence of the problem, the possibility of finding an algorithm with significantly better time complexity applicable generally for arbitrary graphs is unlikely even if $P \neq NP$.

The analysis of the trade-off between the approximation ratio and the time complexity of the algorithm leads to the surprising and partially counterintuitive conclusions.

As an instance of open problems, we mention the generalization for the k-domination sets in the sense of work published in [2]. Another interesting direction for the future research is to examine relationship between the approximation ratio and the time complexity of the algorithms for other graph-theoretical problems so far from the author is currently working on it.

Acknowledgement. The author gratefully acknowledge prof. J. Hromkovič and M. Demetrian for their valuable comments on the manuscript.

This research is supported by the MESRS of the Slovak Republic under the grants KEGA 047STU-4/2016 and VEGA 1/0026/16, respectively.

References

1. Chen, Y.P., Liestman, A.L.: Approximating minimum size weakly-connected dominating sets for clustering mobile ad hoc networks. In: Proceedings of the 3rd ACM International Symposium on Mobile Ad Hoc Networking & Computing MobiHoc 2002, pp. 165–172. ACM, New York (2002)
2. Cooper, C., Klasing, R., Zito, M.: Lower bounds and algorithms for dominating sets in web graphs. Internet Math. **2**(3), 275–300 (2005)
3. Downey, R.G., Fellows, M.R.: Parameterized Complexity. Springer, New York (1999)
4. Garey, M.R., Johnson, D.S.: Computers and Intractability. Freeman, New York (1979)
5. Gast, M., Hauptmann, M., Karpinski, M.: Inapproximability of dominating set in power law graphs. Theor. Comput. Sci. **562**, 436–452 (2015)
6. Gomes, C.P., Williams, R.: Approximation algorithms. In: Burke, E., Kendall, G. (eds.) Introduction to Optimization, Decision Support and Search Methodologies, pp. 557–585. Kluwer, Dordrecht (2005)

7. Haynes, T.W., Hedetniemi, S.T., Slater, P.J.: Fundamentals of Domination in Graphs. Marcel Dekker Inc., New York (1998)
8. Hooker, J.N., Garfinkel, R.S., Chen, C.K.: Finite dominating sets for network location problems. Oper. Res. **39**(1), 100–118 (1991). http://www.jstor.org/stable/171492. INFORMS
9. Hromkovič, J.: Design and Analysis of Randomized Algorithms: Introduction to Design Paradigms. Springer, Heidelberg (2005)
10. Kelleher, L., Cozzens, M.: Dominating sets in social network graphs. Math. Soc. Sci. **16**, 267–279 (1988)
11. Kratsch, D., Fomin, F.V., Grandoni, F.: Exact algorithms for dominating set. In: Kao, M.-Y. (ed.) Encyclopedia of Algorithms, pp. 1–5. Springer, New York (2008)
12. Motwani, R., Raghavan, P.: Randomized Algorithms. Cambridge University Press, New York (1995)
13. Nacher, J.C., Akutsu, T.: Analysis on critical nodes in controlling complex networks using dominating sets. In: Proceedings of the International Conference on Signal-Image Technology & Internet-Based Systems, pp. 649–654. IEEE Computer Society Press (2013)
14. Nehéz, M., Bernát, D., Klaučo, M.: Comparison of algorithms for near-optimal dominating sets computation in real-world networks. In: Proceedings of the 16th International Conference on Computer Systems and Technologies, pp. 199–206. ACM, New York (2015)
15. Nikoletseas, S.E., Spirakis, P.G.: Near-optimal dominating sets in dense random graphs in polynomial expected time. In: van Leeuwen, J. (ed.) WG 1993. LNCS, vol. 790, pp. 1–10. Springer, Heidelberg (1994)
16. Raz, R., Safra, S.: A sub-constant error-probability low-degree test, and a sub-constant error-probability PCP characterization of NP. In: Proceedings of the 29th Annual ACM Symposium on the Theory of Computing STOC 1997, pp. 475–484. ACM, New York (1997)
17. van Rooij, J.M.M., Bodlaender, H.L.: Exact algorithms for dominating set. Discrete Appl. Math. **159**(17), 2147–2164 (2011)
18. Wang, H., Zheng, H., Browne, F., Wang, C.: Minimum dominating sets in cell cycle specific protein interaction network. In: Proceedings of the IEEE International Conference on Bioinformatics and Biomedicine, pp. 25–30. IEEE Computer Society Press (2014)
19. Wuchty, S.: Controllabilty in protein interaction networks. Proc. Natl. Acad. Sci. USA **111**(9), 7156–7160 (2014)

A Multivariate Approach for Checking Resiliency in Access Control

Jason Crampton, Gregory Gutin, and Rémi Watrigant[✉]

Royal Holloway University of London, Egham, UK
remi.watrigant@rhul.ac.uk

Abstract. In recent years, several combinatorial problems were introduced in the area of access control. Typically, such problems deal with an authorization policy, seen as a relation $UR \subseteq U \times R$, where $(u, r) \in UR$ means that user u is authorized to access resource r. Li, Tripunitara and Wang (2009) introduced the RESILIENCY CHECKING PROBLEM (RCP), in which we are given an authorization policy, a subset of resources $P \subseteq R$, as well as integers $s \geq 0$, $d \geq 1$ and $t \geq 1$. It asks whether upon removal of any set of at most s users, there still exist d pairwise disjoint sets of at most t users such that each set has collectively access to all resources in P. This problem possesses several parameters which appear to take small values in practice. We thus analyze the parameterized complexity of RCP with respect to these parameters, by considering all possible combinations of $|P|, s, d, t$. In all but one case, we are able to settle whether the problem is in FPT, XP, W[2]-hard, para-NP-hard or para-coNP-hard. We also consider the restricted case where $s = 0$ for which we determine the complexity for all possible combinations of the parameters.

1 Introduction

1.1 Context and Definition of the Problem

Access control is a fundamental aspect of the security of any multi-user computing system. Typically, it is based on the idea of specifying and enforcing an authorization policy, identifying which interactions between a set of users U and a set of resources R are to be allowed by the system [11]. More formally, an authorization policy is defined as a relation $UR \subseteq U \times R$, where $(u, r) \in UR$ means that user u is authorized to access resource r. Quite recently, we have seen the introduction of resiliency policies, whose satisfaction indicates that a system will continue to function as intended in the absence of some number of authorized users [1, 10, 12]. Li, Tripunitara and Wang's seminal work [10] introduces a number of problems associated with the satisfaction of a resiliency policy. One of their motivating examples concerns the emergency response to a natural disaster, where teams of users must perform the same critical operation(s) at multiple (distinct) geographical locations. Thus the members of each team must

This research was partially supported by EPSRC grant EP/K005162/1. Gutin's research was also supported by Royal Society Wolfson Research Merit Award.

© Springer International Publishing Switzerland 2016
R. Dondi et al. (Eds.): AAIM 2016, LNCS 9778, pp. 173–184, 2016.
DOI: 10.1007/978-3-319-41168-2_15

be authorized collectively to perform the operation(s). In addition, we may wish to impose an upper bound on the size of the teams because, for example, of constraints on transportation.

For a user $u \in U$ and a set of users $V \subseteq U$, we define $N_{UR}(u) = \{r \in R : (u, r) \in UR\}$ the *neighborhood* of u and, by extension, $N_{UR}(V) = \bigcup_{u \in V} N_{UR}(u)$ the *neighborhood* of V, omitting the subscript UR if the authorization policy is clear from the context. Given an authorization policy $UR \subseteq U \times R$, an instance of the RESILIENCY CHECKING PROBLEM (RCP) is defined by a resiliency policy $\mathsf{res}(P, s, d, t)$, where $P \subseteq R$, $s \geq 0$, $d \geq 1$ and $t \geq 1$. We say that UR *satisfies* $\mathsf{res}(P, s, d, t)$ if and only if for every subset $S \subseteq U$ of at most s users, there exist d pairwise disjoint subsets of users $V_1, \ldots, V_d$ such that for all $i \in \{1, \ldots, d\}$:

$$V_i \cap S = \emptyset, \tag{1}$$

$$|V_i| \leq t, \tag{2}$$

$$N(V_i) \supseteq P. \tag{3}$$

We are now ready to define the main problem we study in this paper:

Resiliency Checking Problem (RCP)
Input: $UR \subseteq U \times R$, $P \subseteq R$, $s \geq 0$, $d \geq 1$, $t \geq 1$.
Question: Does UR satisfy $\mathsf{res}(P, s, d, t)$?

Furthermore, we will adopt the bracket notation $\mathrm{RCP}\langle\rangle$ used by Li *et al.* [10] to denote some restrictions of the problem, in which one or more parameters (among s, d and t) are fixed. In particular, we will consider the cases where s and d are respectively set to 0 and/or 1 (or other fixed positive values), while t might be set to ∞, meaning that there is no constraint on the size of the sets (which is actually equivalent to $t = |P|$, implying that we may assume in the remainder that $t \leq |P|$). For instance, $\mathrm{RCP}\langle s = 0\rangle$ denotes the variant in which s is fixed to 0, *i.e.* we ask for the satisfaction of $\mathsf{res}(P, 0, d, t)$. In the remainder of the paper, we set $p = |P|$.

Given an instance of $\mathrm{RCP}\langle\rangle$, we say that a set of d pairwise disjoint subsets of users $V = \{V_1, \ldots, V_d\}$ satisfying conditions (2) and (3) is a *set of teams*. For such a set of teams, we define $\mathcal{U}(V) = \bigcup_{i=1}^{d} V_i$. Given $U' \subseteq U$, the *restriction* of UR to U' is defined by $UR|_{U'} = UR \cap (U' \times R)$. Finally, a set of users $S \subseteq U$ is called a *blocker set* if for every set of teams $V = \{V_1, \ldots, V_d\}$, we have $\mathcal{U}(V) \cap S \neq \emptyset$. Equivalently, observe that S is a blocker set if and only if $UR|_{U \setminus S}$ does not satisfy $\mathsf{res}(P, 0, d, t)$. Throughout the paper, we write $[d]$ to denote $\{1, \ldots, d\}$ for any integer $d \geq 1$, and we will often make use of the $O^*(.)$ notation, which omits polynomial factors and terms.

1.2   Parameters

An instance of $\mathrm{RCP}\langle\rangle$ contains several parameters (namely s, d and t) which may be used for the complexity analysis of the problem. An interesting point of the work of Li *et al.* [10] is that the number of users in an organization will typically be large in comparison to the other parameters (s, d, t, and even p) in practice. In their experiments, the maximum values used are $n = 100$, $p = 10$ and $d = 7$ (they only run experiments on the variant where $t = \infty$, but, as we observed previously, we may set $t = p$). With this in mind, we exploit the theory of fixed-parameter tractability in order to settle the parameterized complexity of the problem.

Given an instance x (of size $|x|$) of a decision problem, with some parameter[1] k, we are interested in algorithms deciding whether x is positive or negative in polynomial time when k is bounded above by a constant. More precisely, if such an algorithm has running time $O(f(k)|x|^{O(1)})$ for some computable function f, then we will say that this algorithm is fixed-parameter tractable (FPT), while if its running time is $O(|x|^{f(k)})$ for some computable function f, we will say that this algorithm is XP (an FPT algorithm is thus an XP algorithm). By extension, FPT (resp. XP) gathers all problems for which an FPT (resp. XP) algorithm exists. Proving the NP-hardness of a problem in the case where a parameter k is bounded above by a constant immediately forbids the existence of any XP (and thus FPT) algorithm unless P $=$ NP. In this case, we will say that this parameterized problem is para-NP-hard. A similar definition can be given using coNP-hard and coNP instead of NP-hard and NP, respectively, leading to the para-coNP-hard complexity class (and thus, if a problem is shown to be para-coNP-hard, then it does not belong to XP unless P $=$ coNP). In the following, para-(co)NP-hard denotes the union of para-NP-hard and para-coNP-hard. Finally, the W[i]-hierarchy of parameterized problems is typically used to rule out the existence of an FPT algorithm, under the widely believed conjecture that FPT $\neq$ W[1]. For more details about fixed-parameter tractability, we refer the reader to the recent monographs [3,5].

1.3 Related Work

As one might expect, the RCP$\langle\rangle$ problem is strongly related to some known combinatorial problems. Indeed, one can observe that RCP$\langle s = 0, d = 1\rangle$ is equivalent to the SET COVER problem, while RCP$\langle s = 0, t = \infty\rangle$ can be reduced in a straightforward way from the DOMATIC PARTITION problem (in the DOMATIC PARTITION problem, one asks whether a given graph admits k pairwise disjoint dominating sets). Li et al. [10] obtained several (mainly negative) results for RCP$\langle\rangle$ in some restricted cases which can be summarized by the following theorem.

Theorem 1 [10]. *We have the following:*

- RCP$\langle\rangle$, RCP$\langle d = 1\rangle$ and RCP$\langle t = \infty\rangle$ are NP-hard and are in[2] coNP^{NP};

---

[1] Note that one can aggregate several parameters $p_1, \ldots, p_m$ by defining $k = p_1 + \cdots + p_m$, in which case we will say the parameter is $(p_1, \ldots, p_m)$.

[2] coNP^{NP} is the set of problems whose complement can be solved by a nondeterministic Turing machine having access to an oracle to a problem in NP.

- $\text{RCP}\langle s = 0, d = 1 \rangle$, $\text{RCP}\langle s = 0, t = \infty \rangle$ *are NP-hard;*
- $\text{RCP}\langle d = 1, t = \infty \rangle$ *can be solved in linear time.*

In addition, they developed and implemented an algorithm for $\text{RCP}\langle\rangle$ which consists of (i) enumerating all subsets of at most s users, and (ii) for each such subset S, determining the satisfaction of $\text{res}(P, 0, d, t)$ for $UR|_{U \setminus S}$. Step (ii) is achieved by a SAT formulation of the problem and the use of an off-the-shelf SAT solver, while they develop a pruning strategy in order to avoid the entire enumeration of all subsets of users of size at most s, resulting in an efficient speed-up of step (i). Quite surprisingly, they observe that the bottleneck of their algorithm lies in the second step, where an instance of $\text{RCP}\langle s = 0 \rangle$ has to be solved. This motivated us to focus on the parameterized complexity of $\text{RCP}\langle s = 0 \rangle$ separately.

1.4 Contribution and Organization of the Paper

Our goal in this paper is thus to determine the parameterized complexity of $\text{RCP}\langle\rangle$ and $\text{RCP}\langle s = 0 \rangle$ with respect to parameters p, s, d, t, by considering every possible combination of them. In each case, we aim at determining whether the problem is (i) in FPT, (ii) in XP but $W[i]$-hard for some $i \geq 1$, or (iii) para-(co)NP-hard.

Figure 1 summarizes the (already known and) obtained results for $\text{RCP}\langle\rangle$ and $\text{RCP}\langle s = 0 \rangle$ with respect to all possible combinations of the parameters

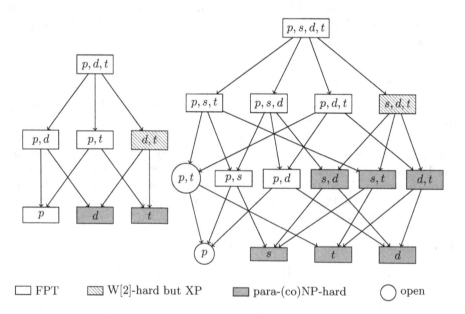

□ FPT   ▧ W[2]-hard but XP   ■ para-(co)NP-hard   ○ open

Fig. 1. Schemas of the complexity of $\text{RCP}\langle s = 0 \rangle$ (left) and $\text{RCP}\langle\rangle$ (right) after the results obtained in this paper (see the end of this section for the difference between old and new results).

specified previously. An arrow $A \longrightarrow B$ means that A is a larger parameter than B, in the sense that an FPT algorithm parameterized by B implies an FPT algorithm parameterized by A, and, conversely, any negative result parameterized by A implies the same negative result parameterized by B. Since (under classical complexity assumptions) a decision problem is either in one of the previous cases (i), (ii) or (iii), one can observe that the parameterized complexity of $\text{RCP}\langle s = 0 \rangle$ is now completely determined with respect to all possible combinations of parameters p, d and t. Concerning the more general case $\text{RCP}\langle \rangle$, only the parameterization by p only remains unknown (recall that as we mentioned earlier, we may assume in any instance that $t \leq p$, implying that adding t in the parameter list is of no importance concerning the membership in these complexity classes, both for positive or negative results).

The next section gathers all our results for the general case $\text{RCP}\langle \rangle$, namely:

- membership in XP parameterized by (s, d, t) (Theorem 2),
- membership in FPT parameterized by (p, d) or (p, s) (Theorem 3),
- para-coNP-hardness parameterized by (d, t) (Theorem 4),
- para-NP-hardness parameterized by (s, t) (Theorem 5).

Note that the para-NP-hardness for (s, d) was already known (Theorem 1), as well as the W[2]-hardness for (s, d, t) (see explanation in Sect. 2.2).

Section 3 gathers all our results for the restricted case $\text{RCP}\langle s = 0 \rangle$, namely:

- an FPT algorithm parameterized by (d, p) with an optimal running time (under ETH) when d is fixed (Theorems 6 and 7),
- membership in FPT parameterized by p only (Theorem 8).

Note that the W[2]-hardness for (d, t) is inherited from $\text{RCP}\langle \rangle$, while the XP membership results from a brute-force enumeration of all subsets of users of size dt. We also investigate in this section the question of data (user) reductions and present positive and negative kernelization results depending on the considered variant: $\text{RCP}\langle s = 0 \rangle$ or $\text{RCP}\langle s = 0, t = \infty \rangle$ (Theorem 10). We finally conclude the paper in Sect. 4.

Due to space restrictions, some proofs (for theorems marked with a ⋆) were omitted and can be found in [1].

2 The General Case

2.1 Positive Results

First, observe that there exists a simple XP algorithm for $\text{RCP}\langle \rangle$ parameterized by (s, d, t). Indeed, recall that the problem actually aims to check whether there is a set $S \subseteq U$ of size at most s such that for any set of teams $V = \{V_1, \ldots, V_d\}$ we have $S \cap \mathcal{U}(V) \neq \emptyset$, and note that finding a set of teams is exactly the $\text{RCP}\langle s = 0 \rangle$ problem, which is in XP parameterized by (d, t), as said in Sect. 1.4. Hence, since $|\mathcal{U}(V)| \leq dt$, by finding iteratively a set of teams and branching on each element to be removed from it (and included in the future blocker set), one can determine whether there exists a blocker set of size at most s in XP time parameterized by (s, d, t):

Theorem 2. RCP$\langle\rangle$ *is in XP when parameterized by* (s, d, t)

Despite its simplicity, this result is actually somehow tight. First, as we will see later (Sect. 2.2), RCP$\langle\rangle$ is W[2]-hard with this parameterization. In addition, considering a strict subset of $\{s, d, t\}$ as a parameter makes the problem para-(co)NP-hard (Theorems 1, 4 and 5). A way of going further is to "replace" t by p (since we may assume $t \leq p$). With this modification, we show in the next result how to get rid of the parameter s or d by designing an FPT algorithm parameterized by (p, d) or (p, s).

Theorem 3. RCP$\langle\rangle$ *is FPT when parameterized by* $(p, \min\{s, d\})$.

Proof. Without loss of generality, we may assume $P = R$ as well as $N(u) \neq \emptyset$ for all $u \in U$. For all $C \subseteq P$, let $U_C = \{u \in U : N(u) = C\}$ (notice that we might have $U_C = \emptyset$ for some $C \subseteq P$). Let $S \subseteq U$ be a blocker set of size at most s, *i.e.* a set whose removal makes $\mathsf{res}(P, 0, d, t)$ unsatisfiable. Moreover, assume that S is a *minimal blocker set*, meaning that there does not exist $S' \subsetneq S$ such that the removal of S' makes $\mathsf{res}(P, 0, d, t)$ unsatisfiable.

Claim. For all $C \subseteq P$, $U_C \cap S \neq \emptyset$ implies that $|U_C \setminus S| < d$.

Before proving the claim, notice that for all $u \in U_C \cap S$, there exists a set of teams $V = \{V_1, \ldots, V_d\}$ such that (i) $\mathcal{U}(V) \cap S = \{u\}$, and (ii) $|\mathcal{U}(V) \cap U_C| \leq d$. Condition (i) comes from the minimality of S, while Condition (ii) comes from the fact that otherwise, there would exist $i \in [d]$ such that $|V_i \cap U_C| \geq 2$, and removing one user from V_i, arbitrarily chosen in $(V_i \cap U_C) \setminus \{u\}$, produces another set of teams V' with $\mathcal{U}(V') \subsetneq \mathcal{U}(V)$ (with exactly one element less) and still such that $V \cap S = \{u\}$. Applying this strategy iteratively, we can get a set of teams V as desired.

Proof (of the claim). To do so, let $u \in U_C \cap S$ and $V = \{V_1, \ldots, V_d\}$ defined as previously. If we have $|U_C \setminus S| \geq d$, then there exists $v \in U_C \setminus S$ such that $v \notin \mathcal{U}(V)$ (since $|\mathcal{U}(V) \cap U_C| \leq d$, and since $u \in S \cap U_C$, it follows that $|(U_C \setminus S) \cap \mathcal{U}(V)| \leq d - 1$), in which case we have that $(\mathcal{U}(V) \setminus \{u\}) \cup \{v\}$ is the union of a set of teams which does not intersect S (recall that $\mathcal{U}(V) \cap S = \{u\}$), and satisfies $\mathsf{res}(P, 0, d, t)$ (since $N(u) = N(v)$), a contradiction. □

We now define a reduced set of users $U^r \subseteq U$ composed of $d_C = \min\{|U_C|, d\}$ users from U_C chosen arbitrarily, for all $C \subseteq P$. By construction, observe that $|U^r| \leq d2^p$. We also define, for all $C \subseteq P$, $U_C^r = U_C \cap U^r$. Finally, consider an algorithm which outputs that $\mathsf{res}(P, s, d, t)$ is unsatisfiable if and only if there exists a blocker set $S \subseteq U^r$ of the instance induced by U^r (*i.e.* with authorization policy $UR|_{U^r}$), and such that $\sum_{C \subseteq P} \zeta_S(C) \leq s$, where

$$\zeta_S(C) = \begin{cases} |S \cap U_C^r| + |U_C| - d_C & \text{if } S \cap U_C^r \neq \emptyset \\ 0 & \text{otherwise.} \end{cases}$$

in which case we will say that S is a *reduced blocker set*. We will prove that this algorithm is FPT parameterized by $(p, \min\{s, d\})$, and is correct.

Concerning the running time, observe first that the construction of U^r as well as the evaluation of ζ_S, given $S \subseteq U^r$, takes $O^*(2^p)$ time. Then, for any reduced blocker set $S \subseteq U^r$, notice that $|S \cap U^r_C| \leq \min\{s, d\}$ for all $C \subseteq P$, and that any set $S' \subseteq U^r$ such that $|S' \cap U^r_C| = |S \cap U^r_C|$ for all $C \subseteq P$ is also a reduced blocker set (because $N(u) = N(v)$ for all $u, v \in C$, for all $C \subseteq P$). Hence, instead of enumerating every possible subset S of U^r, it is sufficient to enumerate the sizes of each intersection with U^r_C for all $C \subseteq P$, and pick the right number of users in U^r_C in an arbitrary way. Since its intersection is of size at most $\min\{s, d\}$, the number of sets to enumerate is $O((\min\{s, d\} + 1)^{2^p})$. Then, for each obtained set $S \subseteq U^r$, we can check whether it is a blocker set of $UR|_{U^r}$ by solving the $\text{RCP}\langle s = 0 \rangle$ problem on the instance $UR|_{U^r \setminus S}$ in FPT time parameterized by p (using, $e.g.$, Theorem 8).

It now remains to prove its correctness, by proving that there exists a reduced blocker set if and only if $\text{res}(P, s, d, t)$ is unsatisfiable. If such a set S exists, then define, for each $C \subseteq P$, a set $S_C \subseteq U_C$ composed of $S \cap U^r_C$ plus all users in $U_C \setminus U^r_C$. By construction, $|S_C| = \zeta_S(C)$, and thus $S^* = \bigcup_{C \subseteq P} S_C$ contains at most s users. We now prove that S^* is a blocker set: suppose by contradiction that there exists a set of teams $V = \{V_1, \ldots, V_d\}$ such that $\mathcal{U}(V) \cap S^* = \emptyset$. As we saw previously, we may assume that $|V_i \cap U_C| \leq 1$ for all $i \in [d]$ and all $C \subseteq P$. Let $I_V = \{i \in [d] : V_i \cap (U_C \setminus U^r_C) \neq \emptyset\}$. We show that we can turn V into another set of teams V' such that $\mathcal{U}(V') \subseteq U^r$ ($i.e.$ such that $I_{V'} = \emptyset$), implying that S is not a reduced blocker set, a contradiction. If $I_V = \emptyset$, then we are done. Otherwise let $i \in I_V$ and $u \in V_i \cap (U_C \setminus U^r_C)$. By construction of U^r, there exists $v \in U^r_C$, and thus $(V \setminus \{u\}) \cup \{v\}$ is the union of a set of teams V' (recall that $N(u) = N(v)$) such that $i \notin I_{V'}$. Repeating this transformation at most d times, we naturally obtain a set of teams V' such that $I_{V'} = \emptyset$ as desired.

Conversely, suppose that $\text{res}(P, s, d, t)$ is unsatisfiable, $i.e.$ there exists a blocker set of users $S \subseteq U$ of size at most s. As previously, we may assume that S is a minimal blocker set. We now use the previous Claim, and thus for all $C \subseteq P$, $|S \cap U_C| \geq \max\{0, |U_C| - d + 1\}$. Thus, we may assume, without loss of generality (since, again, $N(u) = N(v)$ for all $u, v \in U_C$) that $U_C \setminus U^r_C \subseteq S$. Then, we define $S^r = S \setminus (\bigcup_{C \in \wp(C)} U_C \setminus U^r_C)$. Observe that for all $C \subseteq P$, we have:

$$\zeta_{S^r}(C) = |S^r \cap U^r_C| + |U_C| - d_C$$
$$= |S^r \cap U^r_C| + |U_C \setminus U^r_C|$$
$$= |S \cap U_C|$$

and thus $\sum_{C \subseteq P} \zeta_{S_r}(C) = \sum_{C \subseteq P} |S \cap U_C| = |S| \leq s$. Finally, S^r is indeed a blocker set of the instance induced by U^r, since otherwise, there would exist a set of teams $V = \{V_1, \ldots, V_d\}$ with $\mathcal{U}(V) \subseteq U^r$ such that $\mathcal{U}(V) \cap S^r = \emptyset$, which would imply that $\mathcal{U}(V) \cap S = \emptyset$ as well, a contradiction.   □

2.2 Negative Results

It is worth pointing out that the reduction of [10, Lemma 3] proving the NP-hardness of $\text{RCP}\langle s = 0, d = 1 \rangle$ actually proves the W[2]-hardness of this problem parameterized by t (from SET COVER parameterized by the size of the solution [5]). Another implication of this reduction is the para-NP-hardness of RCP$\langle\rangle$ when parameterized by (s, d). We now complement this result by showing that RCP$\langle d = 1, t = \tau\rangle$ is coNP-hard for every fixed $\tau \geq 3$, implying para-coNP-hardness of RCP$\langle\rangle$ parameterized by (d, t). The result is obtained by a reduction from the δ-HITTING SET problem for every $\delta \geq 2$.

Theorem 4 ($\star$). RCP$\langle d = 1, t = \tau\rangle$ is coNP-hard for every fixed $\tau \geq 3$.

We also settle the case of RCP$\langle\rangle$ parameterized by (s, t) (and thus RCP$\langle s = 0\rangle$ parameterized by t). The result is obtained by a reduction from the 3-DIMENSIONAL MATCHING problem.

Theorem 5 ($\star$). RCP$\langle s = 0, t = 4\rangle$ is NP-hard.

3   Refined Positive Results for the Case $s = 0$

We now turn to the particular case where $s = 0$. As said in Sect. 1, one motivation for studying this case is that it is the bottleneck of the algorithm of Li $et~al.$ [10] for RCP$\langle\rangle$. Hence, we believe that designing efficient algorithms for this sub-case might help us solve much larger instances of RCP$\langle\rangle$ than is currently possible. To this end, we now provide a complete characterization of the complexity when considering all possible combinations of parameters among p, d and t. We also investigate the question of reduction rules within the framework of kernelization, highlighting a difference of behavior between RCP$\langle s = 0\rangle$ and RCP$\langle s = 0, t = \infty\rangle$.

3.1   FPT Algorithms

The first algorithm is a dynamic programming-based approach similar to the one for SET COVER [5], in order to obtain an FPT algorithm for RCP$\langle s = 0\rangle$ parameterized by (p, d). While this result was already known, given that RCP$\langle\rangle$ is itself FPT with this parameterization (and that RCP$\langle s = 0\rangle$ is actually FPT parameterized by p only, as we will see in Theorem 8), we provide for RCP$\langle s = 0\rangle$ a better running time. In particular, as we will see later, a previous known reduction of Li $et~al.$ [10] actually proves that when d is fixed, the obtained running time is the best we can hope for, under the Exponential Time Hypothesis (ETH)[3].

Theorem 6 ($\star$). RCP$\langle s = 0\rangle$ can be solved in $O^*(2^{dp})$ time.

Li $et~al.$ [10] showed that RCP$\langle s = 0, t = \infty, d = 3\rangle$ is NP-hard, by a reduction from 3-DOMATIC PARTITION, which transforms a graph of n vertices into an instance $(U, R, UR, \text{res}(P, 0, 3, \infty))$ with $|P| = n$. Since a $2^{o(n)}$ algorithm for 3-DOMATIC PARTITION would violate the ETH (by a linear reduction from SAT [2]), we have the following:

---

[3] The ETH claims that SAT cannot be solved in $O^*(2^{o(n)})$, where n is the number of variables in the CNF formula [7].

Theorem 7. RCP$\langle s = 0, t = \infty, d = 3\rangle$ *cannot be solved in* $2^{o(p)}$ *time unless the ETH fails.*

Hence, for fixed d, the algorithm described in Theorem 6 has an optimal running time. We continue our quest for a better understanding of the frontier between tractable and intractable cases of the RCP$\langle s = 0\rangle$ problem. Given the positive result parameterized by (p, d), a natural question is to consider each parameter separately. The question can well be answered negatively concerning the parameter d, since, as we saw before, RCP$\langle s = 0, d = 3, t = \infty\rangle$ is NP-hard [10], and thus RCP$\langle s = 0\rangle$ is para-NP-hard parameterized by d. However, we are able to give a different answer for the parameter p only.

Theorem 8. RCP$\langle s = 0\rangle$ *is FPT when parameterized by* p.

Proof. The result makes use of Lenstra's celebrated algorithm [9] for Integer Linear Programming Feasibility (ILPF) parameterized by the number of variables.

Theorem 9 (Lenstra [9]). *Whether a given ILP has a non-empty solution set can be decided in* $O^*(f(n))$ *time for some computable function* f, *where* n *denotes the number of variables of the ILP.*

Note that this algorithm has been improved by Kannan [8], with $f(n) = n^{O(n)}$ (but exponential space), and by Frank and Tardos [6] so that the algorithm runs in polynomial space, and with $f(n) = O(n^{2.5n+o(n)})$.

We thus give an ILPF formulation of the problem with a number of variables depending on p and t. As we saw previously, since we may assume that $t \leq p$ in any positive instance, the result will follow (by Lenstra's result) for the parameterization by p only.

Let $(U, R, UR, \mathsf{res}(P, 0, d, t))$ be the input instance of RCP$\langle s = 0\rangle$. For any $N \subseteq P$, let U_N denote the set of users having neighborhood exactly N in P, or, formally: $U_N = \{u \in U : N(u) = N\}$. Moreover, we define the following set called *configurations*:

$$\mathcal{C} = \left\{ \{N_1, \ldots, N_b\} : b \leq t, N_i \subseteq P, i \in [b], \bigcup_{i=1}^{b} N_i = P \right\}.$$

For any $N \subseteq P$, we note

$$\mathcal{C}_N = \{c = \{N_1, \ldots, N_{b_c}\} \in \mathcal{C} : N = N_i \text{ for some } i \in [b_c]\}$$

the set of configurations involving N. Informally, a configuration $\{N_1, \ldots, N_b\}$ represents a way to dominate P, by picking one user in U_{N_i}, for each $i \in [b]$.

The variables of our ILP are in one-to-one correspondence with elements of $\mathcal{C}$, and will be denoted by $\{x_c : c \in \mathcal{C}\}$. Since $\mathcal{C}$ is of size bounded by $O(\sum_{b=1}^{t} 2^{bp})$, the number of variables is bounded by a function of p and t only. Then, we define the following two sets of constraints:

1. $\sum_{c \in C} x_c = d$,
2. $\sum_{c \in C_N} x_c \leq |U_N|$ for all $N \subseteq P$.

We now explain the idea of the ILP. Observe that in a positive instance, there always exists a set of teams in which in each set, each user has a different neighborhood. For any $T \subseteq U$, define $\phi(T) = \{N(u) : u \in T\}$, the set of neighborhoods of users in T. Then, by definition of the problem, for any set of teams $V = \{T_1, \ldots, T_d\}$, we have $\Phi(T_i) \in C$ for all $i \in [d]$. Notice that we might have $\Phi(T_i) = \Phi(T_j)$ for $i, j \in [d]$, $i \neq j$. We can associate, with each such set of teams, a vector $X^V = \{x_c^V\}_{c \in C}$, where x_c^V is the number of sets of V having configuration $c \in C$. By the remark above, we might have $X^V = X^{V'}$ for two different sets of teams V and V', in which case we will say that these two sets of teams are *configuration-equivalent*. Observe that given a vector $X = \{x_c\}_{c \in C}$ such that $X = X^{V^*}$ for a fixed set of teams V^*, we can construct in polynomial time a set of teams V that is configuration-equivalent to V^*; constraints (1) and (2) aim to find such a vector. Suppose that there exists a set of teams $V^* = \{T_1, \ldots, T_d\}$ of the problem. It is clear that X^{V^*} fulfills constraints (1) and (2). Conversely, constraints in (1) ensure that the set of teams will contain d sets, while constraints in (2) ensure that when constructing a set of configuration $c = \{N_1, \ldots, N_{b_c}\}$, there must exist a new user having neighborhood exactly N_i for all $i \in [b_c]$ and that has not been already assigned to another set. $\square$

3.2   User Reductions

We now focus on reduction rules which can be performed in polynomial time and result in an equivalent instance having a smaller number of users. More formally, we say that a (decision) problem has a *kernel* [5] of size f, for some computable function $f : \mathbb{N} \to \mathbb{N}$, if there exists a polynomial algorithm which, given an instance x with parameter k, outputs an instance x' of size $|x'|$ with parameter k' such that: (i) $k' \leq k$, (ii) x is positive if and only if x' is positive, and (iii) $|x'| \leq f(k)$. In the case of RCP$\langle s = 0 \rangle$ our aim is thus to obtain an equivalent instance with a number of users bounded by a function of d and t.

While the role of t was so far of less interest for the complexity of the problem, we show that the problem behaves differently from the kernelization point of view, depending on whether $t = \infty$ or not. We first show that when $t = \infty$, the problem admits a kernel with at most dp users. To do so, we will make use of the following:

Lemma 1 (d-expansion Lemma [3]). *Let $d \geq 1$ be a positive integer and $G = (A, B, E)$ be a bipartite graph with bipartition (A, B) and $E \subseteq A \times B$ such that for all $b \in B$, $N(b) \neq \emptyset$. If $|B| \geq d|A|$, then there exist non-empty vertex sets $X \subseteq A$ and $Y \subseteq B$ which can be found in time polynomial in the size of G, such that:*

(i) $N(Y) \subseteq X$, and

(ii) there is a d-expansion of X into Y: a collection $M \subseteq E \cap (X \times Y)$ such that every vertex of X is incident to exactly d edges of M, and exactly $d|X|$ vertices of Y are incident to an edge of M.

Theorem 10. $\mathrm{RCP}\langle s = 0, t = \infty \rangle$ *admits a kernel with at most dp users.*

Proof. Suppose we are given an instance of $\mathrm{RCP}\langle s = 0, t = \infty \rangle$. We present two reduction rules which are used to decrease the number of users. For each of these rules, we will prove that the instance is positive iff the reduced instance is positive, in which case we will say that the rule is *safe*.

Reduction Rule 1: if there exists $u \in U$ with $N(u) = \emptyset$, then delete u.

Proof (of safeness). Simply observe that such a user cannot participate in any set of teams if the instance is positive, and, conversely, cannot turn a negative instance into a positive one if it is deleted.   □

Reduction Rule 2: if there exist $X \subseteq P$, $Y \subseteq U$ such that $N(Y) \subseteq X$ and there is a d-expansion of X into Y, then delete X from P, Y from U, and $(Y \times X) \cap UR$ from UR.

Proof (of safeness). If the instance is a positive one, then there exists a set of teams $\{V_1, \ldots, V_d\}$. Then, for all $r \in P \setminus X$, there does not exist $u \in Y$ such that $(u, r) \in UR$, since $N(Y) \subseteq X$. Hence, $N(V_i \setminus Y) \supseteq P \setminus X$, and thus $\{V_1 \setminus Y, \ldots, V_d \setminus Y\}$ is a set of teams for the reduced instance, which is thus a positive one.

Conversely, suppose that the reduced instance is a positive one: there exist $V_1, \ldots, V_d$, disjoints sets of users from $U \setminus Y$ such that $N(V_i) \supseteq P \setminus X$. Since there is a d-expansion of X into Y, for all $r \in X$, there exist $u_1^r, \ldots u_d^r \in Y$ such that $(u_i^r, r) \in UR$ for all $i \in [d]$, where $u_i^r \neq u_{i'}^{r'}$ for all $r \neq r'$ and $i \neq i'$. Hence, for all $i \in [d]$, if we set $V_i' = V_i \cup \{u_i^r : r \in X\}$, we have $V_i' \cap V_j' = \emptyset$ for all $1 \leq i < j \leq d$, and $N(V_i') \supseteq P$ for all $i \in [d]$, and thus we have a positive instance as well, which proves that the rule is safe.   □

Since each reduction rule can be applied in polynomial time, and since each of them decreases the number of users by at least one, the algorithm runs in polynomial time. Finally, by Lemma 1, if none of the previous reduction rules applies, then $|U| \leq dp$, and we thus have a kernel with at most dp users, as desired.   □

As Li *et al.* [10] point out, $\mathrm{RCP}\langle s = 0, d = 1 \rangle$ is equivalent to the SET COVER PROBLEM. Known kernel lower bounds for this problem [4] lead to the following theorem, which is in sharp contrast to the previous case.

Theorem 11. $\mathrm{RCP}\langle s = 0, d = 1 \rangle$ *(and thus $\mathrm{RCP}\langle s = 0 \rangle$) does not admit a kernel with $(p + t)^{O(1)}$ users, unless $coNP \subseteq NP/poly$.*

4   Conclusion and Future Work

We considered $\mathrm{RCP}\langle \rangle$, a problem introduced recently in the area of access control to analyze the resiliency of a system. Given the large number of natural parameters in an instance of this problem, and given that these parameters are likely to take small values in practice, our goal was to provide a systematic analysis of the complexity of the problem using the framework of parameterized complexity. For all but one possible combination of the parameters, we were able to obtain either a positive or negative result. We also considered a restricted variant of the problem for which we settled the parameterized complexity of all possible combinations of the parameters. A first obvious idea of future work is thus to fill the remaining hole of Fig. 1, namely to decide whether $RCP\langle\rangle$ is in FPT, XP, W[1]-hard or para-(co)NP-hard parameterized by p.

Another interesting further line of research would be to study resiliency aspects with respect to other problems. In the context of graphs for instance, we could define the problem of determining whether upon removal of at most s vertices, a given graph still satisfies some property given by another combinatorial problem, *e.g.* having a vertex cover of size k. We believe that considering structural parameterizations (together with s) might lead to interesting new results. As in our case, the complexity of such a new problem will certainly depend on the complexity of the considered underlying problem (*i.e.* the case $s = 0$).

References

1. Crampton, J., Gutin, G., Watrigant, R.: A multivariate approach for checking resiliency in access control. CoRR: 1604.01550 (2016)
2. Creignou, N.: The class of problems that are linearly equivalent to satisfiability or a uniform method for proving NP-completeness. Theoret. Comput. Sci. **145**(1–2), 111–145 (1995)
3. Cygan, M., Fomin, F.V., Kowalik, L., Lokshtanov, D., Marx, D., Pilipczuk, M., Pilipczuk, M., Saurabh, S.: Parameterized Algorithms. Springer, Switzerland (2015)
4. Dom, M., Lokshtanov, D., Saurabh, S.: Incompressibility through colors and IDs. In: Albers, S., Marchetti-Spaccamela, A., Matias, Y., Nikoletseas, S., Thomas, W. (eds.) ICALP 2009, Part I. LNCS, vol. 5555, pp. 378–389. Springer, Heidelberg (2009)
5. Downey, R.G., Fellows, M.R.: Fundamentals of Parameterized Complexity. Texts in Computer Science. Springer, London (2013)
6. Frank, A., Tardos, É.: An application of simultaneous diophantine approximation in combinatorial optimization. Combinatorica **7**(1), 49–65 (1987)
7. Impagliazzo, R., Paturi, R., Zane, F.: Which problems have strongly exponential complexity? J. Comput. Syst. Sci. **63**(4), 512–530 (2001)
8. Kannan, R.: Minkowski's convex body theorem and integer programming. Math. Oper. Res. **12**(3), 415–440 (1987)
9. Lenstra, H.W.: Integer programming with a fixed number of variables. Math. Oper. Res. **8**(4), 538–548 (1983)
10. Li, N., Tripunitara, M.V., Wang, Q.: Resiliency policies in access control. ACM Trans. Inf. Syst. Secur. **12**(4), 113–137 (2009)
11. Sandhu, R.S., Coyne, E.J., Feinstein, H.L., Youman, C.E.: Role-based access control models. IEEE Comput. **29**(2), 38–47 (1996)
12. Wang, Q., Li, N.: Satisfiability and resiliency in workflow authorization systems. ACM Trans. Inf. Syst. Secur. **13**(4), 40 (2010)

Efficient Algorithms for the Order Preserving Pattern Matching Problem

Simone Faro[1]([✉]) and M. Oğuzhan Külekci[2]

[1] Department of Mathematics and Computer Science,
Università di Catania, Catania, Italy
faro@dmi.unict.it
[2] Informatics Institute, Istanbul Technical University, Istanbul, Turkey
kulekci@itu.edu.tr

Abstract. Given a pattern x of length m and a text y of length n, both over an ordered alphabet, the *order-preserving pattern matching* problem consists in finding all substrings of the text with the same relative order as the pattern. The OPPM, which might be viewed as an approximate variant of the well known *exact pattern matching* problem, has gained attention in recent years. This interesting problem finds applications in a lot of fields as from time series analysis, like share prices on stock markets or weather data analysis, to musical melody matching. In this paper we present two new filtering approaches which turn out to be much more effective in practice than the previously presented methods by reducing the number of false positives up to 99 %. From our experimental results it turns out that our proposed solutions are up to 2 times faster than the previous solutions.

Keywords: Approximate text analysis · Experimental algorithms · Filtering algorithms · Text processing

1 Introduction

Given a pattern x of length m and a text y of length n, both over a common alphabet Σ, the *exact string matching problem* consists in finding all occurrences of the string x in y. The *order-preserving pattern matching problem* [2,3,8,9] (OPPM in short) is an approximate variant of the exact pattern matching problem which has gained attention in recent years. In this variant the characters of x and y are drawn from an ordered alphabet Σ with a total order relation defined on it. The task of the problem is to find all substrings of the text with the same relative order as the pattern.

For instance the relative order of the sequence $x = \langle 6, 5, 8, 4, 7 \rangle$ is the sequence $\langle 3, 1, 0, 4, 2 \rangle$ since 6 has rank 3, 5 as rank 1, and so on. Thus x occurs in the string $y = \langle 8, 11, 10, 16, 15, 20, 13, 17, 14, 18, 20, 18, 25, 17, 20, 25, 26 \rangle$ at position 3, since x and the subsequence $\langle 16, 15, 20, 13, 17 \rangle$ share the same relative order. An other occurrence of x in y is at position 10 (see Fig. 1).

A preliminary version of this paper appeared in a technical report [8].

© Springer International Publishing Switzerland 2016
R. Dondi et al. (Eds.): AAIM 2016, LNCS 9778, pp. 185–196, 2016.
DOI: 10.1007/978-3-319-41168-2_16

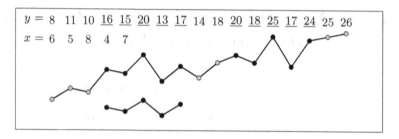

Fig. 1. Example of a pattern x of length 5 over an integer alphabet with two order preserving occurrences in a text y of length 17, at positions 3 and 10.

The OPPM problem finds applications in the fields where we are interested in finding patterns affected by relative orders, not by their absolute values. For example, it can be applied to time series analysis like share prices on stock markets, weather data or to musical melody matching of two musical scores.

In the last few years some solutions have been proposed for the order-preserving pattern matching problem. The first solution was presented by Kubica et al. [12] in 2013. They proposed a $\mathcal{O}(n+m\log m)$ solution over generic ordered alphabets based on the Knuth-Morris-Pratt algorithm [11] and a $\mathcal{O}(n+m)$ solution in the case of integer alphabets. Some months later Kim et al. [10] presented a similar solution running in $\mathcal{O}(n+m\log m)$ time based on the KMP approach. Although Kim et al. stressed some doubts about the applicability of the Boyer-Moore approach [2] to order-preserving matching problem, in 2013 Cho et al. [7] presented a method for deciding the order-isomorphism between two sequences showing that the Boyer-Moore approach can be applied also to the order-preserving variant of the pattern matching problem. In addition in [1] the authors showed that the Aho-Corasik approach can be applied to the OPPM problem for searching a set of patterns.

Chhabra and Tarhio in 2014 presented a new practical solution [5,6] based on filtration. Their algorithm translates the input sequences in two binary sequences and then use any standard exact pattern matching algorithm as a filtration procedure. In particular in their approach a sequence s is translated in a binary sequence β of length $|s|-1$ according to the following position

$$\beta[i] = \begin{cases} 1 & \text{if } s[i] \geq s[i+1] \\ 0 & \text{otherwise} \end{cases} \tag{1}$$

for each $0 \leq i < |s|-1$. This translation is unique for a given sequence s and can be performed on line on the text, requiring constant time for each text character.

Thus when a candidate occurrence is found during the filtration phase an additional verification procedure is run in order to check for the order-isomorphism of the candidate substring and the pattern. Despite its quadratic time complexity, this approach turns out to be simpler and more effective in practice than earlier solutions. It is important to notice that any algorithm for exact string matching can be used as a filtration method. The authors also proved that if the underlying filtration algorithm is sublinear and the text is translated on line, the complexity of the algorithm is sublinear on average. Experiments conducted in [5] show that the approach is faster than the algorithm by Cho *et al.*

In 2015 other efficient filtration algorithms have been presented. Specifically, in [3] the authors presented an efficient solution which combines the Skip-Search approach for searching and a multiple hash function technique for reducing the number of false positives during each bucket inspection. Additionally, in [4] the authors proposed two filtration online solutions based on SSE and AVX instructions, respectively. It turns out from experimental results conducted in [3] and in [4] that such solutions are faster than the previous algorithms in most cases.

In this paper we present two new families of filtering approaches which turn out to be much more effective in practice than the previously presented methods. While the technique proposed by Chhabra and Tarhio translates the input strings in binary sequences, our methods work on sequences over larger alphabets in order to speed up the searching process and reduce the number of false positives. From our experimental results it turns out that our proposed solutions are up to 2 times faster than the previous solutions reducing the number of false positives up to 99 % under suitable conditions.

The paper is organized as follows. In Sect. 2 we give preliminary notions and definitions relative to the order-preserving pattern matching problem. Then we present our new solutions in Sect. 3 and evaluate their performances against the previous algorithms in Sect. 4.

2   Notions and Basic Definitions

A string x over an ordered alphabet Σ, of size σ, is defined as a sequence of elements in Σ. We shall assume that a total order relation "$\leq$" is defined on it.

By $|x|$ we denote the length of a string x. We refer to the i-th element in x as $x[i]$ and use the notation $x[i .. j]$ to denote the subsequence of x from the element at position i to the element at position j (including the extremes), where $0 \leq i \leq j < |x|$. We say that two (nonnull) sequences x, y over Σ are order-isomorphic if the relative order of their elements is the same. More formally:

Definition 1 (Order-isomorphism). *Two nonnull sequences x, y of the same length, over a totally ordered alphabet $(\Sigma, \leq)$, are said to be* order-isomorphic, *and we write $x \approx y$, if the following condition holds*

$$x[i] \leq x[j] \quad \Longleftrightarrow \quad y[i] \leq y[j], \quad for \ \ 0 \leq i, j < |x|.$$

From a computational point of view, it is convenient to characterize the order of a sequence by means of two functions: the *rank* and the *equality* functions.

Definition 2 (Rank function). *Let x be a nonnull sequence over a totally ordered alphabet $(\Sigma, \leq)$. The* rank function *of x is the bijection from $\{0, 1, \ldots, |x| - 1\}$ onto itself defined, for $0 \leq i < |x|$, by*

$$rk_x(i) = \big|\{k : x[k] < x[i] \ or \ (x[k] = x[i] \ and \ k < i)\}\big|.$$

The following property is a trivial consequence of the Definition 2.

Corollary 1. *Let x be a non-null sequence drawn over a totally ordered alphabet $(\Sigma, \leq)$. Then we have $x[rk_x^{-1}(i)] \leq x[rk_x^{-1}(i+1)]$, for $0 \leq i < |x| - 1$.* ∎

Given any non-null sequence x, we shall refer to the corresponding non-null sequence $\langle rk_x^{-1}(0), rk_x^{-1}(1), \ldots, rk_x^{-1}(|x| - 1) \rangle$ as the *relative order* of x (see Example 1). From Corollary 1, it follows that the relative order of x can be computed in time proportional to the time required to (stably) sort x.

The rank function alone allows one to characterize order-isomorphic sequences only when characters are pairwise distinct. To handle the more general case in which multiple occurrences of the same character are permitted, we also need the *equality function*, which is defined next.

Definition 3 (Equality function). *Let x be a sequence of length $m \geq 2$ over a totally ordered alphabet $(\Sigma, \leq)$. The equality function of x is the binary map $eq_x: \{0, 1, \ldots, m - 2\} \to \{0, 1\}$ where, for $0 \leq i \leq m - 2$,*

$$eq_x(i) = \begin{cases} 1 & \text{if } x[rk_x^{-1}(i)] = x[rk_x^{-1}(i+1)] \\ 0 & \text{otherwise.} \end{cases}$$

Lemma 1. *For any two sequences x and y of the same length $m \geq 2$, over a totally ordered alphabet, $x \approx y$ if and only if $rk_x = rk_y$ and $eq_x = eq_y$.* ∎

Example 1. Let $x = \langle 6, 3, 8, 3, 10, 7, 10 \rangle$, $y = \langle 2, 1, 4, 1, 5, 3, 5 \rangle$, and let also $z = \langle 6, 3, 8, 4, 9, 7, 10 \rangle$. They have the same rank function $\langle 2, 0, 4, 1, 5, 3, 6 \rangle$ and, therefore, the same relative order $\langle 1, 3, 0, 5, 2, 4, 6 \rangle$. However, x and y are order-isomorphic, whereas x and z (as well as y and z) are not. Notice that we have $eq_x = eq_y = \langle 1, 0, 0, 0, 0, 1 \rangle$ and $eq_z = \langle 0, 0, 0, 0, 0, 0 \rangle$.

Thus in order to establish whether two given sequences of the same length m are order-isomorphic, it is enough to compute their rank and equality functions. The cost of the test is dominated by the cost $\mathcal{O}(m \log m)$ of sorting the sequences.

Lemma 2. *Let x and y be two sequences of the same length $m \geq 2$, over a totally ordered alphabet. Then $x \approx y$ if and only if the following conditions hold:*

(i) $y[rk_x^{-1}(i)] \leq y[rk_x^{-1}(i+1)]$, for $0 \leq i < m - 1$
(ii) $y[rk_x^{-1}(i)] = y[rk_x^{-1}(i+1)]$ if and only if $eq_x(i) = 1$, for $0 \leq i < m - 1$. ∎

Based on Lemma 2, the procedure ORDER-ISOMORPHIC verifies correctly whether a sequence y is order-isomorphic to a sequence x of the same length as y. It receives as input the functions rk_x and eq_x and the sequence y, and returns true if $x \approx y$, false otherwise. A mismatch occurs when one of the three conditions of lines 2, 3, or 4 holds. Notice that the time complexity of the procedure ORDER-ISOMORPHIC is linear in the size of its input sequence y.

Definition 4 (Order-preserving pattern matching). *Let x and y be two sequences of length m and n, respectively, with $n > m$, both over an ordered alphabet $(\Sigma, \leq)$. The order-preserving pattern matching problem consists in finding all positions i, with $0 \leq i \leq n - m$, such that $y[i .. i + m - 1] \approx x$.*

If $y[i .. i + m - 1] \approx x$, we say that x has an *order-preserving occurrence* in y at position i.

3   New Efficient Filter Based Algorithms

In this section we present two new general approaches for the OPPM problem. Both of them are based on a filtration technique, as in [5], but we use information extracted from groups of integers in the input string, as in [7], in order to make the filtration phase more effective in terms of efficiency and accuracy.

In our approaches we make use of the following definition of q-neighborhood of an element in an integer string.

Definition 5 (q-neighborhood). *Given a string x of length m, we define the q-neighborhood of the element $x[i]$, with $0 \leq i < m - q$, as the sequence of $q + 1$ elements from position i to $i+q$ in x, i.e. the sequence $\langle x[i], x[i+1], \ldots, x[i+q] \rangle$.*

The *accuracy* of a filtration method is a value indicating how many false positives are detected during the filtration phase, i.e. the number of candidate occurrences detected by the filtration algorithm which are not real occurrences of the pattern. The *efficiency* is instead related with the time complexity of the procedure we use for managing grams and with the time efficiency of the overall searching algorithm.

When using q-grams, a great accuracy translates in involving greater values of q. However, in this context, the value of q represents a trade-off between the computational time required for computing the q-grams for each window of the text and the computational time needed for checking false positive candidate occurrences. The larger is the value of q, the more time is needed to compute each q-gram. On the other hand, the larger is the value of q, the smaller is the number of false positives the algorithm finds along the text during the filtration.

3.1   The Neighborhood Ranking Approach

Given a string x of length m, we can compute the relative position of the element $x[i]$ compared with the element $x[j]$ by querying the inequality $x[i] \geq x[j]$. For brevity we will write in symbol $\beta_x(i,j)$ to indicate the boolean value resulting from the above inequality, extending the formal definition given in Eq. (1). Formally we have

$$\beta_x(i,j) = \begin{cases} 1 & \text{if } x[i] \geq x[j] \\ 0 & \text{otherwise} \end{cases} \tag{2}$$

It is easy to observe that if $\beta_x(i,j) = 1$ we have that $rk_x^{-1}(i) \geq rk_x^{-1}(j)$ ($x[j]$ precedes $x[i]$ in the ordering of the elements of x), otherwise $rk_x^{-1}(i) < rk_x^{-1}(j)$.

The neighborhood ranking (NR) approach associates each position i of the string x (where $0 \leq i < m - q$) with the sequence of the relative positions between $x[i]$ and $x[i+j]$, for $j = 1, \ldots, q$. In other words we compute the binary sequence $\langle \beta_x(i,i+1), \beta_x(i,i+2), \ldots, \beta_x(i,i+q) \rangle$ of length q indicating the relative positions of the element $x[i]$ compared with other values in its q-neighborhood. Of course, we do not include in the sequence the relative position of $\beta(i,i)$, since it doesn't give any additional information.

Neighborhood Ranking	Example	NR seq.	$\chi^3_x[i]$
$x[i] \leq x[i+1], x[i+2], x[i+3]$		$\langle 0,0,0 \rangle$	0
$x[i+3] \leq x[i] \leq x[i+1], x[i+2]$		$\langle 0,0,1 \rangle$	1
$x[i+2] \leq x[i] \leq x[i+1], x[i+3]$		$\langle 0,1,0 \rangle$	2
$x[i+2], x[i+3] \leq x[i] \leq x[i+1]$		$\langle 0,1,1 \rangle$	3
$x[i+1] \leq x[i] \leq x[i+2], x[i+3]$		$\langle 1,0,0 \rangle$	4
$x[i+1], x[i+3] \leq x[i] \leq x[i+2]$		$\langle 1,0,1 \rangle$	5
$x[i+1], x[i+2] \leq x[i] \leq x[i+3]$		$\langle 1,1,0 \rangle$	6
$x[i+1], x[i+2], x[i+3] \leq x[i]$		$\langle 1,1,1 \rangle$	7

Fig. 2. The 2^3 possible 3-neighborhood ranking sequences associated with element $x[i]$, and their corresponding NR value. In the leftmost column we show the ranking position of $x[i]$ compared with other elements in its neighborhood $\langle x[i], x[i+1], x[i+2], x[i+3] \rangle$.

Since there are 2^q possible configurations of a binary sequence of length q the string x is converted in a sequence χ^q_x of length $m - q$, where each element $\chi^q_x[i]$, for $0 \leq i < m - q$, is a value such that $0 \leq \chi^q_x[i] < 2^q$ (Fig. 2).

More formally we have the following definition

Definition 6 (q-NR sequence). *Given a string x of length m and an integer $q < m$, the q-NR sequence associated with x is a numeric sequence χ^q_x of length $m - q$ over the alphabet $\{0, \ldots, 2^q\}$ where*

$$\chi^q_x[i] = \sum_{j=1}^{q} \left(\beta_x(i, i+j) \times 2^{q-j} \right), \quad \text{for all } 0 \leq i < m - q$$

Example 2. Let $x = \langle 5, 6, 3, 8, 10, 7, 1, 9, 10, 8 \rangle$ be a sequence of length 10. The 4-neighborhood of the element $x[2]$ is the subsequence $\langle 3, 8, 10, 7, 1 \rangle$. Observe that $x[2]$ is greater than $x[6]$ and less than all other values in its 4-neighborhood. Thus the ranking sequence associated with the element of position 2 is $\langle 0, 0, 0, 1 \rangle$ which translates in a NR value equal to 1. In a similar way we can observe that the NR sequence associated with the element of position 3 is $\langle 0, 1, 1, 0 \rangle$ which translates in a NR value equal to 6. The whole 4-NR sequence of length 6 associated to x is $\chi^4_x = \langle 4, 8, 1, 6, 15, 8 \rangle$.

The following Lemma 3 and Corollary 2 prove that the NR approach can be used to filter a text y in order to search for all order preserving occurrences of a pattern x, i.e. $\{i \mid x \approx y[i \ldots i + m - 1]\} \subseteq \{i \mid \chi^q_x = \chi^q_y[i \ldots i + m - k]\}$.

Lemma 3. *Let x and y be two sequences of length m and let χ^q_x and χ^q_y the q-ranking sequences associated to x and y, respectively. If $x \approx y$ then $\chi^q_x = \chi^q_y$.*

Proof. Let rk be the rank function associated to x and suppose by hypothesis that $x \approx y$. Then the following statements hold

1. by Definition 2 we have $x[rk_x^{-1}(i)] \leq x[rk_x^{-1}(i+1)]$, for $0 \leq i < m-1$;
2. by hypothesis and Definition 1, $y[rk_x^{-1}(i)] \leq y[rk_x^{-1}(i+1)]$, for
    $0 \leq i < m-1$;
3. then by *1* and *2*, $x[i] \leq x[j]$ iff $y[i] \leq y[j]$, for $0 \leq i, j < m-1$;
4. the previous statement implies that $x[i] \geq x[i+j]$ iff $y[i] \geq y[i+j]$
    for $0 \leq i < m-q$ and $1 \leq j < q$;
5. by statement *4* we have that $\beta_x(i, i+j) = \beta_y(i, j+j)$
    for $0 \leq i < m-q$ and $1 \leq j < q$;
6. finally, by statement *5* and Definition 6, $\chi_x^q[i] = \chi_y^q[i]$, for $0 \leq i < m-q$.

This last statement proves the thesis.   ∎

The following corollary proves that the NR approach can be used as a filtering. It trivially follows from Lemma 3.

Corollary 2. *Let x and y be two sequences of length m and n, respectively. Let χ_x^q and χ_y^q the q-ranking sequences associated to x and y, respectively. If $x \approx y[j \ldots j+m-1]$ then $\chi_x^q[i] = \chi_y^q[j+i]$, for $0 \leq i < m-q$.*   ∎

Figure 4 (on the left) shows the procedure for computing the NR value associated with the element of the string x at position i. The time complexity of the procedure is $\mathcal{O}(q)$. Thus, given a pattern x of length m, a text y of length n and an integer value $q < m$, we can solve the OPPM problem by searching χ_y^q for all occurrences of χ_x^q, using any algorithm for the exact string matching problem. During the preprocessing phase we compute the sequence χ_x^q and the functions rk_x and eq_x. When an occurrence of χ_x^q is found at position i the verification procedure ORDER-ISOMORPHIC($inv\text{-}rk, eq, y[i \ldots i+m-1]$) (shown in Fig. 3) is run in order to check if $x \approx y[i \ldots i+m-1]$.

Since in the worst case the algorithm finds a candidate occurrence at each text position and each verification costs $\mathcal{O}(m)$, the worst case time complexity of the algorithm is $\mathcal{O}(nm)$, while the filtration phase can be performed with a $\mathcal{O}(nq)$ worst case time complexity. However, following the same analysis of [5], we easily prove that verification time approaches zero when the length of the pattern grows, so that the filtration time dominates. Thus if the filtration algorithm is sublinear, the total algorithm is sublinear.

3.2 The Neighborhood Ordering Approach

The neighborhood ranking approach described in the previous subsection gives partial information about the relative ordering of the elements in the neighborhood of an element in x. The binary sequence used to represent each element $x[i]$ is not enough to describe the full ordering information of a set of $q+1$ elements.

The q-neighborhood ordering (NO) approach, which we describe in this section, associates each element of x with a binary sequence which completely describes the ordering disposition of the elements in the q-neighborhood of $x[i]$. The number of comparisons we need to order a sequence of $q + 1$ elements is between q (the best case) and $q(q + 1)/2$ (the worst case). In this latter case it is enough to compare the element $x[j]$, where $i \leq j < i + q$, with each element $x[h]$, where $j < h \leq i + q$. Thus each element of position i in x, with $0 \leq i < m - q$, is associated with a binary sequence of length $q(q + 1)/2$ which completely describes the relative order of the subsequence $x[i, \ldots, i + q]$. Since there are $(q + 1)!$ possible permutations of a set of $q + 1$ elements, the string x is converted in a sequence φ_x^q of length $m - q$, where each element $\varphi_x^q[i]$ is a value such that $0 \leq \varphi_x^q[i] < q(q + 1)/2$. More formally we have the following definition

Definition 7 (q-NO sequence). *Given a string x of length m and an integer $q < m$, the q-NO sequence associated with x is a numeric sequence φ_x^q of length $m - q$ over the alphabet $\{0, \ldots, q(q + 1)/2\}$ where*

$$\varphi_x^q[i] = \sum_{k=1}^{q} \left(\chi_x^k[i + q - k] \times 2^{(k)(k-1)/2} \right), \quad \text{for all } 0 \leq i < m - q \tag{3}$$

Thus the q-NO value associated to $x[i]$ is the combination of q different NR sequences $\chi_x^q[i]$, $\chi_x^{q-1}[i + 1]$, $\ldots$, $\chi_x^1[i + q - 1]$.

For instance the 4-NO value associated to $x[i]$ is computed as

$$\varphi_x^4[i] = \chi_x^4[i] \times 2^6 + \chi_x^3[i + 1] \times 2^2 + \chi_x^2[i + 2] \times 2 + \chi_x^1[i + 3]$$

Neighborhood Ordering	Example	NO seq.	$\varphi_x^4[i]$
$\langle x[i], x[i + 1], x[i + 2] \rangle$		$\langle 0, 0, 0 \rangle$	0
$\langle x[i], x[i + 2], x[i + 1] \rangle$		$\langle 0, 0, 1 \rangle$	1
$\langle x[i + 2], x[i], x[i + 1] \rangle$		$\langle 0, 1, 1 \rangle$	3
$\langle x[i + 1], x[i], x[i + 2] \rangle$		$\langle 1, 0, 0 \rangle$	4
$\langle x[i + 1], x[i + 2], x[i] \rangle$		$\langle 1, 1, 0 \rangle$	6
$\langle x[i + 2], x[i + 1], x[i] \rangle$		$\langle 1, 1, 1 \rangle$	7

Fig. 3. The 3! possible ordering of the sequence $\langle x[i], x[i + 1], x[i + 2] \rangle$ and the corresponding binary sequence $\langle \beta_x(i, i + 1), \beta_x(i, i + 2), \beta_x(i + 1, i + 2) \rangle$.

Example 3. As in *Example 2*, let $x = \langle 5, 6, 3, 8, 10, 7, 1, 9, 10, 8 \rangle$ be a sequence of length 10. The 3-neighborhood of the element $x[3]$ is the subsequence $\langle 8, 10, 7, 1 \rangle$. The NO sequence of length 6 associated with the element of position 2 is therefore $\langle 0, 1, 1, 1, 1, 1 \rangle$ which translates in a NO value equal to $\varphi_x[3] = 31$. In a similar way we can observe that the NR sequence associated with the element of position 2 is $\langle 0, 0, 0, 0, 1, 1 \rangle$ which translates in a NO value equal to $\varphi_x^4[2] = 3$. The whole sequence of length 7 associated to x is $\varphi_x^4 = \langle 20, 32, 3, 31, 60, 32, 3 \rangle$.

COMPUTE-NR-VALUE(x, i, q)   COMPUTE-NO-VALUE(x, i, q)
1. $\delta \leftarrow 0$ 5. $\delta \leftarrow 0$
2.   for $j \leftarrow 1$ to q do   6.   for $k \leftarrow q$ downto 1 do
3. $\delta = (\delta \ll 1) + \beta_x(i, i+j)$ 7.   for $j \leftarrow 1$ to k do
4.   return δ 8. $\delta = (\delta \ll 1) + \beta_x(i+q-k, i+q-k+j)$
                                          9.   return δ

Fig. 4. The two functions which compute the q-neighborhood ranking value (on the left) and the q-neighborhood ranking value (on the right).

The following Lemma 4 and Corollary 3 prove that the NO approach can be used to filter a text y in order to search for all order preserving occurrences of a pattern x. In other words they prove that $\{i \mid x \approx y[i \ldots i+m-1]\} \subseteq \{i \mid \varphi_x^q = \varphi_y^q[i \ldots i+m-k]\}$. The proof easily follows from Definition 7 and Lemma 3.

Lemma 4. *Let x and y be two sequences of length m and let φ_x^q and φ_y^q the q-ranking sequences associated to x and y, respectively. If $x \approx y$ then $\varphi_x^q = \varphi_y^q$.*

The following corollary proves that the NR approach can be used as a filtering. It trivially follows from Lemma 4.

Corollary 3. *Let x and y be two sequences of length m and n, respectively. Let χ_x^q and χ_y^q the q-ranking sequences associated to x and y, respectively. If $x \approx y[j \ldots j+m-1]$ then $\chi_x^q[i] = \chi_y^q[j+i]$, for $0 \leq i < m-q$.* ∎

Figure 4 (on the right) shows the procedure used for computing the NO value associated with the element of the string x at position i. The time complexity of the procedure is $\mathcal{O}(q^2)$. Thus, given a pattern x of length m, a text y of length n and an integer value $q < m$, we can solve the OPPM problem by searching φ_y^q for all occurrences of φ_x^q, using any algorithm for the exact string matching problem. During the preprocessing phase we compute the sequence φ_x^q and the functions rk_x and eq_x. When an occurrence of φ_x^q is found at position i the verification procedure ODER-ISOMORPHIC(inv-rk, eq, $y[i \ldots i+m-1]$) is run in order to check if $x \approx y[i \ldots i+m-1]$. Also in this case, if the filtration algorithm is sublinear on average, the NO approach has a sublinear behavior on average.

4   Experimental Evaluations

In this section we present experimental results in order to evaluate the performances of our new filter based algorithms presented in this paper. In particular we tested our filter approaches against three algorithms: the filter approach of Chhabra and Tarhio [5]; the SkSop(k, q) algorithm [3], using k hash functions and q-grams. We implemented it using $1 \leq k \leq 5$ and $3 \leq q \leq 8$; the algorithm based on SSE instructions presented in [4].

According to the experimental evaluations conducted in [5] and in [4] in our experimental evaluation we use in all cases the SBNDM2 algorithm [9]. In our dataset we use the following names to identify the tested algorithms: FCT to identify the SBNDM2 based algorithm by Chhabra and Jorma Tarhio [5]; NRq to identify the SBNDM2 algorithm based on the NR approach (Sect. 3.1); and NOq: to identify the SBNDM2 algorithm based on the NO approach (Sect. 3.2).

We evaluated our filter based solutions in terms of efficiency, i.e. the running times. In particular for the FCT algorithm we will report the average running times, in milliseconds. Instead, for all other algorithms in the set, we will report the speed up of the running times obtained when compared with the time used by the FCT algorithm. In the case of the SkSop(k, q) algorithm we reported only the best speed-up among all different implementations, indicating in brackets the best values of k and q.

We tested our solutions on sequences of short integer values (each element is an integer in the range $[0 \ldots 256]$), long integer values (where each element is an integer in the range $[0 \ldots 10.000]$) and floating point values (each element is a floating point in the range $[0.0 \ldots 10000.99]$). However we don't observe sensible differences in the results, thus in the following tables we report for brevity the results obtained on short integer sequences. All texts have 1 million of elements. In particular we tested our algorithm on two sequences: a RAND-δ sequence of random integer values varying around a fixed mean equal to 100 with a variability of δ; a PERIOD-δ sequence of random integer values varying around a periodic function with a period of 10 elements with a variability of δ.

Table 1. Experimental results on a RAND-δ short integer sequence.

δ	m	FCT	SkSop	SSE	NR2	NR3	NR4	NR5	NR6	NO2	NO3	NO4
	8	44.29	$1.27^{(5,4)}$	1.45	1.16	1.28	1.25	1.25	1.24	<u>1.89</u>	1.71	1.11
	12	28.39	$1.37^{(4,5)}$	1.38	1.16	1.37	1.37	1.33	1.19	1.64	<u>2.00</u>	1.64
	16	20.65	$1.52^{(4,5)}$	1.23	1.15	1.30	1.43	1.34	1.14	1.42	<u>2.01</u>	1.83
5	20	16.29	$1.58^{(4,5)}$	1.12	1.15	1.30	1.45	1.41	1.14	1.39	<u>2.00</u>	1.93
	24	13.64	$1.63^{(5,4)}$	1.05	1.16	1.29	1.42	1.44	1.12	1.34	1.91	<u>2.01</u>
	28	11.48	$1.62^{(5,4)}$	0.99	1.16	1.28	1.44	1.45	1.11	1.31	1.88	<u>1.96</u>
	32	10.34	$1.60^{(5,4)}$	0.83	1.18	1.30	1.40	1.46	1.12	1.30	1.83	<u>2.05</u>
	8	42.34	$1.22^{(3,4)}$	1.68	1.13	1.27	1.25	1.26	1.22	<u>1.92</u>	1.68	1.08
	12	27.93	$1.40^{(4,5)}$	1.44	1.17	1.40	1.37	1.32	1.21	1.71	<u>2.04</u>	1.63
	16	20.05	$1.46^{(4,5)}$	1.32	1.15	1.32	1.41	1.33	1.15	1.48	<u>2.04</u>	1.81
20	20	15.85	$1.51^{(4,5)}$	1.21	1.15	1.29	1.42	1.37	1.11	1.38	<u>2.00</u>	1.90
	24	13.31	$1.55^{(5,6)}$	1.12	1.17	1.31	1.47	1.42	1.12	1.36	1.99	<u>2.02</u>
	28	11.38	$1.58^{(5,6)}$	1.01	1.17	1.31	1.42	1.45	1.09	1.35	1.94	<u>2.07</u>
	32	9.96	$1.58^{(3,7)}$	0.97	1.16	1.29	1.45	1.46	1.09	1.29	1.87	<u>2.09</u>
	8	42.62	$1.19^{(3,4)}$	1.91	1.16	1.28	1.28	1.25	1.25	<u>1.94</u>	1.70	1.09
	12	28.35	$1.39^{(4,5)}$	1.82	1.19	1.41	1.39	1.36	1.21	1.75	<u>2.06</u>	1.65
	16	20.37	$1.43^{(4,5)}$	1.63	1.18	1.32	1.44	1.37	1.17	1.49	<u>2.09</u>	1.83
40	20	16.12	$1.52^{(5,6)}$	1.45	1.15	1.29	1.46	1.39	1.12	1.39	<u>2.04</u>	1.95
	24	13.35	$1.57^{(5,6)}$	1.21	1.18	1.30	1.46	1.44	1.13	1.36	1.97	<u>1.99</u>
	28	11.60	$1.57^{(5,6)}$	1.17	1.18	1.32	1.47	1.50	1.14	1.37	1.96	<u>2.06</u>
	32	10.06	$1.58^{(5,7)}$	0.99	1.16	1.29	1.45	1.48	1.10	1.33	1.89	<u>2.07</u>

For each text in the set we randomly select 100 patterns extracted from the text and compute the average running time over the 100 runs. We also computed the average number of false positives detected by the algorithms during the search. All the algorithms have been implemented using the C programming language and have been compiled on an MacBook Pro using the gcc compiler Apple LLVM version 5.1 (based on LLVM 3.4svn) with 8 Gb Ram. During the compilation we use the -O3 optimization option. In the following table running times are expressed in milliseconds. Best results have been underlined.

Experimental results on RAND-δ numeric sequences have been conducted with values of $\delta = 5$, 20, and 40 (see Table 1). The results show as the No approach is the best choice in all cases, achieving a speed up of 2.0 if compared with the FCT algorithm. Also the NR approach achieves always a good speed up which is between 1.15 and 1.50. In addition we can observe that in all cases the best speed-up achieved by the new algorithms are greater then that achieved by the SkSop and SSE algorithms. For the sake of completeness we report also that the gain in number of detected false positives in most cases between 90 % and 100 %.

Table 2. Experimental results on a PERIOD-δ short integer sequence.

δ	m	FCT	SkSop	SSE	NR2	NR3	NR4	NR5	NR6	No2	No3	No4
	8	41.08	$0.83^{(3,4)}$	1.10	0.99	<u>1.05</u>	0.88	0.79	0.90	0.88	0.73	0.60
	12	36.42	$0.88^{(4,6)}$	1.03	<u>1.06</u>	1.02	0.94	0.86	0.91	0.81	0.67	0.69
	16	34.03	$0.90^{(4,6)}$	0.97	<u>1.04</u>	0.86	0.78	0.74	1.00	0.77	0.64	0.60
5	20	35.31	$0.97^{(4,7)}$	0.90	<u>0.98</u>	0.89	0.88	0.84	0.92	0.73	0.60	0.55
	24	37.90	$1.05^{(4,7)}$	0.82	<u>1.34</u>	1.33	1.30	1.18	1.15	0.99	0.82	0.76
	28	36.26	$1.11^{(4,7)}$	0.75	<u>1.17</u>	1.09	1.10	1.04	0.97	0.78	0.64	0.56
	32	35.38	$1.16^{(4,8)}$	0.69	1.10	<u>1.15</u>	1.05	0.95	0.94	0.82	0.65	0.59
	8	42.35	$0.93^{(3,4)}$	1.12	0.98	<u>1.18</u>	0.91	0.81	0.89	1.02	0.83	0.68
	12	39.09	$0.97^{(4,5)}$	1.05	1.11	<u>1.14</u>	1.06	0.98	1.00	1.02	0.88	0.93
	16	34.25	$1.05^{(4,6)}$	1.01	<u>1.11</u>	1.01	1.02	1.01	1.08	0.96	0.87	0.87
20	20	35.41	$1.11^{(4,6)}$	0.96	1.10	1.09	1.21	<u>1.21</u>	1.07	0.97	0.89	0.89
	24	35.15	$1.18^{(3,7)}$	0.89	1.31	1.51	<u>1.67</u>	1.60	1.14	1.15	1.10	1.18
	28	32.23	$1.24^{(3,7)}$	0.81	1.23	1.40	<u>1.56</u>	1.36	1.07	1.04	1.08	1.15
	32	30.34	$1.36^{(3,7)}$	0.75	1.43	<u>1.60</u>	1.53	1.43	1.22	1.19	1.11	1.07
	8	45.07	$1.10^{(3,4)}$	1.15	0.93	<u>1.18</u>	0.94	0.81	0.89	1.12	0.91	0.78
	12	37.91	$1.14^{(4,5)}$	1.08	1.08	1.12	1.03	0.93	1.03	<u>1.13</u>	1.03	1.08
	16	32.41	$1.19^{(4,5)}$	1.04	1.11	1.04	1.06	<u>1.13</u>	1.07	1.07	1.02	1.10
40	20	28.63	$1.22^{(4,6)}$	1.00	1.05	1.09	1.24	<u>1.35</u>	1.08	1.04	1.04	1.15
	24	27.25	$1.28^{(4,6)}$	0.97	1.18	1.39	<u>1.59</u>	1.53	1.10	1.12	1.14	1.40
	28	24.91	$1.32^{(4,6)}$	0.95	1.20	1.51	<u>1.67</u>	1.41	1.05	1.17	1.30	1.50
	32	23.63	$1.38^{(4,6)}$	0.91	1.39	<u>1.63</u>	1.55	1.31	1.20	1.27	1.41	1.41

Experimental results on PERIOD-δ problem have been conducted on a periodic sequence with a period equal to 10 and with $\delta = 5$, 20 and 40 (see Table 2). The results show as the NR approach is the best choice in most of the cases, achieving a speed up of 1.3 in suitable conditions. However in some cases the FCT algorithm turns out to be the best choice especially on short patterns. The No approach is always less efficient of the FCT algorithm. When the size of δ increases the performances of the No approach get better achieving a speed up of 1.4 in the best cases. However the NR approach turns out to be always the best solutions with a speed up close to 1.7 for long patterns. Also in these cases the best speed-up achieved by the new algorithms are greater then that achieved by the SkSop and SSE algorithms.

References

1. Belazzougui, D., Pierrot, A., Raffinot, M., Vialette, S.: Single and multiple consecutive permutation motif search. In: Cai, L., Cheng, S.-W., Lam, T.-W. (eds.) Algorithms and Computation. LNCS, vol. 8283, pp. 66–77. Springer, Heidelberg (2013)
2. Boyer, R.S., Moore, J.S.: A fast string searching algorithm. Commun. ACM **20**(10), 762–772 (1977)
3. Cantone, D., Faro, S., Külekci, M.O.: An efficient skip-search approach to the order-preserving pattern matching problem. In: Holub and Zdárek [10], pp. 22–35
4. Chhabra, T., Külekci, M.O., Tarhio, J.: Alternative algorithms for order-preserving matching. In: Holub and Zdárek [10], pp. 36–46
5. Chhabra, T., Tarhio, J.: Order-preserving matching with filtration. In: Gudmundsson, J., Katajainen, J. (eds.) SEA 2014. LNCS, vol. 8504, pp. 307–314. Springer, Heidelberg (2014)
6. Chhabra, T., Tarhio, J.: A filtration method for order-preserving matching. Inf. Process. Lett. **116**(2), 71–74 (2016)
7. Cho, S., Na, J.C., Park, K., Sim, J.S.: Fast order-preserving pattern matching. In: Widmayer, P., Xu, Y., Zhu, B. (eds.) COCOA 2013. LNCS, vol. 8287, pp. 295–305. Springer, Heidelberg (2013)
8. Faro, S., Külekci, M.O.: Efficient algorithms for the order preserving pattern matching problem. CoRR abs/1501.04001 (2015). http://arxiv.org/abs/1501.04001
9. Holub, J., Durian, B.: Talk: fast variants of bit parallel approach to suffix automata. In: International Stringology Research Workshop (2005). http://www.cri.haifa.ac. il/events/2005/string/presentations/Holub.pdf
10. Kim, J., Eades, P., Fleischer, R., Hong, S., Iliopoulos, C.S., Park, K., Puglisi, S.J., Tokuyama, T.: Order-preserving matching. Theor. Comput. Sci. **525**, 68–79 (2014)
11. Knuth, D.E., Morris Jr., J.H., Pratt, V.R.: Fast pattern matching in strings. SIAM J. Comput. **6**(1), 323–350 (1977)
12. Kubica, M., Kulczynski, T., Radoszewski, J., Rytter, W., Walen, T.: A linear time algorithm for consecutive permutation pattern matching. Inf. Process. Lett. **113**(12), 430–433 (2013)

Computing the Line-Constrained k-center in the Plane for Small k

Albert Jhih-Heng Huang[1], Hung-Lung Wang[2(✉)], and Kun-Mao Chao[1]

[1] Department of Computer Science and Information Engineering,
National Taiwan University, Taipei, Taiwan
[2] Institute of Information and Decision Sciences,
National Taipei University of Business, Taipei, Taiwan
hlwang@ntub.edu.tw

Abstract. In this paper, we study the line-constrained k-center problem in the Euclidean plane. Given a set of demand points and a line L, the problem asks for k points, called center facilities, on L, such that the maximum of the distances from the demand points to their closest center facilities is minimized. For any fixed k, we propose an algorithm with running time linear to the number of demand points.

1 Introduction

Geometric facility location optimization is an important research topic that asks for appropriate locations of facilities to serve a number of demand points. In general, given a positive integer k and a set of n demand points in $\mathbb{R}^d$, the *k-center problem* is to find k points, to deploy the center facilities, such that the maximum of the distances from the demand points to their closest center facilities is minimized. There is an example for the line-constrained k-center problem in Fig. 1, where the given line is x-axis and $k = 4$. When k is not fixed, the k-center problem is NP-hard [10], even in $\mathbb{R}^2$. However, for any fixed k, the problem can be solved in polynomial time, and there are many elegant algorithms tackling related problems.

For the Euclidean 1-center problem in $\mathbb{R}^2$, Megiddo [9] gave an $O(n)$-time algorithm, based on the technique of prune-and-search. This technique was then widely applied in dealing with the center facility location problems [2,3,6]. For the Euclidean 2-center problem in $\mathbb{R}^2$, Chan [5] proposed an $O(n \log^2 n (\log \log n)^2)$ time deterministic algorithm and an $O(n \log^2 n)$ randomized algorithm. If all demand points are in convex position, Kim and Shin [8] proposed an algorithm which runs in $O(n \log^3 n \log \log n)$ time. They also gave an $O(n \log^2 n)$-time algorithm for the discrete version where the center facilities have to be deployed at the demand points. If the demand points are in general position, Agarwal *et al.* [1] proposed an $O(n^{4/3} \log^5 n)$-time algorithm for the discrete 2-center problem.

In this paper, we are concerned with a variant of the k-center problem, called the *line-constrained k-center problem*. In this problem, the requested k points

© Springer International Publishing Switzerland 2016
R. Dondi et al. (Eds.): AAIM 2016, LNCS 9778, pp. 197–208, 2016.
DOI: 10.1007/978-3-319-41168-2_17

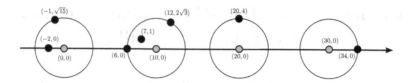

Fig. 1. An example for line-constrained k-center problem.

have to be on a given line. Some results were proposed recently. In 2011, Brass *et al.* [4] gave an $O(n \log^2 n)$-time algorithm. Later, in 2013, Karmakar *et al.* [7] improved the time complexity to $O(n \log n)$. In 2014, Wang and Zhang [11] proposed an $O(n \log n)$-time algorithm for a weighted version of this problem. They also investigated the line-constrained k-center problem under different distance metrics, like L_1- and L_∞-norms.

The main result of this paper is an $O(n)$-time algorithm for the line-constrained k-center problem in $\mathbb{R}^2$, for any fixed k. To solve the problem, we formulate it in a more general way so that we can solve it recursively. This general problem is called the *line-constrained conditional k-center problem*. We organize this paper as follows. First, we formally define the line-constrained conditional k-center problem in Sect. 2. In Sect. 3, we introduce an $O(n^2 \log n)$-time algorithm for the conditional k-center problem. In Sect. 4, we propose a linear-time algorithm for any fixed k. Finally, concluding remarks are given in Sect. 5.

2   Preliminaries

Without loss of generality, we assume that the given line is the x-axis, denoted by X. The x and y coordinates of a point u are denoted by x_u and y_u, respectively. For succinctness, a point u on X is also referred to as x_u, and the distance from a point v to a subset C of X is defined as

$$d(v, C) = \min\{d(v, c) \colon c \in C\},$$

where $d(v, c)$ is the distance between v and c. The line-constrained conditional k-center problem is defined as follows.

Problem 1 (line-constrained conditional k-center problem). Given a nonnegative integer k and two finite sets Δ and Σ with $\Delta \subseteq \mathbb{R}^2$ and $\Sigma \subseteq X$, the problem asks for a subset Γ of X such that $|\Gamma| = k$ and

$$\max\{ d(u, \Gamma \cup \Sigma) \colon u \in \Delta \}$$

is minimized.

In the following, an instance is denoted by (Δ, Σ, k). The requested minimum with respect to (Δ, Σ, k) is called the *radius* of the instance and is denoted by $r(\Delta, \Sigma, k)$. In the discussion below, a solution is alternatively represented as a k-tuple $(c_1, c_2, \ldots, c_k)$, where each c_i is called a *center facility*. When this ordered representation is adopted, we assume that

$$\forall_{i<j\in[k]}, c_i \le c_j,$$

See Fig. 2 for an illustration.

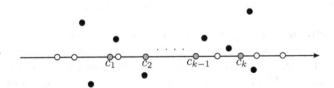

Fig. 2. An instance (Δ, Σ, k) and a set of center facilities. The set Δ consists of black points, and Σ consists of white points. Gray points are the center facilities.

Given a solution $(c_1, c_2, \ldots, c_k)$ of (Δ, Σ, k), the *range* of c_i consists of the demand points closer to c_i, i.e. $\{u \in \Delta: \forall_{j\in[k]} \quad d(u, c_i) \le d(u, c_j)\}$, denoted by $R(c_i)$. For simplicity, the line-constrained conditional k-center problem is referred to as the *conditional k-center problem* in the following. Obviously, the conditional k-center problem is a generalization of the line-constrained k-center problem, but is much more general than what we need. To solve the line-constrained k-center problem, it suffices to consider a restricted version of the conditional k-center problem, given as follows. Let $\Delta = \{v_1, v_2, \ldots, v_n\}$ and $\Sigma = \{c_0, c_{k+1}\}$. We assume

- $\forall_{i<j}, -\infty < x_{v_i} < x_{v_j} < \infty$
- $-\infty \le c_0 < c_{k+1} \le \infty$
- $c_0 \le x_{v_1} < x_{v_n} \le c_{k+1}$

With this assumption, for any instance, we may assume that $|\Sigma| = 2$. In the following, we let $|\Delta| = n$.

Remark 1. Clearly, by letting $\Sigma = \{-\infty, \infty\}$, the conditional k-center problem on the instance (Δ, Σ, k) is equivalent to the line-constrained k-center problem with demand set Δ. Moreover, it can be easily verified that there is an optimal solution Γ such that

$$x_{v_1} \le \min\{x_u: u \in \Gamma\} \le \max\{x_u: u \in \Gamma\} \le x_{v_n}.$$

3   An $O(n^2 \log n)$-Time Algorithm

The line-constrained k-center problem was investigated by Karmakar *et al.* [7], and some algorithms, of various time and space complexity, were proposed. They solved the problem by considering a decision version of the line-constrained

k-center problem, which asks whether $r(\Delta, \Sigma, k) \leq r$ for a given value r. For any given r, it was shown that this decision problem can be solved in $O(n)$ time. To solve the line-constrained k-center problem, they identified a set consisting of $O(n^2)$ candidate values of r and performed binary search on the set. The candidate set consists of

- y_u, where $u \in \Delta$
- d_{uv}, where $d_{uv} = d(u, w) = d(v, w)$ for some w on X.

The idea of Karmakar et $al.$ [7] can be easily extended to solve the conditional k-center problem. First, for a given nonnegative value r, by removing all demand points within distance r to some point in Σ, the decision version of the conditional k-center problem is reduced to that of the line-constrained k-center problem. Second, for the candidate set, beside y_u and d_{uv} mentioned above, the following values have to be included.

- $d(u, v)$, where $u \in \Sigma$ and $v \in \Delta$.

With this modification, one can easily solve the conditional k-center problem. A naive implementation takes $O(n^2 \log n)$ time. Instead of revising the algorithm as Karmakar et $al.$ did [7], we end this section with this straightforward result[1].

Theorem 1. *The conditional k-center problem can be solved in $O(n^2 \log n)$ time.*

4   An $O(n)$-Time Algorithm for Any Fixed k

The technique we apply to derive a linear-time algorithm is prune-and-search. Roughly speaking, for any instance, we efficiently remove a fixed portion of the demand points and then recursively solve the problem. We note here that the idea was extended from the algorithm of Bhattacharya et $al.$ [3], who considered the k-center problem in $\mathbb{R}^1$.

The algorithm we propose is called CONDITIONALCENTER, whose details are given in Sect. 4.2. The procedure outputs the radius of an instance. Before going into the details of our algorithm, we give a simple but essential observation, which provides the information of bounds on the radius as well as the location of some center facilities. For any point α on the x-axis, the bipartition (Δ', Δ'') of Δ with respect to α is defined as

$$\Delta' = \{u \in \Delta \colon x_u \leq \alpha\} \quad \text{and} \quad \Delta'' = \Delta \setminus \Delta'.$$

Lemma 1. *Given an instance (Δ, Σ, k) with $\Sigma = \{c_0, c_{k+1}\}$ and a point α on the x-axis, let (Δ', Δ'') be the bipartition of Δ with respect to α. For any $i \in [k]$, we have*

$$\min\{r_1, r_2\} \leq r(\Delta, \Sigma, k) \leq \max\{r_1, r_2\},$$

---

[1] As indicated in Sect. 4, what we need is an algorithm for solving small instances.

where $r_1 = r(\Delta', \{c_0, \alpha\}, i - 1)$ and $r_2 = r(\Delta'', \{\alpha, c_{k+1}\}, k - i)$. Moreover, if $r_1 \geq r_2$, there is an optimal solution $(c_1, c_2, \ldots, c_k)$ such that $c_i \leq \alpha$.

Proof. Suppose first that $r_1 < r_2$ (resp., $r_1 > r_2$). By considering α as the ith center facility, with the solutions of $(\Delta', \{c_0, \alpha\}, i-1)$ and $(\Delta'', \{\alpha, c_{k+1}\}, k - i)$, the inequality $r_1 < r(\Delta, \Sigma, k) \leq r_2$ (resp., $r_2 < r(\Delta, \Sigma, k) \leq r_1$) holds immediately.

Next, consider the case where $r_1 = r_2$. We claim that $r(\Delta, \Sigma, k) = r_1$. Obviously,

$$r(\Delta, \Sigma, k) \leq r_1. \tag{1}$$

For any optimal solution $(c_1, c_2, \ldots, c_k)$ of (Δ, Σ, k), consider the bipartition (Δ^*, Δ^{**}) of Δ with respect to c_i. We have either $\Delta' \subseteq \Delta^*$ (see Fig. 3(a)) or $\Delta'' \subseteq \Delta^{**}$ (see Fig. 3(b)), which leads to the fact

$$r_1 \leq \max\{r(\Delta^*, \{c_0, c_i\}, i - 1), r(\Delta^{**}, \{c_i, c_{k+1}\}, k - i)\} = r(\Delta, \Sigma, k). \tag{2}$$

By inequalities (1) and (2), we get the requested upper and lower bounds on $r(\Delta, \Sigma, k)$. The existence of an optimal solution $(c_1, c_2, \ldots, c_k)$ such that $c_i \leq \alpha$ can also be guaranteed with the arguments given above. $\qquad\square$

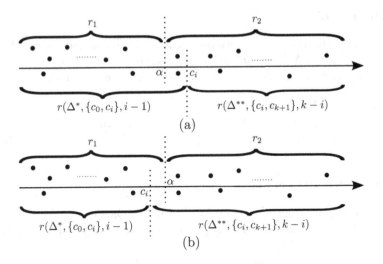

Fig. 3. (a) $\Delta' \subseteq \Delta^*$ and (b) $\Delta'' \subseteq \Delta^{**}$.

With Lemma 1, we can determine that the ith center facility is on the left or right of a point x, as CENTERLOCATION.

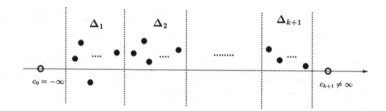

Fig. 4. The balanced partition. In the partition, $\Delta_0 = \emptyset$ since $c_0 = -\infty$. Similarly, Δ_{k+1} is nonempty since $c_{k+1} \neq \infty$. This balanced partition of Δ consists of $k+1$ disjoint subsets, where $\left\lfloor \frac{n}{k+1} \right\rfloor \leq |\Delta_i| \leq \left\lceil \frac{n}{k+1} \right\rceil$ for all $i \in [k+1]$.

CENTERLOCATION(Δ, Σ, k, i, x)
1   $(\Delta', \Delta'') \leftarrow$ BIPARTITION(Δ, x)
    ▷ the bipartition of Δ w.r.t x
2   $r_1 \leftarrow$ CONDITIONALCENTER($\Delta', \{c_0, x\}, i - 1$)
3   $r_2 \leftarrow$ CONDITIONALCENTER($\Delta'', \{x, c_{k+1}\}, k - i$)
4   **if** $r_1 \leq r_2$
5   **then return** RIGHT
6   **else return** LEFT

4.1 Big Region

To recursively solve the problem, we remove a subset of demand points which are from a *big region* of (Δ, Σ, k), defined as follows. Let $k' = k + |\{c \in \Sigma : c \neq \pm\infty\}|$. A *balanced partition* $(\Delta_0, \Delta_1, \ldots, \Delta_{k+1})$ of Δ is a partition satisfying

- $\forall_{i \in [k]}, \left\lfloor \frac{n}{k'} \right\rfloor \leq |\Delta_i| \leq \left\lceil \frac{n}{k'} \right\rceil$
- if $c_0 = -\infty$ (resp. $c_{k+1} = \infty$), then $|\Delta_0| = \emptyset$ (resp. $|\Delta_{k+1}| = \emptyset$)
- if $c_0 \neq -\infty$ (resp. $c_{k+1} \neq \infty$), then $\left\lfloor \frac{n}{k'} \right\rfloor \leq |\Delta_0| \leq \left\lceil \frac{n}{k'} \right\rceil$ (resp. $\left\lfloor \frac{n}{k'} \right\rfloor \leq |\Delta_{k+1}| \leq \left\lceil \frac{n}{k'} \right\rceil$)

An illustration is given in Fig. 4.

Definition 1 (big region). *Given* (Δ, Σ, k), *let* $(\Delta_0, \Delta_1, \ldots, \Delta_{k+1})$ *be a balanced partition of* Δ, *and let* $\Delta^l = \bigcup_{j=0}^{l} \Delta_j$. *The subset* Δ_0 *is a big region if*

$$r(\Delta_0, \{c_0, \infty\}, 0) \leq r(\Delta, \Sigma, k).$$

For $i \in [k+1]$, Δ_i *is a big region if*

$$r(\Delta^{i-1}, \{c_0, \infty\}, i - 1) \geq r(\Delta, \Sigma, k)$$

and

$$r(\Delta^i, \{c_0, \infty\}, i) \leq r(\Delta, \Sigma, k)$$

An essential property of a big region is that all demand points in it belong to a center facility, as shown in Lemma 2.

Lemma 2. *Given (Δ, Σ, k), let $(\Delta_0, \Delta_1, \ldots, \Delta_{k+1})$ be a balanced partition of Δ. If Δ_i is a big region of the instance, then there is an optimal solution $(c_1, c_2, \ldots, c_k)$ such that $\Delta_i \subseteq R(c_i)$.*

Proof. The lemma holds immediately for $i = 0$. For $i \geq 1$, by definition, we have

$$r(\Delta^{i-1}, \{c_0, \infty\}, i - 1) \geq r(\Delta, \Sigma, k)$$

and

$$r(\Delta \setminus \Delta^i, \{-\infty, c_{k+1}\}, k - i) \geq r(\Delta, \Sigma, k),$$

It follows that there is an optimal solution $(c_1, c_2, \ldots, c_k)$ of (Δ, Σ, k) such that

$$\bigcup_{j=0}^{i-1} R(c_j) \subseteq \Delta^{i-1}$$

and

$$\bigcup_{i+1}^{k+1} R(c_j) \subseteq \Delta \setminus \Delta^i.$$

Thus, $\Delta_i \subseteq R(c_i)$. $\square$

To compute a big region of an instance, one can iteratively verify whether

$$r(\Delta^i, \{c_0, \infty\}, i) \leq r(\Delta, \Sigma, k).$$

Notice that this verification can be done in $O(n)$ time using the algorithm of Karmakar *et al.* [7], as mentioned in Sect. 3. The procedure is given in BIGREGION.

BIGREGION$(\Delta, \{c_0, c_{k+1}\}, k)$
1   $(\Delta_0, \Delta_1, \ldots, \Delta_{k+1}) \leftarrow$ a balanced partition of Δ
2   **if** $\Delta_0 = \emptyset$
3   **then** $i \leftarrow 1$
4   **else** $i \leftarrow 0$
5   **if** $\Delta_{k+1} = \emptyset$
6   **then** $End \leftarrow k$
7   **else** $End \leftarrow k + 1$
8   **while** $i < End$
9   **do** $r \leftarrow$ CONDITIONALCENTER$(\Delta^i, \{c_0, \infty\}, i)$
10   **if** $r < r(\Delta, \Sigma, k)$
11   **then return** Δ_i
12             $i \leftarrow i + 1$
13 **return** Δ_i

4.2   Solving the Problem Recursively

In this section, we show how the problem can be solved recursively. More precisely, for an instance (Δ, Σ, k), a set D of demand points in a big region can be discarded so that $r(\Delta, \Sigma, k) = r(\Delta \setminus D, \Sigma, k)$. Moreover, there is a set of center facilities that is a solution of both (Δ, Σ, k) and $(\Delta \setminus D, \Sigma, k)$. For the extremal cases, i.e., taking the big region as Δ_0 or Δ_{k+1}, there is a straightforward result as indicated in Lemma 3.

Lemma 3. *Let $u \in \Delta_i$ such that $d(u, c_i) = \max\{d(v, c_i) \colon v \in \Delta_i\}$. If Δ_i is a big region with $i \in \{0, k+1\}$, then $r(\Delta, \Sigma, k) = r(\Delta \setminus (\Delta_i \setminus \{u\}), \Sigma, k)$.*

For $i \in [k]$, let x^* be the median of $\{x_u \colon u \in \Delta_i\}$. The *balanced bipartition* $(\hat{\Delta}, \check{\Delta})$ is a bipartition of Δ_i with respect to x^*. In the following, we elaborate on the case where $c_i > x^*$ and omit the discussion on $c_i < x^*$ due to the similarity. Note that the two cases can be distinguished by Lemma 1. Let $\hat{\Delta} = \{u_1, u_2, \ldots, u_q\}$, and let α_j be a point on the x-axis such that

$$d(u_{2j-1}, \alpha_j) = d(u_{2j}, \alpha_j),$$

where $x_{u_{2j-1}} \leq x_{u_{2j}}$. For convenience, we assume that $\alpha_1, \ldots, \alpha_{\lfloor q/2 \rfloor}$ are in nondecreasing order. Let α^* be the median, and let

$$\hat{\Delta}_1 = \{u_{2j} \colon 1 \leq j \leq q/4\}$$

and

$$\hat{\Delta}_2 = \{u_{2j-1} \colon q/4 < j \leq q/2\}.$$

We call α^* the *median of pairs* and $(\hat{\Delta}_1, \hat{\Delta}_2)$ the *candidates* to be pruned with respect to $\hat{\Delta}$. An illustration is given in Fig. 5.

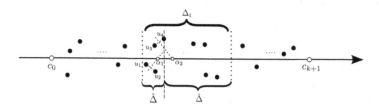

Fig. 5. Generating $\{\alpha_1, \alpha_2, \ldots, \alpha_{q/2}\}$ from $\hat{\Delta}$. In this example, α_1 is α^* of Δ_i, $\hat{\Delta}_1 = \{u_2\}$, and $\hat{\Delta}_2 = \{u_3\}$.

Lemma 4. *Let Δ_i be a big region of (Δ, Σ, k) with $i \in [k]$ and $(\hat{\Delta}, \check{\Delta})$ be a balanced bipartition of Δ_i. Let α^* be the median of pairs and $(\hat{\Delta}_1, \hat{\Delta}_2)$ be the candidates to be pruned with respect to $\hat{\Delta}$. For the bipartition (Δ', Δ'') of Δ with respect to $\max\{x^*, \alpha^*\}$, if $r(\Delta', \Sigma_1, i - 1) \leq r(\Delta'', \Sigma_2, k - i)$, where $\Sigma_1 = \{c_0, \max\{x^*, \alpha^*\}\}$ and $\Sigma_2 = \{\max\{x^*, \alpha^*\}, c_{k+1}\}$, then $r(\Delta, \Sigma, k) = r(\Delta \setminus \hat{\Delta}_1, \Sigma, k)$. Otherwise, $r(\Delta, \Sigma, k) = r(\Delta \setminus \hat{\Delta}_2, \Sigma, k)$.*

Proof. Let $r_1 = r(\Delta', \Sigma_1, i-1)$, $r_2 = r(\Delta'', \Sigma_2, k-i)$, $r^* = r(\Delta, \Sigma, k)$, and $r' = r(\Delta \setminus \hat{\Delta}_1, \Sigma, k)$. For $r_1 \leq r_2$, we prove the lemma by showing $r' \leq r^*$ and $r' \geq r^*$. The first inequality is trivial, and we focus on the second in the following. Suppose to the contrary that $r' < r^*$. Since Δ_i is a big region, by Lemma 1, we have

$$r' < r^* \leq \min\{r(\Delta^{i-1}, \{c_0, \infty\}, i-1), r(\Delta \setminus \Delta^i, \{-\infty, c_{k+1}\}, k-i)\}.$$

It follows that for any optimal solution $(c_1', c_2', \ldots, c_k')$ of $(\Delta \setminus \hat{\Delta}_1, \Sigma, k)$,

$$\Delta_i \setminus \hat{\Delta}_1 \subseteq R(c_i'),$$

which implies

$$\exists_{u \in \hat{\Delta}_1} \quad d(u, c_i') \geq r^*. \tag{3}$$

Since $(\Delta' \setminus \hat{\Delta}_1, \Delta'')$ is a bipartition of $\Delta \setminus \hat{\Delta}_1$ with respect to $\max\{x^*, \alpha^*\}$ and

$$r(\Delta' \setminus \hat{\Delta}_1, \Sigma_1, i-1) \leq r_1 \leq r_2,$$

by Lemma 1 there is an optimal solution of $(\Delta \setminus \hat{\Delta}_1, \Sigma, k)$, say $(c_1', c_2', \ldots, c_k')$, satisfying

$$\max\{x^*, \alpha^*\} \leq c_i'.$$

It follows that

$$\forall_{j \in [\lfloor q/4 \rfloor]} \quad d(u_{2j}, c_i') \leq d(u_{2j-1}, c_i'). \tag{4}$$

Let u_{2l} be the point satisfying predicate (3). It can be derived that

$$r^* \leq d(u_{2l}, c_i') \leq d(u_{2l-1}, c_i') \leq r'. \tag{5}$$

This leads to a contradiction.

For $r_2 < r_1$, it suffices to show

$$r(\Delta' \setminus \hat{\Delta}_2, \Sigma_1, i-1) \geq r_2. \tag{6}$$

Along with the assumption $c_i' > x^*$, we have $c_i' \leq \alpha^*$, and similar arguments as above can be applied. Inequality (6) holds as follows. Let $(c_1'', c_2'', \ldots, c_{i-1}'')$ be an optimal solution of $(\Delta' \setminus \hat{\Delta}_2, \Sigma_1, i-1)$, and let $R'(c_j'')$ be the range of c_j'' on $(\Delta' \setminus \hat{\Delta}_2, \Sigma_1, i-1)$, i.e.,

$$R'(c_j'') = \{u \in \Delta' \setminus \hat{\Delta}_2 : d(u, c_j'') = d(u, \Sigma_1 \cup \bigcup_{l=1}^{i-1}\{c_l''\})\},$$

where $j \in [0 \ldots i-1]$. Recall that Δ_i is a big region. If $R'(\max\{x^*, \alpha^*\}) \subseteq (\Delta' \setminus \Delta^{i-1}) \setminus \hat{\Delta}_2$, then $r(\Delta' \setminus \hat{\Delta}_2, \Sigma_1, i-1) \geq r(\Delta^{i-1}, \{c_0, \infty\}, i-1) > r_2$ (see Fig. 6). Otherwise, we have $r(\Delta' \setminus \hat{\Delta}_2, \Sigma_1, i-1) = r_1$, with which inequality (6) holds. $\square$

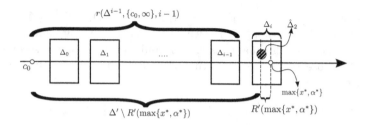

Fig. 6. The case where $R'(\max\{x^*, \alpha^*\}) \subseteq (\Delta' \setminus \Delta^{i-1}) \setminus \hat{\Delta}_2$.

Symmetrically, when $c_i < x^*$, the set to be pruned is a subset of $\check{\Delta}$. One can apply the arguments as above by replacing $\hat{\Delta}$ with $\check{\Delta}$ and defining

$$\check{\Delta}_1 = \{u_{2j} \colon 1 \leq j \leq q/4\}$$

and

$$\check{\Delta}_2 = \{u_{2j-1} \colon q/4 < j \leq q/2\}.$$

For a big region Δ_i with $i \in [k]$, we call the quadruple $(\hat{\Delta}_1, \hat{\Delta}_2, \check{\Delta}_1, \check{\Delta}_2)$ *candidates to be pruned* with respect to Δ_i. The set whose removal results in a smaller instance with the same solution can be obtained by CRITICALREGION, and the pseudocode for the conditional k-center problem is given as CONDITIONALCENTER.

CRITICALREGION$(\Delta, \Sigma, k, i, x, \alpha_1, \alpha_2)$

```
1  if i = 0 or i = k + 1
2      then Find out u ∈ Δᵢ such that d(u, cᵢ) = max{d(v, cᵢ) : v ∈ Δᵢ}
3          Δ* ← Δᵢ \ {u}
4      else
5          lrₓ ← CENTERLOCATION(Δ, Σ, k, i, x)
6          lr_{α₁} ← CENTERLOCATION(Δ, Σ, k, i, α₁)
7          lr_{α₂} ← CENTERLOCATION(Δ, Σ, k, i, α₂)
8          (Â₁, Â₂, Ǎ₁, Ǎ₂) ← candidates to be pruned
       with respect to Δᵢ
9              if lrₓ = RIGHT and lr_{α₁} = RIGHT
10                 then Δ* ← Â₁
11             elseif lrₓ = RIGHT and lr_{α₁} = LEFT
12                 then Δ* ← Â₂
13             elseif lrₓ = LEFT and lr_{α₂} = RIGHT
14                 then Δ* ← Ǎ₁
15                 else  Δ* ← Ǎ₂
16 return Δ*
```

CONDITIONALCENTER(Δ, Σ, k)

 1   **if** $|\Delta| \leq 8k$ or $k = 0$
 2   **then** Compute the radius of the instance immediately
 3   **else**
 4           $\Delta_i \leftarrow$ BIGREGION(Δ, Σ, k)
 5           $x^* \leftarrow$ median of Δ_i
 6           $(\hat{\Delta}, \check{\Delta}) \leftarrow$ BIPARTITION(Δ_i, x^*)
 7           $\alpha_1^* \leftarrow$ median of pairs w.r.t. $\hat{\Delta}$
 8           $\alpha_2^* \leftarrow$ median of pairs w.r.t. $\check{\Delta}$
 9           $\Delta^* \leftarrow$ CRITICALREGION$(\Delta, \Sigma, k, i, x^*, \alpha_1^*, \alpha_2^*)$
10   **return** CONDITIONALCENTER$(\Delta \setminus \Delta^*, \Sigma, k)$

4.3   The Analysis

About the time complexity of this algorithm, we have the following theorem.

Theorem 2. *For any fixed k,* CONDITIONALCENTER *computes the radius of* (Δ, Σ, k) *in $O(n)$ time.*

Proof. We prove this theorem by induction on k. For $k = 0$, it suffices to compute $\max_{u \in \Delta} \min\{d(c_0, u), d(c_{k+1}, u)\}$. This computation can be done in $O(n)$ time. Suppose the theorem holds for $k < k'$. For $k = k'$, since k is fixed, there are a constant number of CONDITIONALCENTERs being executed, within each of which the third argument is less than k. By the induction hypothesis, lines 4 to 9 can be done in $O(n)$ time. Let the running time of CONDITIONALCENTER on (Δ, Σ, k) be $T(n)$. Obviously, Δ^* in line 9 consists of at least $\lfloor n/k \rfloor / 8$ demand points. For $n \geq 8k$, we have

$$T(n) \leq T\left(n - \frac{1}{8}\left\lfloor\frac{n}{k}\right\rfloor\right) + cn.$$

Thus, for any fixed k, the time complexity is $O(n)$. The correctness follows from Lemmas 3 and 4.   □

5   Concluding Remarks

In this paper, we propose a linear-time algorithm for the line-constrained k-center problem, for any fixed k. However, the weakness of this algorithm is that the time complexity is $O(f(k) \cdot n)$, where $f(k)$ is exponential to k. The time complexity could be improved by reducing the order of $f(k)$. Another research work we are interested in is to see if the algorithm can be generalized for higher dimensional Euclidean space.

Acknowledgments. Albert Jhih-Heng Huang and Kun-Mao Chao were supported in part by MOST grants 101-2221-E-002-063-MY3 and 103-2221-E-002-157-MY3, and Hung-Lung Wang was supported in part by MOST grant 103-2221-E-141-004 from the Ministry of Science and Technology, Taiwan.

References

1. Agarwal, P.K., Sharir, M., Welzl, E.: The discrete 2-center problem. Discrete Comput. Geom. **20**(3), 287–305 (1998)
2. Ben-Moshe, B., Bhattacharya, B.K., Shi, Q.: An optimal algorithm for the continuous/discrete weighted 2-center problem in trees. In: Correa, J.R., Hevia, A., Kiwi, M. (eds.) LATIN 2006. LNCS, vol. 3887, pp. 166–177. Springer, Heidelberg (2006)
3. Bhattacharya, B., Shi, Q.: Optimal algorithms for the weighted p-center problems on the real line for small p. In: Dehne, F., Sack, J.-R., Zeh, N. (eds.) WADS 2007. LNCS, vol. 4619, pp. 529–540. Springer, Heidelberg (2007)
4. Brass, P., Knauer, C., Na, H.-S., Shin, C.-S., Vigneron, A.: The aligned k-center problem. IJCGA **21**(2), 157–178 (2011)
5. Chan, T.M.: More planar two-center algorithms. Comput. Geom. **13**, 189–198 (1999)
6. Hoffmann, M.: A simple linear algorithm for computing rectilinear 3-centers. Comput. Geom. **31**(3), 150–165 (2005)
7. Karmakar, A., Das, S., Nandy, S.C., Bhattacharya, B.: Some variations on constrained minimum enclosing circle problem. J. Comb. Optim. **25**(2), 176–190 (2013)
8. Kim, S.K., Shin, C.-S.: Efficient algorithms for two-center problems for a convex polygon. In: Du, D.-Z., Eades, P., Sharma, A.K., Lin, X., Estivill-Castro, V. (eds.) COCOON 2000. LNCS, vol. 1858, pp. 299–309. Springer, Heidelberg (2000)
9. Megiddo, N.: Linear-time algorithms for linear programming in r^3 and related problems. SIAM J. Comput. **12**(4), 759–776 (1983)
10. Megiddo, N., Supowit, K.J.: On the complexity of some common geometric location problem. SIAM J. Comput. **13**(1), 182–196 (1984)
11. Wang, H., Zhang, J.: Line-constrained k-median, k-means, and k-center problems in the plane. In: Ahn, H.-K., Shin, C.-S. (eds.) ISAAC 2014. LNCS, vol. 8889, pp. 3–14. Springer, Heidelberg (2014)

Online k-max Search Algorithms
with Applications to the Secretary Problem

Sizhe Wang[1,2(✉)] and Yinfeng Xu[1,2]

[1] The School of Management, Xi'an Jiaotong University, Xi'an 710049, China
`wang1989sz@stu.xjtu.edu.cn`
[2] The State Key Lab for Manufacturing Systems Engineering,
Xi'an 710049, China

Abstract. This work focuses on a special case of the secretary problem: only a few applicants have access to the interview, and they belong to some class of people (e.g., bachelors, undergraduates or veterans). The overall ability of the applicants is known to the interviewer. This domain occurs when the positions require high professional skills (e.g., pilots, doctors or musicians). This work considers an extension of the classical k-max search model [1,15] to approach this domain. In the new model, we design a case in which the pool of applicants is not large, and introduced the parameter $\bar{S}$ to measure the overall ability of the applicants. First, we modify the *threat based policy* (TBP) algorithm [6], and obtain a new algorithm-TBPS. We prove that TBPS outperforms the optimal k-max search algorithm (KMS) [15] in the new model. Second, we modify KMS and obtain a new algorithm: Modified k-max search algorithm (MKMS). We prove that MKMS outperforms KMS in the new model without respecting to the value of $\bar{S}$. Finally, numerical results are presented to demonstrate the performance of MKMS and TBPS.

Keywords: k-max search · Secretary problem · Online algorithm · Competitive analysis

1 Introduction

The classical secretary problem first appeared in the late 1950's and early 1960's. Gardner [9] presented the classical model of the secretary problem: n applicants come in a random order, and the interviewer is requested to hire the best applicant among them. Once rejected, an applicant cannot later be recalled. Subsequent studies focused on different generalizations of this problem, obtaining many results. In this work, we focus on a special case of the secretary problem: only a few applicants have access to the interview, and the applicants belong to some class of people (e.g., bachelors, undergraduates or veterans). Thus, the ability of the applicant is more likely to be identified before the interview. We introduce the parameter $\bar{S}$ to our new model to measure the average ability of the applicants. This model can be regarded as an extension of the classical *k-max search model* [1,15]. In the following, we would like to introduce several related works before presenting our work.

© Springer International Publishing Switzerland 2016
R. Dondi et al. (Eds.): AAIM 2016, LNCS 9778, pp. 209–221, 2016.
DOI: 10.1007/978-3-319-41168-2_18

1.1   Related Work

Gilbert *et al.* [11] formally presented the optimal algorithm to hire the best applicant in the classical secretary problem, and the corresponding probability to hire the best applicant is $\frac{1}{e}$. A number of variants have subsequently been studied, being with [11] in which multiple choices and various measures of success were considered for the first time. Recent interest in variants of the secretary problem has been motivated by applications in online mechanism design [1,3, 7,12,14]. Babaioff *et al.* [1] studied a generalization of the classical secretary problem called *the matroids secretary problem*. The authors presented a domain of *selling k identical items* which is equal to the *k-max search model* in one-way trading problem [15]. Gharan *et al.* [10] subsequently designed a constant factor competitive algorithm for the *random assignment* model in which the weights are assigned randomly to elements of a matroid.

There is another problem that is similar to the secretary problem: the one-way trading problem. Both problems request the player to choose at least one applicant (or price), and the exact interview score (or the quotation) is unknown. Finally, both of their goals are to maximize the total score (or price) that they have chosen. This work modifies the optimal *k-max search algorithm* (KMS) [15] for application to the secretary problem. Hence, we introduce the related work on one-way trading problems in the following.

El-Yaniv *et al.* [5,6] studied the classical one-way trading problem where the prices vary within an interval $[m, M]$, and the player has one unit asset to sell. They presented the optimal deterministic online algorithm with a competitive ratio of $\sqrt{\frac{M}{m}}$. Lorenz *et al.* [15] applied the optimal *RPP* algorithm to both *k*-max and *k*-min search models in the original one-way trading model. The algorithm for *k*-search seeks to achieve a fixed competitive ratio regardless of how the adversary selects the maximum and minimum. In other words, the algorithm helps bound the risk. Javeria *et al.* [13] studied a *k*-min search model and designed an optimal deterministic algorithm for one-way trading that was less restricted than [15]. Subsequent studies then focused on other forms of one-way trading models: In [2], Chen *et al.* introduced *a planning game* problem, which is similar to the one-way trading problem. They presented an algorithm considering n players with a competitive ratio of $\frac{n\alpha\beta-(n-1)(\alpha+\beta)+(n+2)}{\alpha\beta-1}$ for some fixed $\alpha, \beta > 1$. Chin et al. [1] presented a one-way trading algorithm that did not impose any bounds on market prices and whose performance guarantee depends directly on the input. Xu *et al.* [16] extended the basic one-way trading model by introducing a profit function that decreases over time. Damaschke *et al.* [4] extended the basic model to two new models in which the upper and lower bounds of prices vary with time. Fujiwara *et al.* [8] studied the case that the distribution of the maximum exchange rate in one-way trading is known in advance, and presented an average-case competitive analysis for the one-way trading problem. Zhang *et al.* [17] presented a different case of one-way trading where the variation range of each price depends on its preceding price. They applied the *reservation price policy* (RPP) to this model and thus transformed the problem into a linear programming problem.

In recent studies, there are two forms of the input in the basic one-way trading and the secretary problems: (1) the price or the ability of an applicant is chosen by an adversary and the player has no idea of the future price or the ability. The performance of an algorithm is measured by the worst case competitive ratio. (2) the price or the ability of an applicant arrive in a random order. The performance of an algorithm is measured by the expected competitive ratio.

In this work, we consider a different input: the ability (score) of the applicants is chosen by an adversary, the applicants' ability (score) is unknown to the interviewer unless it arrives. Moreover, we assume that the average score $\bar{S}$ is known in advance. This assumption is able to describe a special domain in the secretary problem: the applicants belong to a class of people (e.g., bachelors, undergraduates or veterans). We present two deterministic online algorithms to this problem, and we measure these two algorithms by worst case competitive ratio.

1.2 Our Results

This work extends the model of one-way trading problem in order to apply to the secretary problem. In this problem, we consider a case in which the interviewer has no idea of each applicant's ability in advance. However, he is able to know the overall ability of the applicants. This work applies the *k-max search model* [5] to this problem by a new parameter $\bar{S}$. We present two algorithms to the new model. First, we design the TBPS algorithm which outperforms KMS if $\bar{S} \geqslant \sqrt{Mm}$. Second, we modify the optimal algorithm (KMS) in the k-max search and obtain another algorithm, MKMS, which outperforms KMS in the new model. Finally, we present several numerical results to demonstrate the performance of MKMS.

The remainder of this paper is organized as follows. In Sect. 2, we set up a new model, $(n, k, \bar{S})$. We present a new deterministic algorithm TBPS in Sect. 3, proving it outperforms KMS in $(n, k, \bar{S})$ if $\bar{S} > \sqrt{Mm}$. In Sect. 4, we present a deterministic algorithm MKMS which is derived from the KMS algorithm, we prove MKMS outperforms KMS. Numerical results of MKMS are presented in Sect. 5. We conclude in Sect. 6 with the main results and possible future work.

2 $(n, k, \bar{S})$ Model

Before we present our new model, it is necessary to repeat the original k-max search model [15]: There are n independent trading periods with k unit asset need to be sold (each asset can be sold only once). The player observes only one selling price at each period. All the prices vary within an fixed interval $[m, M]$. However, the player completely has no idea about the quotation in the future. The goal is maximizing the final return, and the player is forced to sell all the assets before the trading ends. For example, if the player sells all his asset at a price M, his final return is kM. The value of k, m, M and n are known in advance. In following, we call this model (n, k).

In this work, we generalize the (n, k) model above, and obtain a different model-$(n, k, \bar{S})$. In the following, we formally present the $(n, k, \bar{S})$-model with several assumptions.

$(n, k, \bar{S})$-Model: There are n applicants competing for k positions (These positions are all identical). For each applicant, the interviewer is requested to score him and decide whether to hire him immediately. Only one applicant can be interviewed for each time. Job vacancy is not allowed. His goal is to maximize the total score of the k hired applicants. When the interviewer makes a decision, it cannot be canceled. The interviewer must score all applicants unless he has hired k applicants. Unlike the previous work on the secretary problem, we assume that the applicants' score varies within a fixed interval $[m, M], (0 < m < M)$. The interviewer knows the average score of the n applicants in advance, which can be represented by a positive real number $\bar{S}$, $\bar{S} \in [m, M]$. The values of $\bar{S}$, n, m, M and k are known in advance, and we assume $n > k$.

Without loss of generality, we assume that the interview for each applicant is fair, that is, the score of each applicant are assigned correctly. It is necessary to make another important assumptions in this work: *the value of $\frac{n}{k}$ is sufficiently low* (In numerical result, we assume $k = 5, 10$). This assumption accurately describes a situation in which the positions require high qualifications (e.g., pilots, doctors, musicians). In this case, only a few applicants have access to the interview. If $\frac{n}{k} \to \infty$, $(n, k, \bar{S})$-model becomes the (n, k)-model. Another difference is that (n, k)-model allows the player trade (buy or sell) all of his asset (e.g. stock, currency) at one price. However in the $(n, k, \bar{S})$-model, the interviewer can only hire or reject one applicant at each time.

3   TBPS Algorithm

In the (n, k)-model, the *total score* of the applicants varies within an interval $[nm, nM]$. However, the *total score* of the applicants is a fixed number $n\bar{S}$ in the $(n, k, \bar{S})$-model. The conditions in the new model are quite different from those in the (n, k)-model.

3.1   TBPS

In this section, we modify the RPP algorithm in [5] based on *threat based policy* (TBP) which forces the player or the interviewer sells (or hire) the assets (or applicants) as long as the price (or score) is not lower than some preset value by the player or the interviewer. The RPP algorithm is proved to be the optimal algorithm for the classical one-way trading model. We show that the modified algorithm, TBPS, outperforms the original k-max search algorithm in [15] when the value of $\bar{S}$ satisfies the inequality:

$$M > \bar{S} \geqslant (Mm)^{\frac{1}{2}}$$

In the following, we formally present TBPS.

Algorithm 1. TBPS

Step 0: Let S_i be the interview score of the ith applicant. Let j be the number of hired applicants. Set $i = 1$, $j = 0$.

Step 1: At time i, if $S_i \geqslant S_0$, hire the ith applicant, let $i = i + 1, j = j + 1$ go to step 2; If $S_i < S_0$, let $i = i + 1$, go to Step 2.

Step 2: If $n - i + j = k$, hire all the rest of the applicants and the game terminates; If $n - i + j > k$, go to Step 1.

3.2 Competitive Analysis of TBPS

In this section, we present worst case analysis according to the behavior of the TBPS algorithm.

Let S_0 be the lowest acceptable score according to the interviewer, if the one of the applicants' score is lower than S_0, the interviewer will reject him and wait for the next applicant. We set S_0 be a positive number varying within the interval: $[m, M]$. Let ε be an arbitrary positive integer. Let θ be the competitive ratio of TBPS. In the following we present worst cases for the interviewer with respect to the value of S_0 and $\bar{S}$.

Case 1: The interviewer chooses $S_0 < \bar{S}$. Since we do not consider any specific value of $\bar{S}$, the worst sequence made by adversary depends on the value of $\bar{S}$. Without loss of generality, we discuss the two cases based on the value of $\bar{S}$.

(1) If $nM \geqslant n\bar{S} > (n - k)M + kS_0$. Since $n\bar{S} > (n - k)M + kS_0$, the adversary will let the interviewer pick the applicants first. After the interviewer finishes hiring, the adversary will rise the score to M and begin hiring. Therefore, the adversary presents the first k applicants with a total score of $n\bar{S} - (n-k)M$. As a result, the interviewer hired these k applicants. After that, the adversary hires k applicants with a total score of kM (each one with a score of M.). The interview sequence is: $\sigma = \underbrace{(...)}_{k(interviewer)}, \underbrace{(M, ..., M)}_{k(adversary)}, \underbrace{(M, ..., M)}_{n-2k}$. In this case, the competitive ratio is: $\theta = \frac{kM}{n\bar{S} - (n-k)M}$.

(2) If $(n-k)M + kS_0 \geqslant n\bar{S} \geqslant Mk + kS_0 + (n-2k)m$, the adversary will firstly present k applicants with each one's score of S_0. The interviewer hired these k applicants with a total score of kS_0 (each one's score equals S_0). After that, the adversary hires k applicants with a total score of kM. The interview sequence is: $\sigma = \underbrace{(S_0, ..., S_0)}_{k(interviewer)}, \underbrace{(M, ..., M)}_{k(adversary)}, \underbrace{(...)}_{n-2k}$. In this case, the competitive ratio is: $\theta = \frac{M}{S_0}$.

(3) If $nm < n\bar{S} < Mk + kS_0 + (n - 2k)m$, the adversary is unable to present M as the interview score for exactly k times because the value of $\bar{S}$ is small. Instead, the adversary will firstly present k applicants with score of S_0. After that, the adversary presents m as the score of the next $n - 2k$ applicants in order to maximize the total score of the last k applicants. As a result, the adversary hires the last k applicants with a total score of $n\bar{S} - kS_0 - (n - 2k)m$. Hence,

the interview sequence is: $\sigma = \underbrace{(S_0, ..., S_0)}_{k(interviewer)}, \underbrace{(m, ..., m)}_{n-2k}, \underbrace{(...)}_{k(adversary)}$. In this case,

the competitive ratio is: $\theta = \frac{n\bar{S}-kS_0-(n-2k)m}{kS_0}$.

Case 2: If the interviewer chooses $S_0 \geqslant \bar{S}$, there are 5 subcases according to the value of S_0:

(1) If $nM \geqslant n\bar{S} > (n-k)M + kS_0$, the adversary will firstly present k applicants with a total score of $n\bar{S} - (n-k)M$. The interviewer hired these k applicants. After that, the adversary hires k applicants with a total score of kM. Hence, the interview sequence is: $\sigma = \underbrace{(...)}_{k(interviewer)}, \underbrace{(M, ..., M)}_{k(adversary)}, \underbrace{(M, ..., M)}_{n-2k}$

In this case, the competitive ratio is: $\theta = \frac{kM}{n\bar{S}-(n-k)M}$.

(2) If $(n-k)M + kS_0 \geqslant n\bar{S} \geqslant Mk + kS_0 + (n-2k)m$, this case is similar to Case 1, the interview sequence is: $\sigma = \underbrace{(S_0, ..., S_0)}_{k(interviewer)}, \underbrace{(M, ..., M)}_{k(adversary)}, \underbrace{(...)}_{n-2k}$.

In this case, the competitive ratio is: $\theta = \frac{M}{S_0}$.

(3) If $Mk + kS_0 + (n-2k)m > n\bar{S} > 2kS_0 + (n-2k)m$, this case is similar to Case 1, the interview sequence is: $\sigma = \underbrace{(S_0, ..., S_0)}_{k(interviewer)}, \underbrace{(m, ..., m)}_{n-2k}, \underbrace{(...)}_{k(adversary)}$.

In this case, the competitive ratio is: $\theta = \frac{n\bar{S}-kS_0-(n-2k)m}{kS_0}$.

(4) If $kS_0 + (n-k)m \leqslant n\bar{S} \leqslant 2kS_0 + (n-2k)m$, the adversary will present the first k applicants with score of $S_0 - \varepsilon$, and hires all of them. In this case, the interviewer has no choice but hires the last k applicants with a total score of km.

Hence, the interview sequence is: $\sigma = \underbrace{(S_0 - \varepsilon, ..., S_0 - \varepsilon)}_{k(adversary)}, \underbrace{(...)}_{n-2k}, \underbrace{(m, ..., m)}_{k(interviewer)}$.

As $\varepsilon \to 0^+$, $\theta \to \frac{S_0}{m}$.

(5) If $nm \leqslant n\bar{S} < kS_0 + (n-k)m$, the adversary will present the last $n - 2k$ applicants with score of m. The interviewer has no choice but hires the last k applicants with a total score of km.

Hence, the interview sequence is: $\sigma = \underbrace{(...)}_{k(adversary)}, \underbrace{(m, ..., m)}_{n-k}$.

In this case, the competitive ratio is: $\theta = \frac{n\bar{S}-(n-k)m}{km}$.

Now we are ready to prove the competitive ratio of TBPS

Theorem 1. TBPS has a competitive ratio of $\theta \leqslant \sqrt{\frac{M}{m}}$ if $S_0 = \sqrt{Mm}$.

Proof. Without loss of generality, we do not consider any specific value of $\bar{S}$. The TBPS algorithm may achieve different ratios if the values of S_0 and $\bar{S}$ change. Since the value of average score $\bar{S}$ is fixed, the adversary cannot always present the lowest possible scores that the player accepts and reserve the highest scores for himself. Therefore the value of $\bar{S}$ must be considered into the worst case analysis. If we set $S_0 = \sqrt{Mm}$, there are two cases below:

Case 1: $\bar{S} \geqslant S_0$

(1) If $nM \geqslant n\bar{S} > (n-k)M + kS_0, \theta = \frac{kM}{n\bar{S}-(n-k)M}$

(2) If $(n-k)M + kS_0 \geqslant n\bar{S} \geqslant Mk + kS_0 + (n-2k)m, \theta = \frac{M}{S_0}$.

(3) If $nm \leqslant n\bar{S} < Mk + kS_0 + (n-2k)m, \ \theta = \frac{n\bar{S}-kS_0-(n-2k)m}{kS_0}$.

Hence, $\theta = min[\frac{kM}{n\bar{S}-(n-k)M}, \frac{M}{S_0}, \frac{n\bar{S}-kS_0-(n-2k)m}{kS_0}]$.

Case 2: $\bar{S} < S_0$

(1) If $nM \geqslant n\bar{S} > (n-k)M + kS_0, \ \theta = \frac{kM}{n\bar{S}-(n-k)M}$.

(2) If $(n-k)M + kS_0 \geqslant n\bar{S} \geqslant Mk + kS_0 + (n-2k)m, \ \theta = \frac{M}{S_0}$.

(3) If $Mk + kS_0 + (n-2k)m > n\bar{S} > 2kS_0 + (n-2k)m, \theta = \frac{n\bar{S}-kS_0-(n-2k)m}{kS_0}$.

(4) If $kS_0 + (n-k)m \leqslant n\bar{S} \leqslant 2kS_0 + (n-2k)m, \ \theta = \frac{S_0}{m}$.

(5) If $nm \leqslant n\bar{S} < kS_0 + (n-k)m, \ \theta = \frac{n\bar{S}-(n-k)m}{km}$.

Hence, $\theta = min[\frac{kM}{n\bar{S}-(n-k)M}, \frac{M}{S_0}, \frac{n\bar{S}-kS_0-(n-2k)m}{kS_0}, \frac{S_0}{m}, \frac{n\bar{S}-(n-k)m}{km}]$.

Obviously, $\frac{kM}{n\bar{S}-(n-k)M} < \frac{M}{S_0}, \frac{kS_0}{n\bar{S}-(n-k)S_0} < \frac{S_0}{m}, \frac{n\bar{S}-kS_0-(n-2k)m}{kS_0} < \frac{M}{S_0}$,

$\frac{n\bar{S}-(n-k)m}{km} < \frac{S_0}{m}$. When $S_0 = (Mm)^{\frac{1}{2}}$, for an arbitrary $\bar{S}$, $\theta \leqslant \sqrt{\frac{M}{m}}$. $\qquad\square$

4 MKMS Algorithm

Lorenz *et al.* [15] designed the optimal algorithm (in the following, we refer to this algorithm as simply KMS) for the original one-way trading algorithm (we call the (n, k)-model). In this section, we modify KMS algorithm in [15]. Through competitive analysis, we prove that MKMS outperforms KMS in the $(n, k, \bar{S})$-model.

4.1 KMS Algorithm

Now we need to recall the main idea of the optimal k-max search algorithm in [15]. In the following, we recall the KMS algorithm according the secretary back ground: the KMS algorithm seeks a series of scores: $S_1, S_2, ..., S_k$ to reach a constant competitive ratio of $c = \frac{kS_i}{S_1+...+S_{i-1}} > 1$, $(i = 2, ..., k$. If $i = 1, c = \frac{S_1}{m})$. i.e. The KMS algorithm force the interviewer hire the ith applicant only when the interview score is not lower than S_i. Hence, there are k possible worst cases according to the behavior of KMS : $\frac{S_1}{m} = \frac{kS_2}{p_1+(k-1)m} = \frac{kM}{S_1+S_2+...+S_k} = c$.

However, the conditions above may change in the $(n, k, \bar{S})$-model. It is necessary to discuss what conditions make KMS algorithm available in the $(n, k, \bar{S})$-model. For example, if the interviewer applies the KMS algorithm, and the lowest score for the ith hired applicant is S_i, $(i = 1, ..., k)$. We assume the adversary chooses S_i as the maximum score in the interview. The adversary present the first $i - 1$ applicants with each score of:$S_1, ..., S_{i-1}$, then he presented S_i for k times. If the value of S_i is too large to satisfy the inequality $kS_i + S_1 + ... + S_{i-1} + (k-i+1)m + (n-2k)m > n\bar{S}$, though the adversary reduces the rest applicants' scores to m, the total score becomes larger than the actual total score: $n\bar{S}$. Since the value of $\bar{S}$ is fixed in the $(n, k, \bar{S})$-model. The value of S_i must satisfies the inequality: $kS_i + S_1 + ... + S_{i-1} + (k-i+1)m + (n-2k)m > n\bar{S}$, otherwise the rest of the sequence(from S_{i+1} to S_k) makes no sence to the interviewer. Hence, the value of S_i should satisfy the condition below:

Condition A: $kS_i + S_1 + S_2 + ... + S_{i-1} + (n - k - i + 1)m \leqslant n\bar{S}$, $i = 1, ..., k$.

Similarly, if the value of S_i satisfies the inequality- $kS_i + S_1 + ... + S_{i-1} + (k - i + 1)m + (n - 2k)(S_i - \varepsilon) < n\bar{S}$, KMS becomes unavailable because the total score made by adversary will be lower than $n\bar{S}$. So that the value of S_i should satisfy another condition:

Condition B: $kS_i + S_1 + S_2 + ... + S_{i-1} + (k - i + 1)m + (n - 2k)S_i \geqslant n\bar{S}$, $i = 1, ..., k$.

Therefore, KMS is available in the $(n, k, \bar{S})$-model only when the two conditions above are met. Let c_1 be the competitive ratio of KMS. KMS achieves a competitive ratio of $c = max[\frac{S_1}{m}, ..., \frac{kM}{S_1 + S_2 + ... + S_k}]$. When Condition B establishes, it is obvious that $kS_1 + km + (n - 2k)S_1 \geqslant n\bar{S}$. We can conclude that $S_1 > \bar{S}$. Thus we have $c \geqslant \frac{S_1}{m} > \frac{\bar{S}}{m}$, indicating that TBPS outperforms KMS if $\bar{S} \geqslant \sqrt{Mm}$.

4.2   MKMS Algorithm

In this section we modified the KMS algorithm, and obtain a new algorithm-MKMS (Modified KMS algorithm). The main ideas of MKMS and KMS are similar. However, MKMS constructs a special series-$S_1^*, S_2^*, ..., S_k^*$ as the lowest acceptable score for the interviewer, which leads to different possible worst cases from [15]. The details of MKMS and S_k^* are formally presented below.

Algorithm 2. MKMS

Step 0: Let S_t be the score of the tth applicant, $t = 1, ..., n$. Let j be the number of hired applicants.

(1) Construct an arithmetic sequence $S_1^*, S_2^*, ..., S_t^*$ ($S_t^* = \lambda(t - 1) + S_1^*, \lambda > 0$)

(2) Construct a series $S_{t+1}^*, S_{t+2}^*, ..., S_k^*$. $S_{t+1}^* = (M - m)(\frac{k}{k+1})^{k-t} + m, (t = 2, ..., k)$. The value of S_{t+1}^* must satisfy Conditions A and B. Moreover, the value of S_1^*, i and λ must satisfy the inequalities: 1)$S_{t-1}^* < \bar{S}$, 2)$S_t^* > \bar{S}$, 3)$kS_t^* + S_1^* + S_2^* + ... + S_{t-1}^* + (k - t + 1)m + (n - 2k)S_t^* < nS$, $S_{t+1}^* > S_t^*$, 4)$k(M - m)(\frac{k}{k+1})^{k-t} + (t + 1 - k)m + n(\bar{S} - S_1^*) \leqslant [S_1^* + \frac{t(t-1)}{2} + n(t - 1)]\lambda$

Step 1: Set $t = 1, j = 0$.

Step 2: At time t, if $S_t \geqslant S_{j+1}^*$, hire ith applicant, let $t = t + 1, j = j + 1$, go to step 3; If $S_t < S_{j+1}^*$, let $t = t + 1$, go to Step 5.

Step 3: If $j = k$, the game terminates; If $j < k$, go to Step 4.

Step 4: If $n - t + j = k$, hire all the rest of the applicants and the game terminates. If $n - t + j > k$, go to Step 2.

4.3   Competitive Analysis of MKMS

Let S_{max} be the maximum score which is determined by the adversary. Note that $S_{t-1}^* < \bar{S}$, the value of S_{max} satisfies the inequality: $S_{max} \geqslant S_t^*$. There are two cases according to the behavior of the MKMS.

Case 1: If the adversary chooses $S_{max} = S_t^*$.

The adversary will present the first $t-1$ applicants with a total score series: $S_1^*, S_2^*, ..., S_{t-1}^*$. According to the behavior of MKMS, the interviewer hires all the $t-1$ applicants. The adversary then sets the next $n-k$ applicants each with a score of $S_t^* - \varepsilon$. After that, the adversary hires exactly k applicants among these $n-k$ applicants and the corresponding score is $(n-k)(S_t^* - \varepsilon)$. Hence, the interviewer must hire the remaining $k-t$ applicants. The total score achieved by MKMS is $n\bar{S} - (n-k)(S_t^* - \varepsilon)$. Since $\varepsilon \to 0^+$, $n\bar{S} - (n-k)(S_t^* - \varepsilon) \to n\bar{S} - (n-k)S_t^*$. The competitive ratio is: $c_2 = \frac{kS_t^*}{n\bar{S} - (n-k)S_t^*}$.

Case 2: If the adversary chooses $S_{max} = S_{t+1}^*$: In this case, Conditions A and B are both established. The worst case is similar to [15]: In the beginning, the adversary presents the first t applicants with a score series: $S_1^*, S_2^*, ..., S_t^*$. According to the behavior of MKMS, the interviewer hires all t applicants. According to A_t and B_t, the adversary then presents the next k applicants with a score series: $S_{t+1}^* - \varepsilon, S_{t+1}^* - \varepsilon, ..., S_{t+1}^* - \varepsilon$. As $\varepsilon \to 0^+$, the total score score obtained by the adversary obtains is kS_{t+1}^*. However, $S_{t+1}^* - \varepsilon < S_{t+1}^*$, according to MKMS, the interviewer rejects these k applicants and waits for the remainder. After that, the adversary presents the next $n-2k$ applicants with an average score of S, and S holds the equality:

$$(n-2k)S + (k-t)m + S_1^* + ... + S_t^* + k(S_{t+1}^* - \varepsilon) = n\bar{S} \tag{1}$$

Clearly, $S < S_{t+1}^*$ (Condition A_{t+1}). The interviewer rejects these $(n-2k)$ applicants again. According to (1), the last $k-t+1$ applicants come with a total score of $(k-t)m$, and the interviewer has no choice but to hire them. Hence, the total score that the interviewer obtains is $S_1^* + ... + S_t^* + (k-t)m$.

In this case, the worst interview sequence can be written as follows:

$$\sigma = S_1^*, ..., S_t^*, \underbrace{(S_{t+1}^* - \varepsilon, ..., S_{t+1}^* - \varepsilon)}_{k}, \underbrace{(S, ..., S)}_{n-2k}, \underbrace{(m, ..., m)}_{k-t}.$$

Hence, the competitive ratio is $c_2 = \frac{kS_{t+1}^*}{S_1^* + ... + S_t^* + (k-t)m}$. Hence, we have the following possible worst cases:

$$\begin{cases} 1 < \dfrac{kS_{t+1}^*}{S_1^* + ... + S_t^* + (k-t)m} \leqslant c_2, (t = 2, ..., k) \\ ... \\ 1 < \dfrac{kM}{S_1^* + ... + S_k^*} \leqslant c_2 \end{cases}$$

To summarize the two cases above, there are $k - i + 1$ possible worst cases according to the behavior of MKMS:

$$\begin{cases} 1 < \frac{kS_t^*}{n\bar{S} - (n-k)S_t^*} \leqslant c_2 \\ 1 < \frac{kS_{t+1}^*}{S_1^* + ... + S_t^* + (k-t)m} \leqslant c_2 \\ ... \\ 1 < \frac{kM}{S_1^* + ... + S_k^*} \leqslant c_2 \end{cases}$$

Therefore, the competitive ratio of MKMS is $c_2 = max[\frac{kS_t^*}{n\bar{S} - (n-k)S_t^*}, ...,$ $\frac{kM}{S_1^* + ... + S_k^*}]$. In the following, we prove the competitive ratio of MKMS.

Theorem 2. MKMS has a competitive ratio of $\frac{kS_t^*}{n\bar{S} - (n-k)S_t^*}$.

Proof. According to MKMS, we have 4 restrictions on the series $S_1^*, ..., S_k^*$:

$$S_t^* = (M - m)(\frac{k}{k+1})^{k-t+1} + m(t = 2, ..., k) \tag{2}$$

$$K(M-m)(\frac{k}{k+1})^{k-t} + (t+1-k)m + n(\bar{S} - S_1^*) \leqslant [S_1^* + \frac{t(t-1)}{2} + n(t-1)]\lambda \tag{3}$$

$$kS_t^* + S_1^* + S_2^* + ... + S_{t-1}^* + (k-t+1)m + (n-2k)S_t^* < n\bar{S} \tag{4}$$

$$S_1 < \bar{S}, S_t > \bar{S} \tag{5}$$

An improper fraction becomes smaller if the numerator and denominator are coupled with a same positive number. Besides, we have: $\frac{k(S_{t+2}^* - S_{t+1}^*)}{S_{t+1}^* - m} = ... = \frac{kM - kS_k^*}{S_k^* - m} = 1$ according to (2). We can conclude that:

$$\frac{kS_{t+1}^*}{S_1^* + ... + S_t^* + (k-t)m} > \frac{kS_{t+1}^* + k(S_{t+2}^* - S_{t+1}^*)}{S_1^* + ... + S_t^* + (k-t)m + S_{t+1}^* - m}$$

$$max[\frac{kS_{t+1}^*}{S_1^* + ... + S_t^* + (k-t)m}, ..., \frac{kM}{S_1^* + ... + S_k^*}] = \frac{kS_{t+1}^*}{S_1^* + ... + S_t^* + (k-t)m}$$

Therefore, if the adversary want choose S_{t+1}^* or larger ones as the maximum score, he finally chooses S_{t+1}^*. According to MKMS, $S_{t+1}^* = (M - m)(\frac{k}{k+1})^{k-t} + m$, $S_t^* = \lambda(t-1) + S_1^*$, and (3) equals to the inequality: $kS_{t+1}^* - kS_t^* < [S_1^* + ... + S_t^* + (k-t)m] - [n\bar{S} - (n-k)S_t^*]$. Thus we have: $\frac{kS_t^*}{n\bar{S} - (n-k)S_t^*} > \frac{kS_{t+1}^*}{S_1^* + ... + S_t^* + (k-t)m}$, the competitive ratio of MKMS is $c_2 = \frac{kS_t^*}{n\bar{S} - (n-k)S_t^*}$. $\square$

For KMS, there exists: $(n - k)S_1 + km > n\bar{S}$, whereas for MKMS, we have $kS_1 > kS_t^*$ according to (5). Moreover, there exists the inequality: $n\bar{S} - (n - k)S_t^* > km$. Thus, $\frac{kS_t^*}{n\bar{S}-(n-k)S_t^*} < \frac{kS_1}{km} \leqslant c_1$.

Therefore, we conclude that MKMS always outperforms KMS in the $(n, k, \bar{S})$ model without considering the value of $\bar{S}$. We present several numerical results in Sect. 5.

5   Numerical Results

This section presents numerical computations on the performance of MKMS, TBPS and KMS. Without loss of generality, we set $m = 10$, $M = 100$ and $n = 100$. We consider two cases of the number of positions, i.e., k = 5 and 10. We consider the value of $\bar{S}$ to be neither too high nor too low. (If the value of $\bar{S}$ is near to m or M, the competitive ratio is nearly 1.) For convenience in comparing the performance of MKMS,KMS and TBPS, we replace the competitive ratio of KMS (c_1) and TBPS (θ) with $\frac{\bar{S}}{m}$ and $\sqrt{\frac{M}{m}}$ respectively. For example, if $\bar{S} = 40$, we have $c_1 \geqslant \frac{S_1}{m} > \frac{\bar{S}}{m} = 4$.

As shown in Table 1, the competitive ratio of MKMS is lower than 2. MKMS and TBPS outperforms KMS considerably when the value of $\bar{S}$ is neither too high nor too low. If $\bar{S} = 10 + \varepsilon$ and $k = 10$, the overall ability is rather low, the total score that adversary and interviewer achieve are both near 100. Thus, the competitive ratio is almost one.

Table 1. Numerical results of MKMS and KMS

k	$\bar{S}$	S_1^*	t	λ	$KMS(\frac{\bar{S}}{m})$	$MKMS$	$TBPS(\sqrt{\frac{M}{m}})$
5	40	36	3	2.5	4	1.95	3.16
5	45	42	3	2	4.5	1.77	3.16
5	50	47	3	2	5	1.64	3.16
5	55	52	3	2	5.5	1.56	3.16
5	60	57	3	2	6	1.49	3.16
10	40	34	5	2	4	1.91	3.16
10	45	39	5	2	4.5	1.74	3.16
10	50	40	5	2	5	1.64	3.16
10	55	54	5	2	5.5	1.72	3.16
10	60	55	5	2	6	1.91	3.16

6   Conclusion

There are scenarios in which applicants belong to a class of people and the scale of the applicants is small. It is meaningful to find a suitable strategy in this case.

This study has made an attempt in this direction. First, we present a simple algorithm TBPS which outperforms KMS if $\bar{S} \geqslant \sqrt{Mm}$. Second, We modify the optimal algorithm KMS in [15], and obtain an algorithm, MKMS. We prove that MKMS outperforms KMS in the new model. Finally we give several numerical results to show the performance of TBPS and MKMS. Additional assumptions will be considered in our future studies, which will lead to the development of practical applications.

Acknowledgments. This paper was supported by the National Natural Science Foundation of China (No. 61221063), the Program for Changjiang Scholars, Innovative Research Team in University (IRT1173), and China Postdoctoral Science Foundation (No. 2015T81040).

References

1. Babaioff, M., Immorlica, N., Kleinberg, R.: Matroids, secretary problems, and online mechanisms. In: SODA 2007, pp. 434–443 (2007)
2. Chen, G.H., Kao, M.Y., Lyuu, Y.D., Wong, H.K.: Optimal buy-and-hold strategies for financial markets with bounded daily returns. SIAM J. Compt. **31**(2), 447–459 (2001)
3. Chin, F.Y.L., Fu, B., Jiang, M., Ting, H.-F., Zhang, Y.: Competitive algorithms for unbounded one-way trading. In: Gu, Q., Hell, P., Yang, B. (eds.) AAIM 2014. LNCS, vol. 8546, pp. 32–43. Springer, Heidelberg (2014)
4. Damaschke, P., Ha, P.H., Tsigas, P.: Online search with time-varying price bounds. Algorithmica **55**(4), 619–642 (2009)
5. El-Yaniv, R., Fiat, A., Karp, R., Turpin, G.: Competitive analysis of financial games. In: 33rd Symposium on Foundations of Computer Science, pp. 327–323 (1992)
6. El-Yaniv, R., Fiat, A., Karp, R., Turpin, G.: Optimal search and one-way trading online algorithms. Algorithmica **30**(1), 101–139 (2001)
7. Feldman, J., Henzinger, M., Korula, N., Mirrokni, V.S., Stein, C.: Online stochastic packing applied to display ad allocation. In: de Berg, M., Meyer, U. (eds.) ESA 2010, Part I. LNCS, vol. 6346, pp. 182–194. Springer, Heidelberg (2010)
8. Fujiwara, H., Iwama, K., Sekiguchi, Y.: Average case competitive analysis for one-way trading. J. Comb. Optim. **21**(1), 83–107 (2011)
9. Gardner, M.: Mathematical games column. Sci. Am. February 1960
10. Gharan, S.O., Vondrak, J.: On the variants of the matroids secretary problems. Algorithmica **67**, 472–497 (2013)
11. Gilbert, J., Mosteller, F.: Recognizing the maximum of a sequence. J. Am. Stat. Soc. **61**, 35–73 (1966)
12. Hajiaghayi, M.T., Kleinberg, R., Sandholm, T.: Automated online mechanism design and prophet inequalities. In: ICAI 2007, pp. 58–65 (2007)
13. Javeria, I., Iftikhar, A.: Optimal online k-min search. EURO J. Comput. Optim. **3**, 147–160 (2015)
14. Kleinberg, R.: A multiple-choice secretary algorithm with applications to online auctions. In: SODA 2005, pp. 630–631 (2005)

15. Lorenz, J., Panagiotou, K., Steger, A.: Optimal algorithm for k-search with application in option pricing. Algorithmica **55**, 311–328 (2009)
16. Xu, Y.F., Zhang, W.M., Zheng, F.F.: Optimal algorithms for the online time series search problem. Theor. Comput. Sci. **412**, 192–197 (2011)
17. Zhang, W.M., Xu, Y.F., Zheng, F.F., Dong, Y.C.: Optimal algorithms for online time series search and one-way trading with interrelated prices. J. Comb. Optim. **23**, 159–196 (2012)

Author Index

Printed in the United States
By Bookmasters